HANDBOOK FOR FLUE GAS DESULFURIZATION SCRUBBING WITH LIMESTONE

Handbook for Flue Gas Desulfurization Scrubbing with Limestone

by

D.S. Henzel
B.A. Laseke

PEDCo Environmental, Inc.

and

E.O. Smith
D.O. Swenson

Black & Veatch Consulting Engineers

NOYES DATA CORPORATION
Park Ridge, New Jersey, U.S.A.
1982

Copyright © 1982 by Noyes Data Corporation
Library of Congress Catalog Card Number: 82-7926
ISBN: 0-8155-0912-X
ISSN: 0090-516X
Printed in the United States

Published in the United States of America by
Noyes Data Corporation
Mill Road, Park Ridge, New Jersey 07656

10 9 8 7 6 5 4 3 2 1

Library of Congress Cataloging in Publication Data
Main entry under title:

Handbook for flue gas desulfurization
   scrubbing with limestone.

   (Pollution technology review, ISSN 0090-
516X ; no. 94)
     Bibliography: p.
     Includes index.
     1. Flue gases--Desulphurization.
2. Scrubber (Chemical technology) 3. Limestone.
I. Henzel, D. S. II. Series.
TD885.5.S85H36 1982   628.5′32   82-7926
ISBN 0-8155-0912-X   AACR2

Foreword

This handbook, based on a study by PEDCo Environmental, Inc. and Black & Veatch Consulting Engineers, provides guidance for the selection, installation, and operation of limestone flue gas desulfurization (FGD) scrubber systems. The book covers all of the stages of the project from inception, through design, procurement, operation, and maintenance of the system.

Of the many available processes for FGD, the limestone wet scrubbing process is widely used and is continually being improved by numerous technological advances. The book deals extensively with optional process features and recent innovative modifications that enhance the efficiency of a system. Another Noyes publication, *New Developments in Flue Gas Desulfurization Technology,* covers the general field of FGD.

The information in the book is from *Limestone FGD Scrubbers: Users Handbook* (EPA Report 600/8-81-017), prepared by D.S. Henzel and B.A. Laseke of PEDCo Environmental, Inc. and E.O. Smith and D.O. Swenson of Black & Veatch Consulting Engineers for the U.S. Environmental Protection Agency, Industrial Environmental Research Laboratory, April 1981.

The table of contents is organized in such a way as to serve as a subject index and provides easy access to the information contained in the book.

> Advanced composition and production methods developed by Noyes Data are employed to bring this durably bound book to you in a minimum of time. Special techniques are used to close the gap between "manuscript" and "completed book." In order to keep the price of the book to a reasonable level, it has been partially reproduced by photo-offset directly from the original report and the cost saving passed on to the reader. Due to this method of publishing, certain portions of the book may be less legible than desired.

Notice

The material in this book was prepared as an account of work sponsored by the U.S. Environmental Protection Agency. Publication does not signify that the contents necessarily reflect the views and policies of the contracting agency or the publisher, nor does mention of trade names or commercial products constitute endorsement or recommendation for use.

Acknowledgements

This handbook was prepared under the sponsorship of the EPA Industrial Environmental Research Laboratory at Research Triangle Park, North Carolina. The Project Officer who provided overall guidance and coordination was Mr. Robert H. Borgwardt. The prime contractor was PEDCo Environmental, Inc., in Cincinnati, Ohio. The subcontractor was Black & Veatch Consulting Engineers in Kansas City, Missouri. The PEDCo Project Director was Mr. William Kemner and the PEDCo Project Manager was Mr. David Henzel. The Senior Technical Reviewer for PEDCo was Mr. Bernard Laseke. The Project Manager for Black & Veatch was Mr. Earl Smith, and Mr. Donald Swenson was Managing Project Engineer.

A central element of the project was a Review Panel, who provided comment and guidance as to content and emphasis. In addition to those mentioned above, the members of the Review Panel were:

> Dr. Gary Rochelle
> University of Texas at Austin
>
> Dr. Norman Ostroff
> Peabody Process Systems
>
> Dr. Nicholas Stevens
> Research Cottrell, Inc.
>
> Mr. Philip Rader
> Combustion Engineering, Inc.
>
> Mr. Earl Smith
> Black & Veatch Consulting Engineers

In addition to his role as a reviewer, Dr. Rochelle was the primary author of Appendix A, Chemistry of Limestone Scrubbing.

Other authors are Messrs. Henzel, Laseke, and Avi Patkar of PEDCo and Messrs. Swenson and John Noland of Black & Veatch.

Contents and Subject Index

1. **INTRODUCTION** 1
   Purpose, Scope, and Structure of the Handbook 1
      Purpose 2
      Scope 3
      Organization 6
   Overview of the Limestone Scrubbing Process 7
      Basic Limestone FGD Process 7
      Optional Process Features 9
   Process Chemistry and Operational Factors 16
      Type and Grind of Limestone 16
      Stoichiometric Ratio and pH 17
      Stoichiometry and Mist Eliminator Fouling 17
      SO_2 Removal and pH 18
      SO_2 Removal and L/G 18
      L/G Ratio and Liquid Holdup 18
      Liquid Phase Alkalinity and L/G 19
      Scrubber Effluent Hold Tank and Relative Saturation 19
      Prevention of Scale Formation 20
      Degree of Oxidation 20
      Chloride Removal 21
      Equipment Considerations 21
   References 21

2. **OVERALL SYSTEM DESIGN** 23
   Powerplant Considerations 23
      Coal-Related Factors (Properties and Supply) 25
      Steam Generator Design 27
      Power Generation Demand 27
      Site Conditions 28

    Environmental Regulations............................30
  Design Basis...34
    Flue Gas Flow and Composition........................34
    Pollutant Removal Requirements.......................35
    Reagent Stoichiometric Ratio.........................35
    Limestone Composition................................35
    Makeup Water...36
  Material and Energy Balances.........................36
  System Configuration Options........................38
    Particulate Removal..................................38
    Fan Location...40
    Flue Gas Bypass (Versus No Bypass)...................41
    Reheat (Versus No Reheat)............................42
    Sludge Disposal......................................43
    Flexibility and Redundancy...........................45
  Computerized Design Guides..........................45
    TVA Lime/Limestone Scrubbing Computer Model..........45
    PEDCo Flue Gas Desulfurization Information System (Experience Records)..47
    Bechtel-Modified Radian Equilibrium Computer Program.......48
  References..48

3. **THE FGD SYSTEM**....................................50
  Scrubbers..50
    Description and Function.............................50
    Basic Scrubber Types (PEDCo Environmental, Inc. 1981; IGCI 1976; Calvert 1977; Saleem 1980)..........................51
    Design Considerations................................59
    Operational Systems..................................63
  Mist Eliminators...................................63
    Description and Function (Conkle et al. 1976)........63
    Basic Types (Conkle et al. 1976; PEDCo Environmental, Inc. 1981; IGCI 1975)...65
    Design Considerations................................67
    Mist Eliminator Wash.................................73
    Operational Systems (PEDCo Environmental, Inc. 1981).......77
  Reheaters (Choi et al. 1977)........................77
    Description and Function.............................77
    Basic Types..77
    Design Considerations................................84
    Operational Systems..................................92
  Fans...92
    Description and Function.............................92
    Design Parameters....................................93
    Operational Systems..................................95
  Thickeners and Mechanical Dewatering Equipment......98
    Description and Function............................100
    Design Considerations...............................107

Operational Systems 109
Sludge Treatment 112
  Treatment Processes........................... 112
  Regulating Considerations..................... 116
Limestone Slurry Preparation 118
  Slurry Storage................................ 119
  Slurry Feed 119
  Design Considerations 121
Pumps... 122
  Description and Function...................... 124
  Design Considerations 129
  Operational Systems 132
Piping, Valves, and Spray Nozzles................. 135
  Piping.. 135
  Valves.. 137
  Spray Nozzles................................. 138
Ducts, Expansion Joints, and Dampers 139
  Ductwork 139
  Expansion Joints.............................. 141
  Dampers 143
Tanks... 145
  Scrubber Effluent Hold Tank (EHT)............. 146
  Limestone Slurry Feed Tank 147
  Thickener Overflow Tank 147
  Mix Tank (Optional) 147
  Mist Eliminator Wash Tank 147
Agitators... 148
Materials of Construction 148
  Basic Classification of Materials 148
  Performance Characteristics................... 149
  Basis for Selection of Materials.............. 152
Process Control and Instrumentation............... 154
  Limestone FGD System Control Loops 156
  Instrumentation 159
  Control Philosophy 167
  Operational Systems 170
References.. 176

4. **PROCUREMENT OF THE FGD SYSTEM**................. 180
   Competitive Bidding 180
   The Purchase Specification 180
   Procurement Planning 182
     Scope of Supply 182
     Allocation of Procurement Packages 184
     Selection of Bidders.......................... 185
   Preparation of Specifications................. 186
     Bidding Requirements 187
     Contract Requirements......................... 188

        Technical Requirements 188
        Detailed Equipment Specifications 194
    Evaluation of Proposals 202
        Technical Evaluation 202
        Commercial Evaluation. 203
        Economic Evaluation. 204
        Selection of Successful Bidder 206
    Installation, Startup, and Testing 206
        Engineering Data. 207
        System Installation 207
        System Startup. 208
        Performance Testing 208
    References 210

5. OPERATION AND MAINTENANCE 211
    Standard Operations 212
        Varying Inlet SO_2 and Boiler Load 213
        Verification of Flow Rates 213
        Surveillance of Scrubber Operations 214
        Mist Eliminators 214
        Flue Gas Reheat 215
        Fans, Ductwork, and Chimney. 215
        Limestone Receiving, Storage, and Slurry Preparation 216
        Limestone Slurry Feed Control 216
        Pumps, Pipes, and Valves 217
        Thickener 218
        Sludge Disposal. 218
        Process Instrumentation and Controls 219
    Initial Operations 220
        Initial Operational Tests 220
    Startup, Shutdown, Standby, and Outage. 221
        Scrubber Startup. 221
        Scrubber Shutdown. 222
        System Standby 222
        Extended Outage. 223
    System Upsets 223
    Operating Staff and Training 224
    Preventive Maintenance Programs 225
        Scrubber Modules. 225
        Mist Eliminators 226
        Reheat System 226
        Dampers, Fans, Ductwork, and Chimneys 226
        Limestone Slurry Preparation 226
        Limestone Slurry Feed. 227
        Pumps, Pipes, and Valves 227
        Thickeners. 228
        Sludge Disposal Equipment. 228
        Process Instruments and Controls. 228

Housekeeping	229
Unscheduled Maintenance	**229**
Scrubber Modules	230
Mist Eliminator	230
Reheat System	230
Fans	230
Ductwork	230
Limestone Slurry Preparation	231
Pumps, Piping, Valves	231
Thickeners and Sludge Disposal Equipment	232
Process Instrumentation and Controls	232
Troubleshooting Techniques	233
Spare Parts	234
Maintenance Staff Requirements	**234**
References	**235**
APPENDIX A: CHEMISTRY OF LIMESTONE SCRUBBING	**236**
Chemistry Design Objectives	**236**
SO_2 Gas/Liquid Mass Transfer	**239**
Effect of SO_2 Gas Concentration	240
Effects of pH and Excess Limestone	242
Effect of Alkali Additives	242
Effect of Buffer Additives	246
Effect of Forced Oxidation	247
Limestone Dissolution	**251**
Equilibrium	251
Mass Transfer	252
Oxidation	**256**
Mass Transfer	256
Reaction Kinetics	257
Forced Oxidation	257
$CaSO_3/CaSO_4$ Crystallization	**258**
Crystallization Kinetics	259
$CaSO_3$ Scaling	261
$CaSO_4$ Scaling	262
Forced Oxidation	263
Summary	**263**
SO_2 Removal	263
Scale-free Operation	264
References	**265**
APPENDIX B: OPERATIONAL FACTORS	**268**
Liquid-To-Gas Ratio	**268**
Effect on SO_2 Removal	268
Minimum L/G Ratio and Liquid Phase Alkalinity	271
Actual L/G Ratio	273
Gas/Liquid Distribution	**273**
Gas Distribution	273

xiv   Contents and Subject Index

    Gas Inlet Design 274
    Liquid Distribution 274
  Gas Velocity and Pressure Drop 275
    Flooding Potential. 275
    Entrainment Potential 276
    Pressure Drop. 276
  Turndown Capability. 277
  Prevention of Scale Formation 278
    Calcium Sulfite Scaling. 278
    Calcium Sulfate Scaling 280
    Mechanical Considerations 281
  Chloride Control. 281
  Energy Demand 283
  References. 286
  Bibliography. 287

APPENDIX C: COMPUTER PROGRAMS. 288
  Lime/Limestone Economics Computer Model. 289
    Background 289
    Model Description. 290
    Model Usefulness. 307
  Flue Gas Desulfurization Information System. 307
    Background 307
    System Description 308
    System Usefulness. 314
  Bechtel-Modified Radian Equilibrium Program 316
    Background 316
    Model Usefulness. 317
  References. 318

APPENDIX D: INNOVATIONS IN LIMESTONE SCRUBBING 320
  New Types of Scrubbers. 320
    Jet Bubbling Scrubber 320
    Cocurrent Scrubber 322
    Charged Particulate Separator 323
  New Process Modifications 326
    Dowa Process 326
    Limestone Regeneration in Dual-Alkali Systems 328
    Effect of Limestone Type and Grind. 328
    Forced Oxidation 329
    Gypsum Stacking 331
    Adipic Acid Addition. 333
    Magnesium Addition 334
  Economic Evaluation. 335
  References. 339

APPENDIX E: MATERIAL AND ENERGY BALANCES. 341
  Process Description 341

Material Balance Calculations: High-Sulfur Coal Case............347
    SO_2 Removal Requirement (Step 1)......................347
    Limestone Requirement/Slurry Preparation (Step 2)..........349
    Humidification of Flue Gas (Step 3)......................351
    Recirculation Loop and Sludge Production (Step 4)...........354
    Total Makeup Water Required (Step 5)....................359
    Mist Eliminator Wash..................................359
Estimation of Energy Consumption: High-Sulfur Coal Case.......360
    Flue Gas Fans..360
    Slurry Recirculation Pumps.............................363
    Reheat of Scrubber Flue Gas...........................366
    Total Energy for FGD System as Percent of Plant Input........366
Material Balance Calculations: Low-Sulfur Coal Case...........366
    Determination of SO_2 Removal Requirement and Bypass Gas
      Fraction (Step 1)...................................368
    Limestone Requirement/Slurry Preparation (Step 2)..........370
    Humidification of Flue Gas (Step 3)......................370
    Recirculation Loop and Sludge Production (Step 4)...........372
    Total Makeup Water Required (Step 5)....................373
    Mist Eliminator Wash..................................376
Estimation of Energy Consumption: Low-Sulfur Coal Case........376
    Flue Gas Fans..376
    Slurry Recirculation Pumps.............................378
Reference..378

APPENDIX F: LIMESTONE UTILITY FGD SYSTEMS IN THE U.S......380

APPENDIX G: MATERIALS OF CONSTRUCTION.................389
Base Metals...389
    Test Programs.......................................389
    Performance, Economic, and Fabrication Considerations.......396
Protective Linings.....................................400
    Resin and Rubber Linings..............................400
    Plastic Linings (Kensington 1978; Furman 1977)............403
    Bricks (Brova 1977; Mellan 1976).......................406
    Ceramics (Gleekman 1978)............................406
    Concrete (Gleekman 1978)............................407
**Operating Experience (PEDCo Environmental 1981; Rosenberg
1980)**..407
    Prescrubbers..408
    Scrubbers..409
    Mist Eliminators.....................................411
    Reheaters..412
    Fans..413
    Ducts...414
    Dampers...415
    Expansion Joints.....................................416
    Stacks...417

Contents and Subject Index

    Pumps. 419
    Storage Silos. 419
    Ball Mills 420
    Spray Nozzles. 420
    Piping 420
    Spray Headers. 421
    Valves. 421
    Tanks 422
    Pond Linings 422
References. 423

METRIC CONVERSIONS. 425

Section 1

Introduction

PURPOSE, SCOPE, AND STRUCTURE OF THE HANDBOOK

Since the early attempts to control emissions of sulfur dioxide (SO_2) in the flue gas of power plants, there has been a pronounced preference for limestone wet scrubbing systems. About 63 percent of the flue gas desulfurization (FGD) systems that are now operational, under construction, or planned for the next 5 years in the utility industry use limestone as the absorption reagent (Laseke, Devitt, and Kaplan 1979).

The widespread acceptance of limestone systems is due to several factors. Detailed cost studies by TVA (McGlamery et al. 1980) indicate that both the capital and the annual operating costs are competitive with those of other FGD systems designed for high-sulfur coal applications. Additionally, the on-line experience of full-scale limestone units at utility plants has generated a wealth of operational data, which are being used to enhance system reliability. Advances in sludge disposal technology, such as forced oxidation of the sludge, have enabled utility operators to reduce the volume of sludge to be disposed of and to improve its handling and disposal properties. With continuing technological advances and increasingly wide utilization, limestone scrubbers are a major means of compliance with regulations promulgated by the U.S. Environmental Protection Agency (EPA) for control of SO_2 emissions from power plants.

The performance of early commercial systems often did not reach full potential because design and operational experience was limited. Ten years have passed since the first full-scale systems went into operation, and the experience over that period has generated much information. The aim now is to bring together in one volume--clearly and concisely--the best guidance available for achieving optimum performance from limestone scrubber systems.

Since 1972, an intensive research and development program sponsored by the EPA has been under way in prototype scrubbers at the Tennessee Valley Authority's Shawnee Station. That program has produced a large amount of accurate data on scrubber performance, design, operating parameters, and reliability. In this volume the Shawnee work is drawn on, together with the full-scale commercial experience, for information regarding best engineering practice.

Purpose

The information presented here provides guidance to those involved in selecting, installing, and operating a limestone FGD system. For the utility Project Manager, it provides the background needed to select a limestone scrubber, develop the configuration most appropriate for the site, and prepare specifications for procurement of system hardware and services. Additionally, the handbook gives information on installing, testing, operating, and maintaining the system.

Figure 1-1 shows the principal alternative FGD processes. It is assumed throughout the remainder of this handbook that the process selected from among those shown in Figure 1-1 is the limestone wet scrubbing process that generates wet sludge as a "throwaway" product or gypsum as a recoverable byproduct.

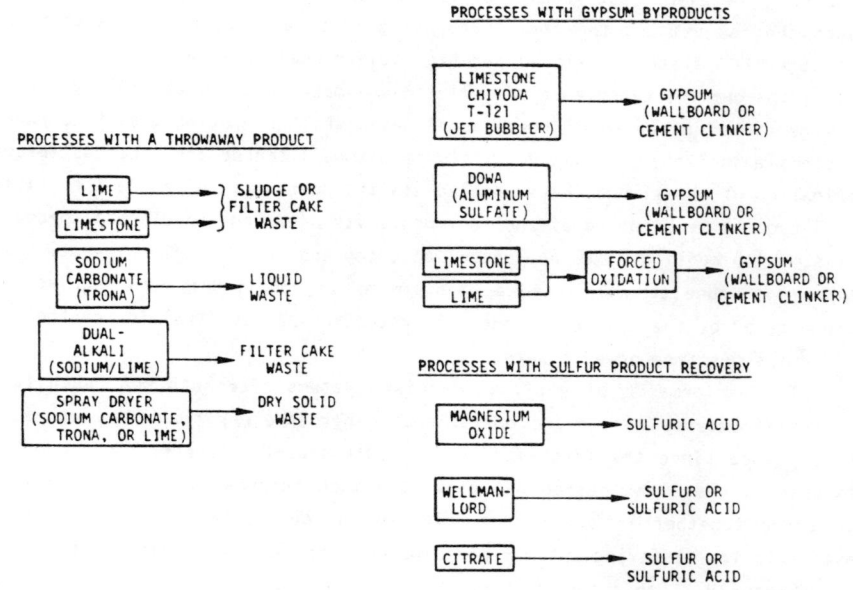

Figure 1-1. Some alternative FGD processes available for commercial application.

The emphasis throughout is on practical applications. For example, the discussion of system design and performance provides the kind of information routinely requested by regulating agencies in permit applications. This information can also be applied in evaluating preliminary studies and recommendations of a consulting architectural/engineering (A/E) firm. Further, it can be used in developing detailed equipment specifications and assessing the performance predictions of various scrubber suppliers.

Scope

The scope of the handbook is the entire limestone FGD system, including the procedures and processes involved in selecting and operating a successful system. Figure 1-2 illustrates a project coordination sequence, from the initial system concept through the purchase of equipment, installation, startup, operation, and maintenance of the limestone scrubbing system. The

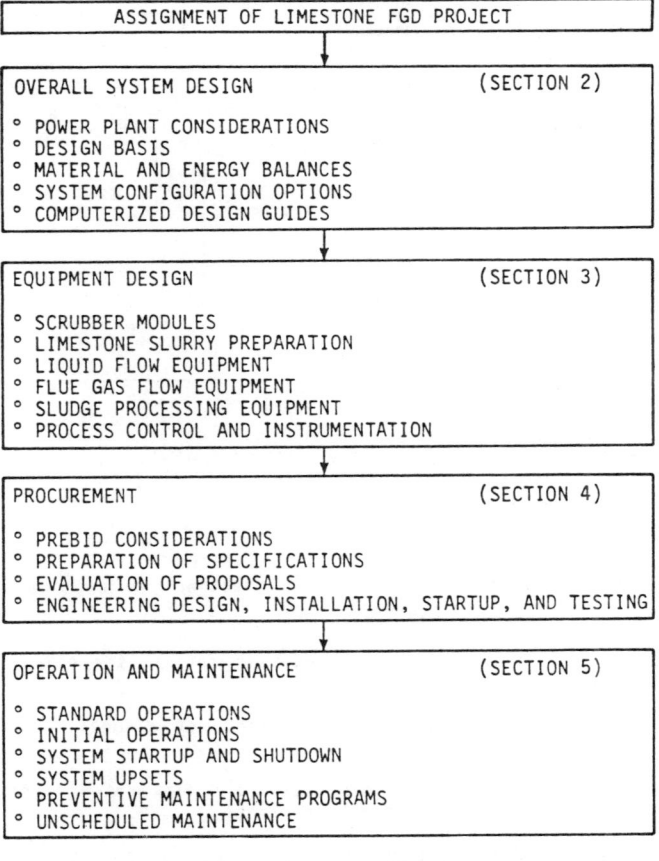

Figure 1-2. FGD project coordination sequence.

goal is successful performance, as measured by compliance with SO_2 emission regulations and by highly reliable day-to-day operation.

Figure 1-3 amplifies the decision factors to be dealt with by a utility Project Manager. Notice that these decision factors are based on an understanding of the basic powerplant design, which affects all of the systems, components, and operations that are to be applied in SO_2 control. Awareness of the optional process features will figure importantly in development of the scrubber system design. The information presented here as an aid to the Project Manager is based in part on current engineering design practices and in part on the records and experience of operational limestone FGD systems. It includes concise guidelines and a detailed review of proven operational procedures and maintenance practices.

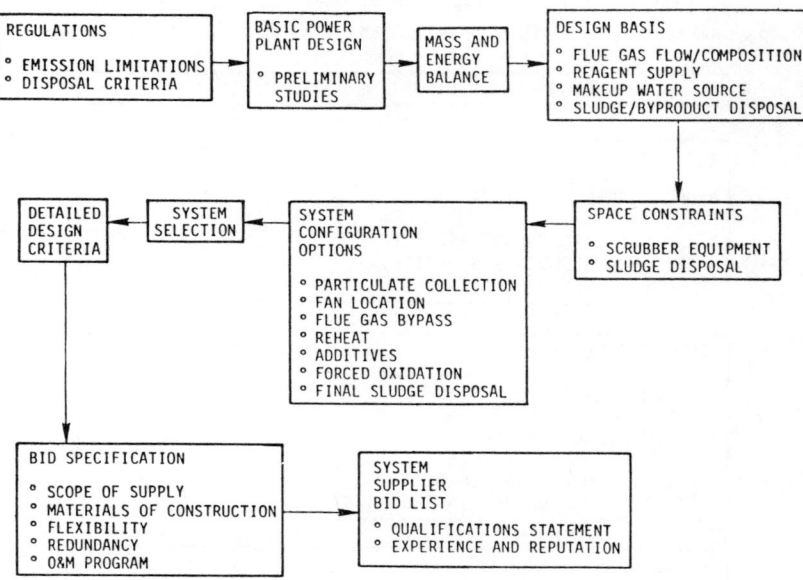

Figure 1-3. Limestone FGD decision sequence.

Following are some of the important questions to be answered by the utility project management team:

- What degree of control is needed to satisfy the emission regulations?

- What criteria should be established for disposal of solids and wastewater?

- What are the flue gas composition and flow rate generated by the "worst case" coal fired at the plant?

- What variability is expected in the coal as received from the supplier?

- What are the availability, chemical composition, and reactivity of the limestone reagent?
- What variability is expected in the limestone as received from the supplier?
- What is the source of the makeup water and what is its expected chemical composition?
- What variability is expected in the makeup water?
- Should the sludge be disposed of as a wet slurry or processed into a dry solid?
- How should wastewater effluent be treated?
- How can the system be designed for a closed-loop water balance?
- Is land available at the site for location of a final sludge disposal area?
- What are the spatial constraints for placement of the scrubber system equipment?
- What should the process design configuration consist of?
- Where will the scrubber system fan be placed?
- How will fan placement affect the final design requirements?
- Should the scrubbing system follow equipment for collection of dry fly ash?
- What type of scrubber and scrubber internals should be used?
- Should the system use reheat?
- Can the system operate as a partial scrubbing unit that bypasses some of the flue gas to provide reheat?
- What provisions can be made for reheat if the environmental regulations will not permit bypassing?
- Should the system use chemical additives (adipic acid or magnesium oxide) to enhance the limestone scrubbing performance?
- Should the system include forced oxidation to yield a more manageable solid material for disposal or a saleable gypsum byproduct?
- What will be the process control philosophy?
- What generic types of equipment have been proven in limestone scrubbing systems?
- What are the best materials of construction for various equipment items?
- How can flexibility be designed into the system to accommodate future needs?
- Which items should be provided as redundant spares?

- What should be the scope of supply of the various contracts?
- Who are appropriate suppliers of major subsystems and equipment?
- What operating and maintenance practices should be established?

These questions and related matters are dealt with in detail in this handbook. The answers to such questions will dictate the configuration of the limestone scrubbing system and its mode of operation.

Organization

The handbook is structured in accordance with the project coordination sequence shown in Figure 1-2. It is assumed that the Project Manager is undertaking his first assignment involving an SO_2 control system. The first step, therefore, is a survey of the major elements of the project. For this purpose this introduction will provide a review of the fundamentals of limestone scrubbing, including recent advances in scrubbing technology, process chemistry, and key operational factors.

In succeeding sections, the subject matter is addressed in greater detail. The handbook format reflects the probable sequence of project events, from planning/design/procurement to installation/operation/maintenance.

Section 2 deals with overall design considerations, starting with factors that affect selection of an SO_2 control system. This discussion includes a detailed account of the options for system configuration and describes some of the tools available to a Project Manager in conducting the preliminary study; e.g., pertinent computer programs and experience records of operational systems.

Section 3 analyzes the FGD system, the process control options, and criteria for selecting ancillary equipment. This section provides detailed guidelines for completion of the engineering effort. It also can serve as reference material for the Project Manager in overseeing the major engineering aspects of the project. The discussion of equipment in Section 3 identifies generic and specific scrubber system components and is supplemented by operational records.

Section 4 describes the purchase documents required to obtain competitive proposals from qualified bidders and the procedures for evaluating the proposals. Guidance is given also on the project activities that immediately follow the award of contract; these include liaison with vendors and consultants during installation, startup, and performance testing of the FGD system.

Section 5 deals with system operation and maintenance (O&M), both of which are critical to successful performance and thus are significant for the entire project staff. Experience to date has shown that many of the major operational problems with limestone FGD scrubbing systems relate directly to inadequate operation and maintenance practices. The discussions encompass establishment of O&M protocols, relationships of preventive and reactive maintenance, troubleshooting operations, staffing, methods of training, and other major facets of an effective O&M program.

The appendixes consist of supplementary reference material, giving more specific details concerning the topics treated in Sections 2 and 3, together with other useful reference materials. Appendix A is a detailed discussion of process chemistry, addressing the theoretical aspects of scrubber performance. Appendix B focuses on process operation, emphasizing the effects of the key operational factors on system reliability. Appendix C describes three pertinent computer programs: the Tennessee Valley Authority (TVA) design and cost-estimating program; the PEDCo FGD Information System (FGDIS), which constitutes an experience record data bank; and the Bechtel-Modified Radian Equilibrium Program for monitoring gypsum relative saturation levels and gypsum scale formation potential. Appendix D discusses further the potential of advanced limestone scrubbing processes and innovative designs; Appendix E gives examples of detailed calculations of mass and energy balance; Appendix F lists the operating limestone FGD systems in the United States; and Appendix G gives detailed guidance on materials of construction.

OVERVIEW OF THE LIMESTONE SCRUBBING PROCESS

A brief description of the basic limestone FGD process is followed by a review of some of the process options that are now available. This overview also briefly summarizes process chemistry and the key operational factors that affect limestone scrubber performance.

Basic Limestone FGD Process

The basic limestone FGD process is shown schematically in Figure 1-4. The process incorporates proven equipment components and utilizes well-established technology. The throwaway process considered here is relatively simple in comparison with the chemically more complex processes that yield a recoverable product or with other advanced SO_2 control processes now available. Limestone scrubbing systems of this basic type have provided the high SO_2 removal efficiencies that are needed to meet the current New Source Performance Standards and have operated reliably at power plants. These systems have successfully treated flue gas generated from the combustion of coals having wide ranges of sulfur and ash contents. Additionally, limestone systems are the least expensive to maintain and operate among all of the commercially available systems.

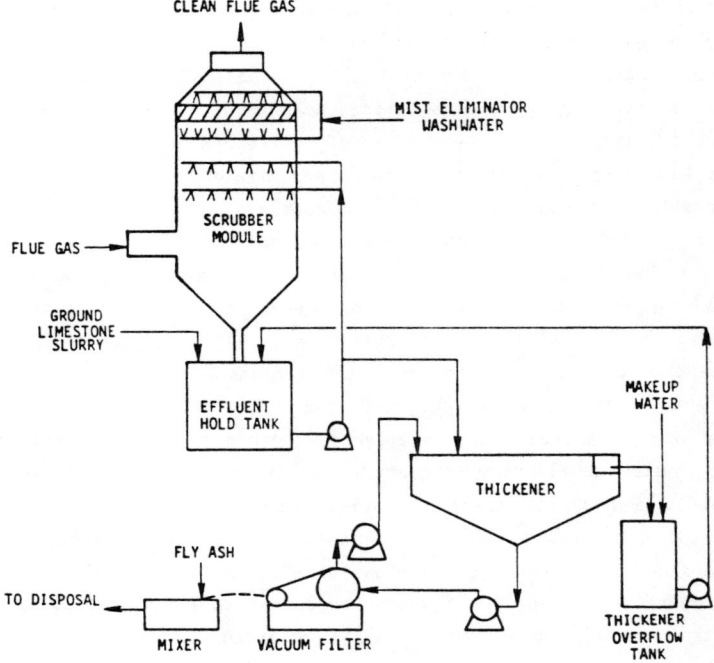

Figure 1-4. Limestone FGD process: basic process flow diagram.

A goal of all wet scrubbing SO_2 control processes is "closed loop" operation, in which fresh water is added to the system only as makeup for water lost by evaporation and that lost with the sludge. EPA personnel have conducted extensive pilot-scale development work on closed-loop systems at the Agency's Industrial Environmental Research Laboratory/Research Triangle Park (IERL/RTP) and have used the Shawnee test facility to demonstrate closed-loop operation of limestone systems. Commercial systems now being installed and planned will operate in a closed-loop mode.

The basic limestone process flow diagram shows incoming flue gas from which fly ash has been removed by treatment in a particulate collection device such as an electrostatic precipitator (ESP) or a fabric filter. The flue gas is brought into contact with the limestone slurry in a simple scrubber tower. Chemical reaction of limestone with SO_2 in the flue gas produces waste solids, which must be removed continuously from the scrubbing loop. These waste solids are concentrated in a thickener and then dewatered in a vacuum filter to produce a filter "cake," which is mixed with fly ash. The resulting stabilized mixture is then transported to a landfill. The limestone scrubbing system is called a "throwaway" process because the product sludge is disposed of rather than regenerated to recover sulfur.

The limestone FGD process has been enhanced with the advent of recent technology improvements. The utility Project Manager can consider these innovations as options to the basic limestone process. Selection of the final system obviously depends greatly on factors specific to an individual project.

Optional Process Features

The ability of optional process features to enhance the limestone scrubbing process is being verified by current commercial experience. The options discussed here are offered by manufacturers of scrubbing systems; they do not constitute an inclusive list but represent a cross section of systems that can be purchased currently. Many have been developed on the basis of research and development at the Shawnee test facility. The introduction of these process features has led to higher SO_2 removal efficiency, greater reliability of operation in a scale-free mode, reduced consumption of limestone reagent, and production of smaller quantities of sludge that can be more easily handled and disposed of. Some of these improvements involve additional equipment items or reagent additives. To the extent possible, schematic diagrams of these optional features are superimposed on the basic process flow diagram to indicate the location and nature of the process change. Most of the equipment items in these figures are discussed in detail in later sections of the manual.

Presaturation is a process feature that is required in some cases to cool the hot flue gas to protect the materials that line the scrubber vessel. Cooling is accomplished by quenching the hot gas with scrubber slurry. Evaporation of the water cools the flue gas to approximately 125°F and saturates it with water vapor. When saturation is effected in an inlet section of the main scrubbing vessel, that portion of the vessel is known as the quencher. When saturation takes place in an external vessel or section of the incoming flue gas ductwork, the external vessel or ductwork section is called the presaturator. Figure 1-5 depicts a presaturator. The presaturator can also serve as a prescrubber for secondary particulate collection and removal of sulfur trioxide and/or hydrogen chloride ahead of the main scrubbing vessel. The prescrubber then can serve as a separate loop to collect the chloride and isolate it from the principal scrubbing loop.

Figure 1-6 shows a venturi scrubber placed ahead of the main scrubber module. The venturi scrubber is an efficient device for removing particulates but not for removing SO_2 from a flue gas stream. Use of the venturi provides a means of removing any particulates remaining in the flue gas after it has passed through the particulate collection system. Installation

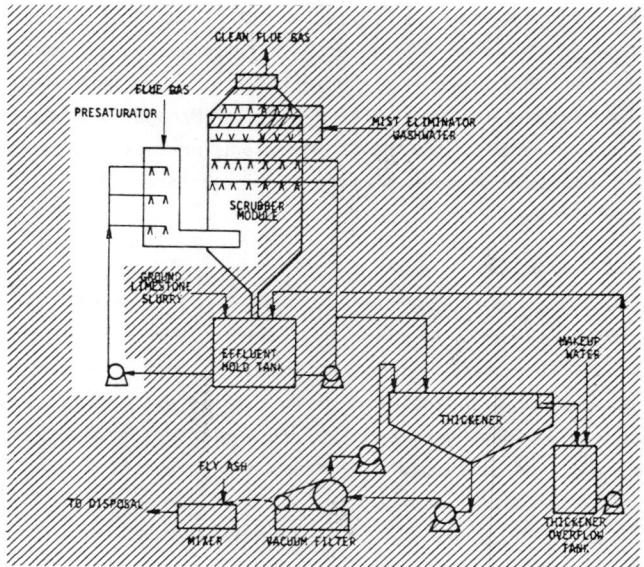

Figure 1-5. Limestone FGD process: external presaturator.

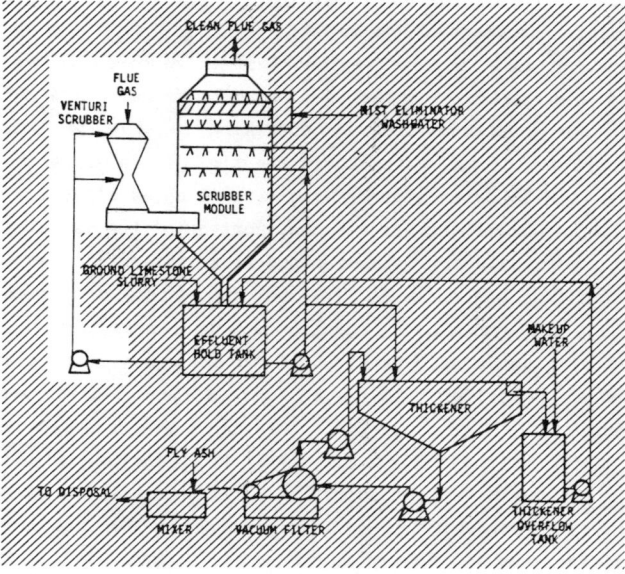

Figure 1-6. Limestone FGD process: venturi scrubber.

of a venturi scrubber is often recommended as best available retrofit technology (BART) for effective control of particulate emissions.

Collection of fine particulates is difficult for a venturi scrubber operating at a reasonable pressure drop. A venturi scrubber can serve efficiently, however, as a primary collector followed by an SO_2 scrubber and a charged particulate separator. Figure 1-7 shows such a system, which is offered by at least two system suppliers (Combustion Engineering and Peabody Process Systems). The low-pressure-drop venturi provides primary collection of particulates (approximately 90 percent removal). The spray tower scrubber then removes the SO_2, and the charged particulate separator provides final collection of particulate and scrubber carryover.

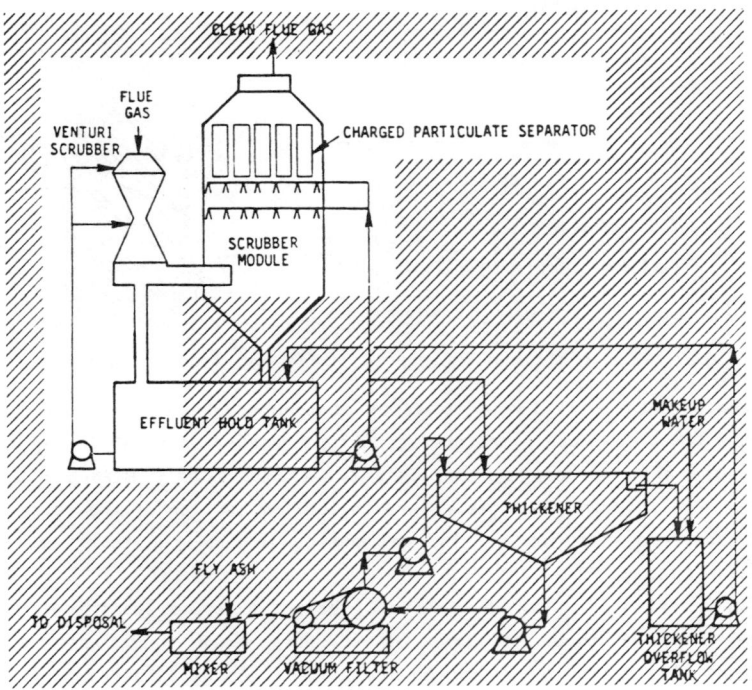

Figure 1-7. Limestone FGD process: venturi spray-tower scrubber with a charged particulate separator.

Experimentation at the Shawnee test facility has shown that utilization of the limestone reagent is greatly improved when an additional tank is placed in the process flow to collect and recycle the venturi scrubber slurry and to keep this slurry separate from the main scrubbing vessel (Head 1977). Figure 1-8 shows the type of installation that constitutes a two-loop limestone FGD system.

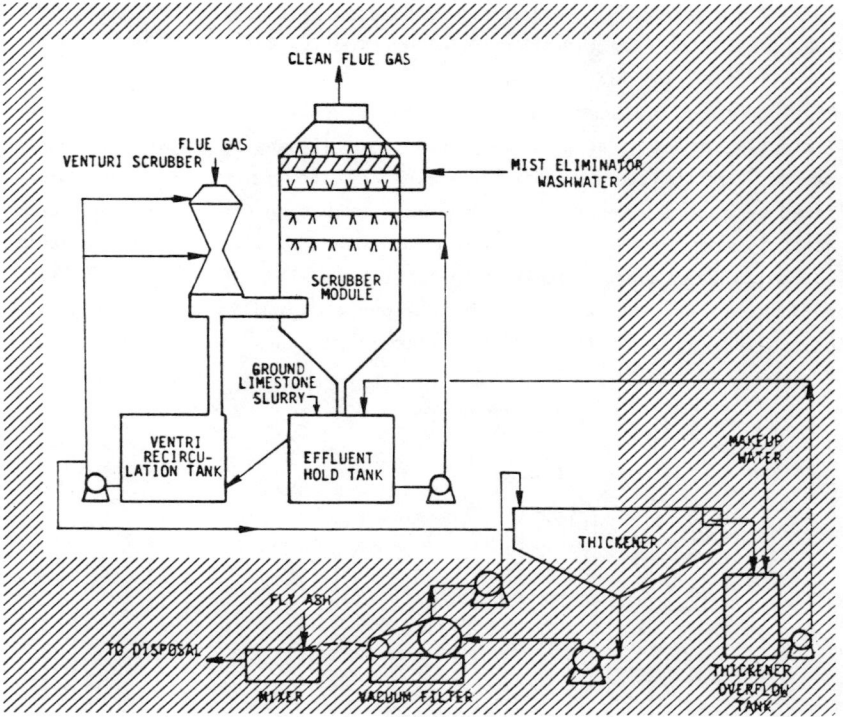

Figure 1-8. Limestone FGD process: EPA two-loop scrubbing.

Concurrently with the experiments at Shawnee, Research-Cottrell developed their Double Loop® Limestone Scrubbing Process, as shown in Figure 1-9. Unlike the EPA two-loop system, this process uses only a single scrubber module. The incoming flue gas enters the bottom section of the tower, where it is quenched; the cooled flue gas passes through the tower, where SO_2 removal occurs.

An important innovative feature of both two-loop systems is separation of the prescrubber/quencher from the scrubber slurry loop, which enhances limestone utilization by promoting low pH in the prescrubber/quencher loop. The two-loop process confines chloride ions to the prescrubber/quencher loop; thus special materials of construction are needed only in this loop. Additionally, separation of the loops permits operation of the scrubber loop in a gypsum-subsaturated mode and enhances oxidation in the prescrubber/quencher loop, thus permitting production of a gypsum byproduct.

The equipment item that permits this separation in the system is the hydrocyclone, which feeds the solids forward to the quencher loop from the scrubber effluent hold tank and returns the clear liquor to the scrubber loop. This procedure applies to high-sulfur applications; in low-sulfur

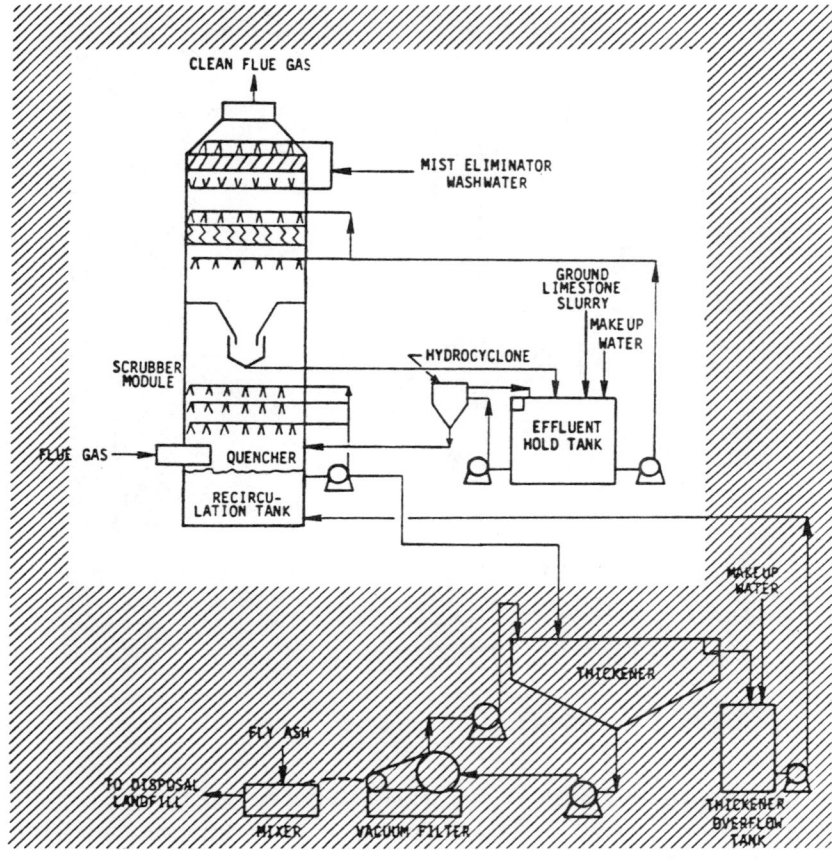

Figure 1-9. Limestone FGD process: RC Double Loop® scrubbing.

applications the discharge mode is reversed. A two-loop system thus maintains a closed-loop water balance while improving limestone utilization and achieving high SO_2 removal efficiency.

Peabody Process Systems has recently applied for a patent on a system that uses a hydrocyclone to increase process stability and reliability by permitting the system to "control" the chemistry in the critical mist elimination area. Figure 1-10 shows the Peabody system. The hydrocyclone separates unreacted limestone in the recycle slurry from the calcium sulfite and calcium sulfate solids. Because no limestone is available to react with SO_2, plugging cannot occur. Thus, recycled liquor can be used for mist eliminator wash without upsetting the closed-loop water balance. Additionally, the hydrocyclone is used to achieve 100 percent limestone utilization (Johnson 1978). Research Cottrell has recently been awarded a patent on a similar system.

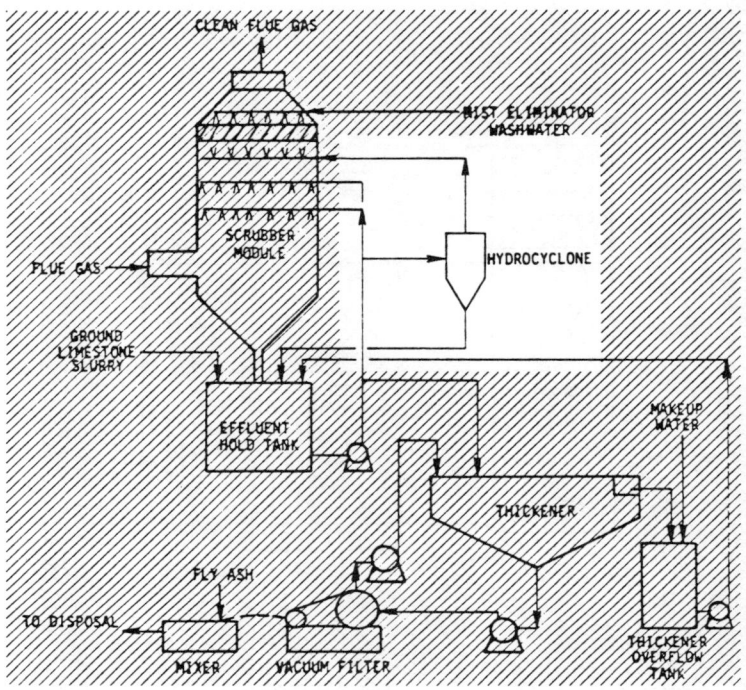

Figure 1-10. Limestone FGD process: hydrocyclone used to provide mist elimination wash liquor.

Other development work at the Shawnee test facility has shown that the volume of sludge is substantially reduced by forced oxidation. In addition, forced oxidation improves the physical properties of the sludge so that it need not be mixed with fly ash for disposal (Head and Wang 1979). Secondary benefits of forced oxidation are increased utilization of limestone and improved control of scaling in single-loop limestone scrubbing systems (Borgwardt 1978). Figure 1-11 shows a scheme for forced oxidation by aerating the scrubber effluent hold tank of a single-loop system.

Forced oxidation can be incorporated into a two-loop scrubbing process to produce gypsum for use in wallboard manufacture. In this system a second hydrocyclone concentrates the solids to produce a wet gypsum underflow and an overflow that is returned to the quencher (Braden 1978).

Limestone scrubbing technology has been improved not only with equipment changes but also by the use of chemical additives. Figure 1-12 shows a process in which either adipic acid [$HOOC(CH_2)_4COOH$] or magnesium oxide (MgO) is added to the scrubber slurry. Recent EPA tests indicate that low concentrations (600 to 1500 ppm) of adipic acid will improve SO_2 absorption (Head et al. 1979). Addition of adipic acid, which is an advanced process

Introduction 15

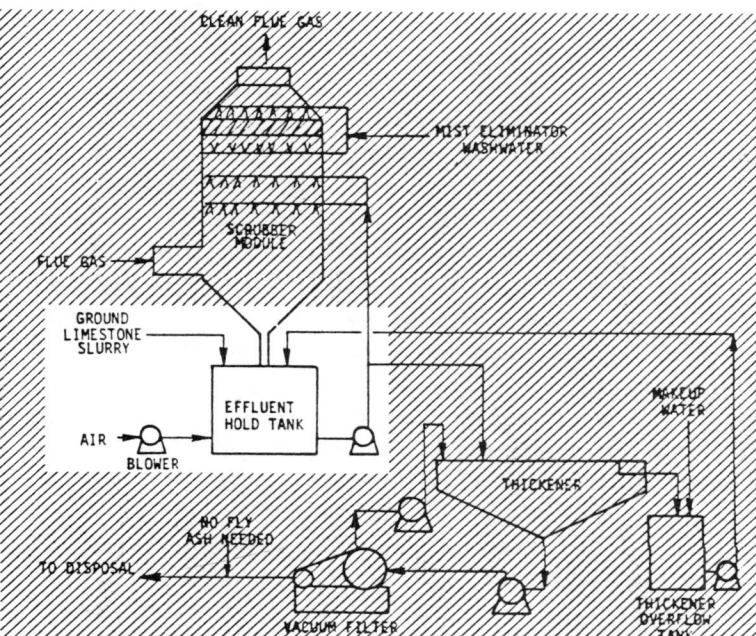

Figure 1-11. Limestone FGD process: forced oxidation of scrubber hold tank in a single-loop system.

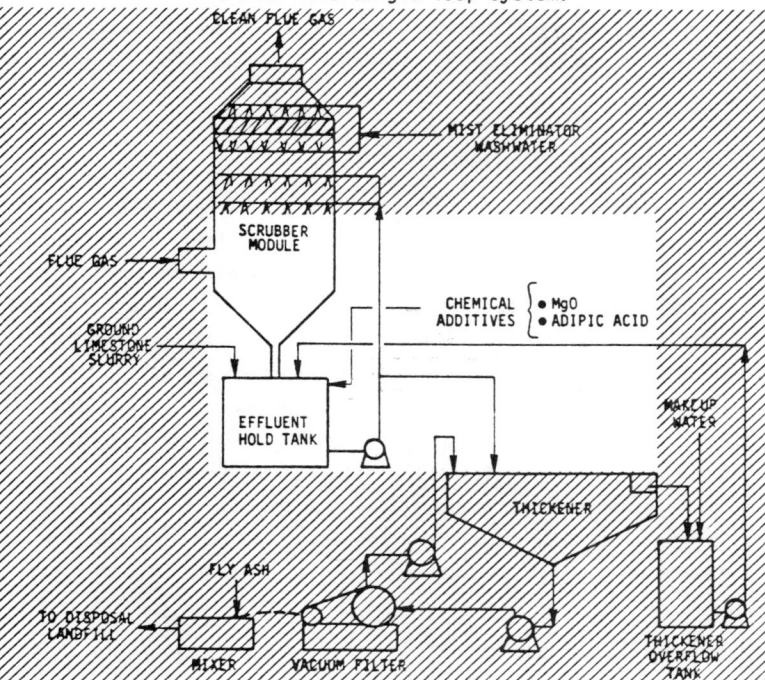

Figure 1-12. Limestone FGD process: chemical additives.

option, has been shown to be effective in conjunction with forced oxidation and in the presence of chloride ions. It improves scrubber operating conditions and can be used with forced oxidation in single-loop scrubbers. Because there is no chloride interference, addition of adipic acid is effective in systems with the most tightly closed loops. To verify the commercial viability of adipic acid as a process additive, the EPA is demonstrating this process on the limestone FGD system in service at the 200-MW Southwest No. 1 Station of City Utilities, Springfield, Missouri. Early results of this demonstration have verified results from the Shawnee test facility. The use of adipic acid has been cost-effective because the higher utilization of limestone and lower sludge volume offset the cost of the additive (Burbank 1980).

Alkali additives like magnesium oxide also enhance the capacity of the scrubbing slurry to absorb SO_2 by buffering the pH. Experiments with magnesium additives at the Shawnee facility have resulted in higher SO_2 removal efficiencies in both single- and two-loop systems (Head et al. 1977). Pullman Kellogg offers commercial, magnesium-buffered limestone FGD systems that use the Weir horizontal crosscurrent spray scrubber.

The optional features discussed here represent some of the more promising of current process modifications. The basic limestone FGD process is, of course, amenable to further variations that may substantially enhance overall operation and reliability. Moreover, the use of several of these features in combination may be practicable, depending on site-specific factors.

PROCESS CHEMISTRY AND OPERATIONAL FACTORS

Process chemistry and operational factors are closely interrelated. For example, the ready availability of dissolved alkaline species (neutralizing capacity) in the scrubbing liquor can yield an energy savings by reducing the required liquid-to-gas (L/G) ratio. In operation at a low stoichiometric ratio, the resulting high utilization of limestone reduces plugging and fouling of mist eliminators. Control of solids concentration in the slurry reduces the potential for scaling in the scrubber vessel. Scaling potential is reduced also by establishment of optimum residence times of slurry in the scrubber effluent hold tank to dissolve the limestone and precipitate the calcium solids. These and other important relationships between process chemistry and operational factors are described briefly in the following discussion, which also provides guidelines for design and operation.

Type and Grind of Limestone

The promotion of good process chemistry begins with selecting the limestone reagent and determining how fine to grind it. The limestone recommended for scrubber applications is a high-purity material containing 90 percent or more calcium carbonate ($CaCO_3$) and less than 5 percent magnesium carbonate ($MgCO_3$). Scrubber operation has demonstrated that the finer the grind, the better is the utilization of limestone, especially when the goal is 90 percent SO_2 removal (Rochelle 1980). The grind should be selected to maximize the limestone utilization, i.e., the portion of the limestone fed to the scrubber that is actually used to neutralize the SO_2 and other acidic species absorbed from the flue gas. Recent work at the Shawnee test facility has shown that a good limestone grind for use in a typical limestone wet scrubbing system would be one in which 90 percent by weight passes through a 325-mesh screen. Decreasing the limestone particle size increases the rate of dissolution of solids in the scrubber and the overall rate of SO_2 removal. For a given degree of SO_2 removal, reduction of the limestone particle size decreases the required pH setpoint of the scrubber recirculation slurry liquor and improves the limestone utilization.

Good limestone utilization ($\geq$ 85 mol percent) is related to the efficiency of SO_2 removal by the limestone scrubbing unit. High limestone utilization is generally attained at relatively low pH's, and scrubbers must therefore be designed for efficient mass transfer.

Stoichiometric Ratio and pH

Stoichiometric ratio (SR) is defined as the ratio of the actual amount of SO_2 absorbing reagent, usually calcium carbonate ($CaCO_3$), in the limestone fed to the scrubber to the theoretical amount required to neutralize the SO_2 and other acidic species absorbed from the flue gas. The neutralizing capacity of the scrubbing liquor, which is directly related to the pH level, can be increased by increasing the SR up to a limit of about 1.2 and maintaining pH at 5.8. Beyond this limit scaling can occur in the scrubber.

Control of pH is essential to reliable scrubber operation. Appendix A explains in detail that the chemical reactions in a scrubbing system must be confined to separate parts of the system. The absorption of SO_2 in the flue gas must take place in the scrubber vessel, whereas the neutralization and precipitation reactions must occur primarily in the effluent hold tank.

Stoichiometry and Mist Eliminator Fouling

Stoichiometric ratio has a great impact on performance of the mist eliminator. Low SR values reflect high limestone utilization; a low SR prevents fouling and plugging of the mist eliminators and thus enhances the system reliability. Testing at the Shawnee facility has demonstrated this relationship. Operation has been most successful at SR values not exceeding 1.18.

SO_2 Removal and pH

Commercial experience has shown that 5.8 is the maximum practical pH level for high SO_2 removal efficiency in a single-loop system. When this pH level is exceeded, the greater quantity of limestone dissolved in the system can lead to problems with process chemistry. Increasing the pH increases the neutralizing capacity of the slurry, but use of too much limestone is a principal cause of scale formation. An operating limestone scrubbing system must achieve proper balance of limestone consumption, SO_2 removal, and pH level.

SO_2 Removal and L/G

The major operational means of achieving the required degree of SO_2 removal is maintenance of proper L/G ratio. The ratio of liquid flow rate to gas flow rate in the SO_2 scrubber is expressed as gallons of slurry flow per 1000 actual cubic feet of flue gas flow at scrubber outlet conditions. The primary effect of higher liquid flow rate, or higher L/G ratio at a given gas flow rate, is to increase the SO_2 removal efficiency. The minimum L/G ratio required for a given degree of SO_2 removal can be decreased by increasing the neutralizing capacity of the recirculation slurry, which can be effected, within limits, by increasing the limestone stoichiometric ratio.

A high L/G ratio is normally needed for efficient SO_2 removal at the low pH control setpoint ($\leq$ 5.8) and low SR ($\leq$ 1.18) that must be maintained to minimize fouling of the mist eliminator.

L/G Ratio and Liquid Holdup

The operational L/G ratio is affected by holdup of liquid in the scrubbing vessel. Liquid holdup is achieved by placement of various internal structures such as packing, rods, or trays in the scrubber. All scrubbers except venturi and open spray towers incorporate some type of internal device to impede the free fall of slurry and thereby prolong the contact of slurry with the gas stream. The term "tower internals" refers to these internal structures.

Increasing the holdup of liquid slurry can increase SO_2 removal. Alternatively, operators can achieve a required level of SO_2 removal efficiency by providing a means for slurry holdup and reducing the liquid flow rate. This procedure, however, increases the gas-side pressure drop and the horsepower requirements of system fans. The gas pressure drop through the scrubber depends on the operating gas velocity and the type of scrubber design. The pressure drops in the scrubber and in the mist eliminator usually constitute most of the total system gas-side pressure drop and are major factors in the energy consumption of the system. A sudden increase in pressure drop across the scrubber, mist eliminator, or reheater usually indicates the deposition of solids on scrubber internals.

The trade-off between gas-side pressure drop and L/G ratios has a major impact on the total energy demand of the scrubber system. Sulfur dioxide removal can be increased by increasing the system power consumption, which is achieved by increasing the liquid flow rate and/or the gas pressure-side drop.

A well-designed limestone FGD system must be able to operate over a wide range of gas flow rates. The ratio of maximum to minimum gas flow that a scrubber can handle without unstable operation or a reduction in SO_2 removal efficiency is known as turndown capability. The use of parallel scrubber modules provides overall stepwise turndown capability. The use of staging pumps and installation of multiple sprays, gas dampers, and baffles can provide turndown capability for individual modules of the system.

Liquid Phase Alkalinity and L/G

The available alkalinity in the scrubbing liquor, which is supplied by the dissolving limestone, continuously neutralizes the absorbed SO_2.
Removal of the SO_2 as a neutralized product allows for the continuing transfer of additional SO_2 from the gas phase to the liquid phase. The available liquid-phase alkalinity (LPA) thus provides a driving force for the absorption and neutralization of SO_2. The higher the LPA, the higher is the driving force for SO_2 removal.

The availability of alkaline species in the liquid phase is critical. In addition to absorbing SO_2 from the flue gas, the scrubbing liquid absorbs hydrogen chloride (HCl), which is formed from combustion of the chloride in the coal. When the scrubber liquid accumulates substantial amounts of chlorides, less alkalinity is then available for SO_2 removal. The impact of HCl absorption on limestone consumption is minimal except when the sulfur content of the coal is very low and the chloride content is high. Addition of soluble alkalis or buffering agents such as MgO or adipic acid increases the available LPA at a given pH level and thus reduces the impact of chlorides on the SO_2 removal process.

The primary operational means of increasing available alkalinity is to increase the liquid flow rate. Increasing the volume of liquid per unit flow of gas allows more alkaline neutralizing species to contact acidic components of the flue gas. Increasing the liquid flow rate, however, also increases the energy demand of the scrubbing system. Therefore, an optimum balance of the available alkalinity, L/G ratio, energy consumption, and SO_2 removal must be determined.

Scrubber Effluent Hold Tank and Relative Saturation

Design of the scrubber effluent hold tank is very important in the control of relative saturation. All of the precipitation chemical reactions should occur in the hold tank, as well as the dissolution of limestone. The hold tank in a single-loop system should be sized to allow a minimum of 8 minutes residence time of the recirculation slurry flow to ensure optimum limestone utilization and precipitation of solids as gypsum. A residence time of 5 to 7 minutes in a double-loop hold tank will achieve the same objective.

In limestone FGD systems, the term "relative saturation" (RS) pertains to the degree of saturation (or approach to the solubility limit) of calcium sulfite and sulfate in the scrubbing liquor; RS is important as an indicator of scaling potential, especially of gypsum scaling, which can become severe. Relative saturation is defined as the ratio of the product of calcium and sulfate ion activities to the solubility product constant. The solution is subsaturated when RS is less than 1.0, saturated when RS equals 1.0, and supersaturated when RS is greater than 1.0. Generally a limestone scrubbing system will operate in a scale-free mode when the RS of gypsum is maintained below a level of 1.4 and the RS of calcium sulfite is maintained below a level of approximately 6. Operation below these levels provides a margin of safety to ensure scale-free operation.

Prevention of Scale Formation

Scale formation can lead to the accumulation of solids on scrubber internals and ultimately can lead to shutdown of a module or of the entire system. Scaling can also cause instrument malfunction and loss of process control.

Scaling can be prevented by close control of the pH of the scrubber inlet liquor, stoichiometric ratio, and relative saturation of calcium sulfate in the scrubber inlet liquor. As indicated earlier, control of these parameters is achieved by providing sufficient residence time in the effluent hold tank, optimizing the L/G ratio, and maintaining a proper solids content in the scrubbing slurry. The solids content of recirculation slurry should be not less than 8 weight percent in a fly-ash-free limestone scrubbing system and 15 weight percent in a scrubbing system that simultaneously removes fly ash.

Degree of Oxidation

The degree of oxidation of sulfite to sulfate in the FGD system affects the precipitation of the solid reaction products. In general, gypsum ($CaSO_4 \cdot 2H_2O$) begins to precipitate in limestone FGD systems when the degree of oxidation exceeds about 16 percent (sulfite to sulfate on a molar basis).

The degree of oxidation generally increases with an increase in the ratio of O_2 to SO_2 in the inlet flue gas, an increase in the concentrations of certain trace metals (such as manganese) in the scrubbing liquor, and a decrease in the pH level of the scrubbing liquor. Various techniques have been developed for forced oxidation of scrubber systems to improve sludge dewatering characteristics and thereby to reduce the total volume of sludge generated.

Chloride Removal

Because of operational and regulatory considerations, scrubber system designers should consider the available options for reducing high chloride concentrations in the scrubbing liquor. The use of a prescrubber or of a two-loop system can confine the chlorides to a single loop, so that the concentrated chlorides can be removed from the scrubbing system. Some methods currently used to permit chloride liquor blowdown are effective but are not recommended "closed-loop" practice; these methods include sluicing the blowdown through the bottom-ash pond, or diluting it in the plant wastewater system, or controlling the dewatering of sludge so as to maximize its liquor content.

In the future, more stringent enforcement of zero-effluent discharge regulations may necessitate the use of feasible but costly control options, such as vapor-compression evaporation. Vapor compression can evaporate the concentrated stream of chlorides and reduce them to a soluble salt product. Appendix B describes this technique, which is currently used in power plants to evaporate cooling tower blowdown and will soon be tried on the blowdown liquor from FGD systems.

Equipment Considerations

Some of the equipment considerations that should be addressed in the design phase to facilitate maintenance of the scrubbing system are access to internals and effluent hold tanks and to the spray headers and nozzles. In addition, the design should incorporate such features as sootblowers near the wet/dry interfaces and in-line reheater tubes, sprays for mist eliminator wash, and strainers in the effluent hold tanks to prevent plugging and erosion of piping and spray nozzles.

REFERENCES FOR SECTION 1

Borgwardt, R. H. 1978. Effect of Forced Oxidation on Limestone/SO_x Scrubber Performance. In: Proceedings of the Symposium on Flue Gas Desulfurization, Hollywood, Florida, November 1977. Vol. I. EPA-600/7-78-058a. NTIS No. PB-282 090.

Braden, H. B. 1978. Double-Loop Operating Offers "Best of Both Worlds" Approach to Sulfur Dioxide Scrubbing. Public Utilities Fortnightly, August 17, 1978.

Burbank, D. A., S. C. Wang, and R. R. McKinsey. 1980. Test Results on Adipic Acid-Enhanced Limestone Scrubbing at the EPA Shawnee Test Facility--Third Report. Presented at the Symposium on Flue Gas Desulfurization, Houston, Texas, October 28-31, 1980.

Head, H. N. 1977. EPA Alkali Scrubbing Test Facility: Advance Program, Third Progress Report. EPA-600/7-77-105. NTIS No. PB-274 544.

Head, H. N., and S. C. Wang. 1979. EPA Alkali Scrubbing Test Facility: Advance Program, Fourth Progress Report. EPA-600/7-79-244a. NTIS No. PB80-117906.

Head, H. N., et al. 1977. Results of Lime and Limestone Testing With Forced Oxidation at the EPA Alkali Scrubbing Test Facility. In: Proceedings of the Symposium on Flue Gas Desulfurization, Hollywood, Florida, November 1977. Vol. I. EPA-600/7-78-058a. NTIS No. PB-282 090.

Johnson, C. 1978. Minimizing Operating Costs of Lime/Limestone FGD Systems. Power Engineering, 82(2):62-65, February 1978.

Laseke, B. A., T. W. Devitt, and N. Kaplan. 1979. Status of Utility Flue Gas Desulfurization in the United States. Presented at the 72nd Annual AIChE Meeting, San Francisco, California, November 1979.

McGlamery, G. G., et al. 1980. FGD Economics in 1980. Presented at the Symposium on Flue Gas Desulfurization, Houston, Texas, October 28-31, 1980.

Rochelle, G. T. 1980. May Monthly Progress Report on Limestone Type and Grind Project. EPA Grant R. 806251 done at University of Texas at Austin.

Section 2

Overall System Design

This section describes a systematic approach to design of a limestone FGD system. Consideration of the major design factors outlined here will enable the utility project team, together with their A/E consultants, to formulate an overall system configuration. Detailed analysis of the system components and related equipment is given in Section 3.

Figure 2-1 shows the major elements to be considered in design of the overall system. The discussion that follows is presented in the order shown in the figure. Analysis of design factors begins with those related to the powerplant: coal properties and supply, steam generator design, power generation demand, site-related factors, and environmental regulations. In establishment of the design basis, the size and configuration of the FGD system are determined by such parameters as flue gas flow and composition, pollutant removal requirements, reagent stoichiometric ratio, limestone composition, and makeup water. Calculation of material and energy balances provides further basis for finalization of the FGD system design. The design details are refined by evaluating and selecting from among a number of system configuration options, such as reheat versus no reheat.

Throughout the design process, the project team may apply some of the computerized design tools that are readily accessible for this purpose. Three major FGD computer programs are outlined briefly in this section and are discussed in detail in Appendix C.

POWERPLANT CONSIDERATIONS

Design of the limestone FGD system will be strongly affected by certain conditions related to the powerplant and its operation. Ideally, an emission control system is planned and developed as an integral part of the powerplant, with the system design based on economic, geographical, governmental, and other factors that affect the overall plant operation. These factors are considered here in five major categories: (1) the powerplant's coal supply, (2) the steam generator design, (3) the power generation demand, (4) the powerplant site, and (5) the applicable environmental regulations. Table 2-1 lists the major powerplant-related considerations.

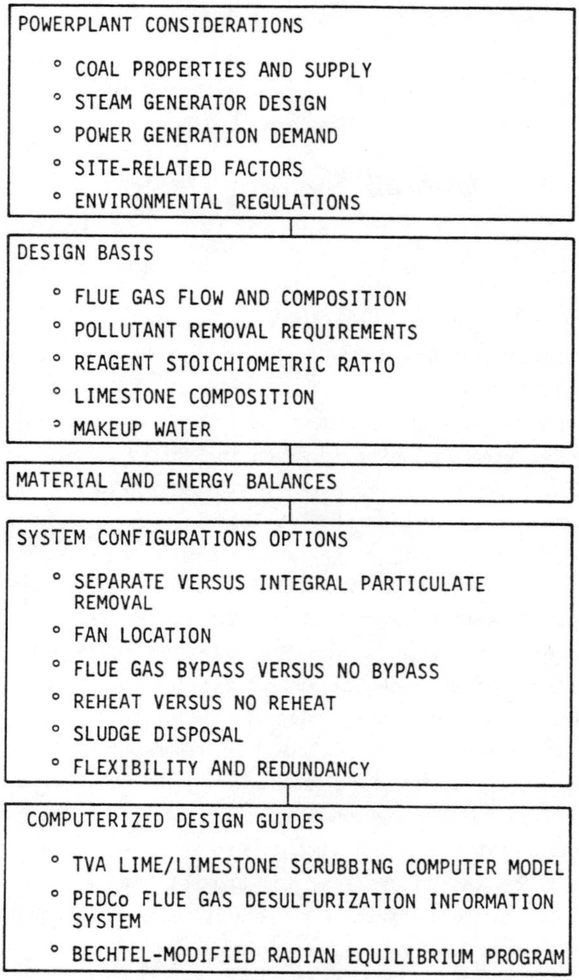

```
┌─────────────────────────────────────────────┐
│ POWERPLANT CONSIDERATIONS                   │
│   ° COAL PROPERTIES AND SUPPLY              │
│   ° STEAM GENERATOR DESIGN                  │
│   ° POWER GENERATION DEMAND                 │
│   ° SITE-RELATED FACTORS                    │
│   ° ENVIRONMENTAL REGULATIONS               │
├─────────────────────────────────────────────┤
│ DESIGN BASIS                                │
│   ° FLUE GAS FLOW AND COMPOSITION           │
│   ° POLLUTANT REMOVAL REQUIREMENTS          │
│   ° REAGENT STOICHIOMETRIC RATIO            │
│   ° LIMESTONE COMPOSITION                   │
│   ° MAKEUP WATER                            │
├─────────────────────────────────────────────┤
│ MATERIAL AND ENERGY BALANCES                │
├─────────────────────────────────────────────┤
│ SYSTEM CONFIGURATIONS OPTIONS               │
│   ° SEPARATE VERSUS INTEGRAL PARTICULATE    │
│     REMOVAL                                 │
│   ° FAN LOCATION                            │
│   ° FLUE GAS BYPASS VERSUS NO BYPASS        │
│   ° REHEAT VERSUS NO REHEAT                 │
│   ° SLUDGE DISPOSAL                         │
│   ° FLEXIBILITY AND REDUNDANCY              │
├─────────────────────────────────────────────┤
│ COMPUTERIZED DESIGN GUIDES                  │
│   ° TVA LIME/LIMESTONE SCRUBBING COMPUTER MODEL │
│   ° PEDCo FLUE GAS DESULFURIZATION INFORMATION  │
│     SYSTEM                                  │
│   ° BECHTEL-MODIFIED RADIAN EQUILIBRIUM PROGRAM │
└─────────────────────────────────────────────┘
```

Figure 2-1. Elements of overall system design.

TABLE 2-1. MAJOR POWERPLANT FACTORS THAT INFLUENCE FGD SYSTEM DESIGN

Coal Properties and Supply
    Sulfur content
    Ash content
    Fly ash composition
    Chloride content
    Moisture content
    Heating value
    Availability of coals
    Transportation considerations
    Flexibility for firing alternative coals

(continued)

Steam Generator Design
> Type of steam generator
> Size of steam generator
> Flue gas (Note that the following items are also related to
>   properties of the coal)
>   weight flow rate
>   volume flow rate
>   temperature
>   dewpoint
>   fly ash loading
> Additional control equipment

Power Generation Demand
> Base load
> Cycling load
> Intermediate load
> Peak load

Site Conditions
> Land availability
> Soil permeability
> Disposal facility
> Climatic and geographic effects
> Quality and availability of limestone and makeup water

Environmental Regulations
> New Source Performance Standards
> Disposal site standards
>   Resource Conservation and Recovery Act (sealed site)
> Effluent discharge standards
>   Clean Water Act (closed loop)

Coal-Related Factors (Properties and Supply)

Properties of the coal to be burned affect the design of components and operational characteristics of a powerplant. The combustion characteristics and physical properties of the coal affect the design of the steam generator, air heaters, primary air fans, pulverizers, and coal-handling system, as well as the maximum volume of flue gas to be treated in the FGD system. The chemical composition of the coal affects the ash-handling system, electrostatic precipitators, limestone scrubbers, limestone handling systems, sludge-handling systems, and miscellaneous equipment (Galluzzo and Davidson 1979).

The ash and sulfur contents of the coal, together with the applicable environmental regulations, determine the required efficiency of particulate and SO_2 control; the coal's chloride content affects the materials of construction and the ability of a limestone scrubber to operate in the closed-loop mode.

Table 2-2 lists the properties of four widely used types of coal from different locations in the United States. This table indicates the broad range of properties characteristic of these coal supplies (note, for

TABLE 2-2. FUEL PROPERTIES OF FOUR REPRESENTATIVE COALS[a]

	Wyoming subbituminous		Montana subbituminous		Illinois (raw) bituminous		Gulf Coast lignite	
	Average	Range	Average	Range	Average	Range	Average	Range
Proximate analysis								
Moisture, %	30	28-32	24	23-25	11	10-14	32	30-35
Volatile matter, %	32	30-33	32	30-32	37	32-42	29	b
Fixed carbon, %	32	30-34	40	40-41	42	37-47	27	b
Ash, %	6	5-8	4	3.5-6.0	10	10-11	12	10-15
Heating value, Btu/lb	8,000	7,800-8,200	9,200	8,300-10,000	11,000	10,800-11,200	6,800	6,200-7,000
Ultimate analysis (as received), %								
Carbon	48	46-49	55	53-56	60	57-63	41	b
Hydrogen	3	3-4	4	3-4	4	4-5	3	b
Nitrogen	0.6	0.5-0.7	0.9	0.8-1.0	1.0	1.1-1.2	0.7	b
Sulfur	0.5	0.3-0.7	0.5	0.4-0.7	3.9	3.7-4.1	0.8	0.6-0.9
Chlorine	0.03	0.00-0.05	0.01	0.00-0.02	0.10	0.04-0.60	0.1	0.0-0.2
Oxygen (difference)	11.87	9-12	11.59	11-12	10.0	10-11	10.4	b
Ash analysis, %								
SiO_2	32	28-36	32	25-40	47	42-51	49	b
Al_2O_3	15	14-17	17	15-20	23	18-24	21	b
Fe_2O_3	5	4-5	6	5-9	16	15-21	7	b
TiO_2	1	0.9-1.4	1	1.0-1.1	1	0.8-1.2	1.5	b
P_2O_5	1	0.2-1.3	1	0.5-1.0	0.1	0.0-0.1	0.3	b
CaO	23	19-27	15	10-20	6	2-6	9	b
MgO	5	4-6	5	3-5	0.1	0.1-0.5	2	b
Na_2O	1	1.0-1.6	7	6-9	1.0	0.9-1.3	2	1-3
K_2O	0.4	0.3-0.6	0.5	0.4-0.6	1.6	1.5-1.7	0.7	b
SO_3	16	14-19	15	10-20	4	3-5	7	b
Undetermined	0.6		0.5		0.2		0.5	b
Grindability, Hardgrove index	64	60-70	48	45-50	56	54-58	55	50-60

[a] Galluzzo and Davidson 1979.
[b] Data not available.

TABLE 2-3. COMBUSTION CHARACTERISTICS FOR A 500-MW CASE[a]

	Wyoming subbituminous	Montana subbituminous	Illinois (raw) bituminous	Gulf Coast lignite
Slagging potential	High	High	High	Severe
Fouling potential	Medium	Severe	High	High
Erosion potential	Low	Low	Low	Medium
Combustion air, 10^6 acfm[b]	1.49	1.57	1.34	1.57
Flue gas, 10^6 acfm[c]	2.24	2.24	2.01	2.39
Maximum ash, tons/h	26.9	17.2	26.1	64.9
Maximum ash, lb/10^6 Btu	10.3	6.7	10.2	24.2
Minimum sulfur, lb/10^6 Btu	0.4	0.4	3.3	0.9
Maximum sulfur, lb/10^6 Btu	0.9	0.6	3.8	1.5

[a] Adapted from Galluzzo and Davidson 1979.
[b] At 30 percent excess air.
[c] At 15 percent air heater leakage.

example, the ranges of heating values and of sulfur, ash, and moisture contents). Table 2-3 lists the combustion characteristics of the four representative coals described in Table 2-2 for a hypothetical 500-MW generating station.

The diversity of coal supply sources for a coal-fired powerplant must be considered in terms of the performance of both the steam generator and the associated limestone FGD system. Building flexibility into the scrubbing system is of utmost importance at the design stage. Although the properties of coals are diverse, designers can incorporate a high degree of flexibility so that substantially different coals may be fired without diminishing the efficiency of the SO_2 control system.

Steam Generator Design

The scope of this handbook does not include guidelines for selection of the most suitable steam generator for a powerplant. The type and size of steam generator are determined by numerous factors specific to the overall service function of the powerplant. The type of coal burned in the generator does, however, affect the arrangement and operation of the limestone FGD system.

The combustion characteristics of a coal impact not only the potential for slagging, fouling, and erosion of the steam generator but also the composition and quantity of flue gas introduced to the scrubbing system. Coal properties influence the weight and volume flow rates, the temperature and dewpoint, and the fly ash loading of the flue gas. They also determine the need for additional control equipment ahead of or after the scrubbing system. Because of these considerations, design of the steam generator should be integrated with selection of the coal supplier and with design of the FGD system.

Power Generation Demand

The powerplant's electric generation demand affects both the basic powerplant and the associated limestone FGD system. The utility Project Manager, working with a consulting A/E firm, may be required to design for operation at the following loads, either individually or in any combination:

- Base load--operation to take all or part of a minimum load over a given period of time; consequently, operation would be essentially at a constant output.

- Cycling load--operation to provide power during extended periods of low power demand (25 percent of station rated capacity), followed by cycling to high power demand (full rated capacity).

- Intermediate load--operation to provide power for a portion of a day at full rated capacity as a base load and for the rest of the day at a reduced (50 to 67 percent range) constant load.

- Peak load--operation to provide power during periods of maximum demand.

Additionally, future power demand situations could necessitate other modes of operation of both the powerplant and the associated limestone FGD system. These situations could include daily startup and shutdown with operation at station rated capacity, or weekly startup (following weekend shutdown) with operation at full rated capacity. Because of the possibilities for diverse power generation demands, it is important that flexibility of response be built into the limestone FGD system. The following are the kinds of design provisions that will enhance system flexibility:

- Provision of space in the process layout for future installation of additional equipment.

- Provision for fitting of the individual scrubber modules with additional internal gas-liquid contacting devices to increase SO_2 removal efficiency.

- Provision for fitting of an open spray tower scrubber with additional spray bank headers.

- Provision for handling of additional scrubber slurry by the slurry recirculation pumps, pipe headers, and spray nozzles to increase the slurry flow rate.

- Provision of additional horsepower capabilities for the fans to move the flue gas through the scrubbing system at a higher pressure drop or to handle a greater gas flow.

- Provision for future use of chemical additives.

- Provision for future use of forced oxidation to reduce sludge volumes.

- Provision for future incorporation of a chloride removal system.

Site Conditions

Several site-specific factors will strongly influence the FGD system design. Among the most significant are those related to sludge handling and disposal. The availability of land for disposal and the permeability of the soil are two key parameters. Containing untreated sludge onsite could be the preferred sludge disposal option where land is available, where the soil properties tend to prevent leaching, or where any leaching that occurs will not pollute local ground waters.

In many instances, local requirements might dictate such sludge treating operations as forced oxidation, chemical stabilization, and the like, either singly or in combination, for preparation of a nonpolluting, ecologically acceptable landfill material. When some form of sludge treatment is needed, the utility should consider the proprietary sludge fixation technologies that are available.

Final decisions regarding sludge disposal are affected by economic considerations, local regulations, and provisions of the Resource Conservation and Recovery Act of 1976 (RCRA), which establishes guidelines for disposal of solid wastes. Where direct disposal is desired but the permeability of the soil could allow ground water pollution, direct disposal into a pond is a possibility. Guidelines for pond disposal indicate that a coefficient of permeability of 1×10^{-7} cm/sec (0.1 ft/yr) is necessary to adequately isolate the disposal site from the surrounding environment. The effective permeability of an area can be reduced by the use of a liner. The choice of liner materials includes a polymeric or elastomeric sheet covering the entire area of the disposal facility, or an impermeable shell of natural clay or chemically stabilized scrubber sludge.

Climatic and geographic conditions at the site also affect design of the FGD system. The designer must anticipate the effects of such factors as mean evaporation rate and rainfall, temperature range, maximum wind velocity, and seismic phenomena. Local atmospheric conditions, together with local regulations, often dictate the need to reheat the flue gas prior to venting. These factors can directly affect the cost, operation, and maintenance of the planned system. Table 2-4 shows site evaluation factors that were used as a design basis for a limestone scrubber at a 500-MW powerplant (Smith et al. 1980).

Another site-specific design consideration is development of a workable layout. The designer should attempt to minimize the distance between major process operations and at the same time allow adequate space at ground level and overhead for access to the equipment for maintenance, inspection, and

TABLE 2-4. SITE EVALUATION FACTORS FOR A 500-MW CASE[a]

	Site conditions
Land availability	
Total area available for site development, acres	965
Disposal area, acres	480
Scrubber module area, ft	~ 200 x 200
Sludge preparation area, ft	~ 250 x 150
Guidelines for sealing disposal site, gal/ft^2/yr	Ensure that leakage does not exceed 0.75 (equivalent to 1×10^{-7} cm/sec permeability)
Natural clay availability	Available near site
Grade elevation, ft	540
Barometric pressure, in. Hg	29.4
Ambient temperature, °F	
Minimum	-5
Maximum	95
Mean avg. rainfall, in./yr	46.5
Mean avg. evaporation, in./yr	36
Maximum wind velocity, mph	100
Seismic activity	-

[a] Smith et al. 1980.

testing. These factors must be considered by the A/E consultants as they work with the utility project team to develop preliminary design studies.

Additional site-related factors that may affect system design are the quality and availability of limestone and of makeup water. Maintaining consistent and uninterrupted supplies of these materials is a key to successful long-term performance of the FGD system. An FGD system may be able to accept cooling tower blowdown water as process makeup. If the blowdown water is compatible with the FGD process chemistry, then the FGD system can dispose of at least a portion of this contaminated water that would otherwise require expensive treatment prior to discharge. Details regarding the analysis of limestone and makeup water supplies are outlined later in this section (Design Basis).

Environmental Regulations

The applicable environmental regulations exert an obvious influence on FGD system design, particularly with respect to particulate and SO_2 removal capabilities and the sludge disposal system. The following discussion places primary emphasis on Federal regulations. Where state or local regulations are more stringent, they would govern the system design.

New Source Performance Standards. New Source Performance Standards (NSPS) were issued on June 11, 1979, for implementation by coal-fired electric utility steam generating units capable of firing more than 250 million Btu/h heat input (about 25 MW electric output), and upon which construction commenced after September 18, 1978. Following is a brief summary of these standards:

(1) SO_2 emissions are subject to the following:

   (a) If the uncontrolled emissions are less than 2 lb/million Btu of heat release, the SO_2 removal efficiency of the system may not be less than 70 percent.

   (b) If the uncontrolled emissions are between 2 and 6 lb/million Btu, the emission from the scrubbing system may not exceed 0.6 lb/million Btu, and SO_2 removal must be between 70 and 90 percent, depending on the uncontrolled emissions.

   (c) If the uncontrolled emissions are between 6 and 12 lb/million Btu, the SO_2 removal efficiency of the scrubber system must be 90 percent.

   (d) If the uncontrolled emissions are greater than 12 lb/million Btu, the emission from the scrubber system may not exceed 1.2 lb/million Btu.

These regulations are coal-related in that SO_2 removal rate requirements are affected by sulfur content and heating value. Note that at no time may the overall system SO_2 removal efficiency be less than 70 percent, nor may the actual emission exceed 1.2 lb/million Btu. Compliance is determined on a continuous basis by using continuous monitors to obtain a 30-day rolling average. Figure 2-2 delineates the NSPS requirements relating to SO_2 emissions as required removal efficiency.

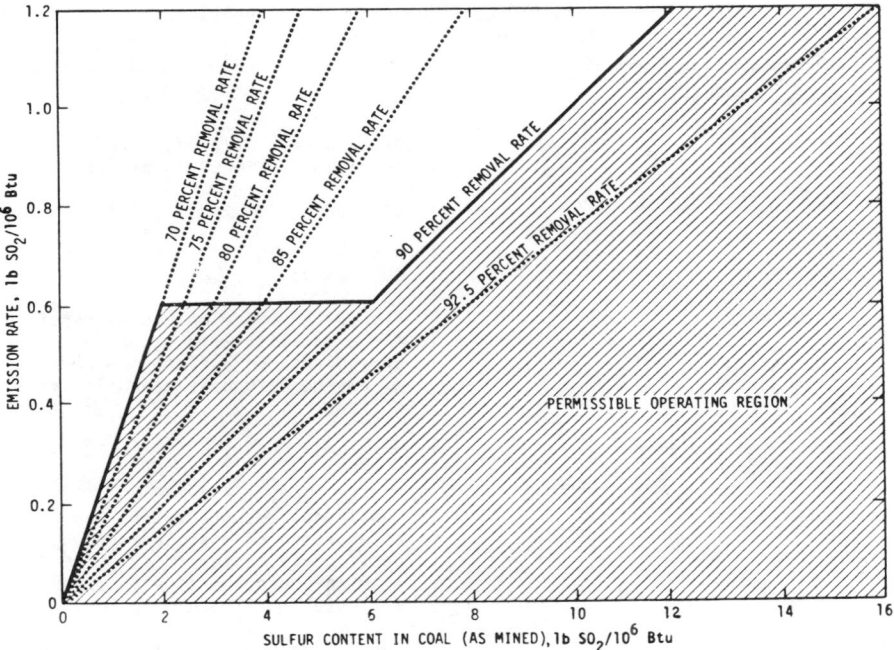

Figure 2-2. Influence of sulfur in coal on required SO_2 removal efficiency.

(2) The NSPS particulate standard limits emissions to 0.03 lb/million Btu heat input. The opacity standard limits the opacity of emissions to 20 percent (6-minute average).

(3) Instrumentation installed to verify compliance must meet performance specifications and must exhibit general characteristics as given in Appendix B of Part 60, Title 40, Code of Federal Regulations.

(4) A spare scrubber module must be provided if emergency bypass capabilities are desired at the plant without a commitment to load reduction. A spare scrubber module is required for any facility whose capacity is greater than 125 MW electric.

Table 2-5 shows the NSPS parameters applicable to a hypothetical coal of specified properties. The design of a limestone FGD system can be affected by the SO_2 limitations and by the degree of redundancy required.

Owners of new powerplants are required under NSPS to notify regulatory officials of the state in which the proposed plant will be located before beginning construction. They are also required to give notification before startup and to submit operating data after startup.

TABLE 2-5. NSPS AND EMISSION REGULATIONS APPLICABLE TO A COAL[a] OF SPECIFIED PROPERTIES[b]

Emission limitations:	
Particulates	0.03 lb particulate/10^6 Btu (corresponds to 99.7% removal)[c]
Opacity	Limited to 20%
SO_2	0.6 lb SO_2/10^6 Btu (corresponds to 90% removal)[d]
Redundancy	A spare scrubber module is required for a facility with capacity exceeding 125 MW electrical output if emergency bypass capabilities are desired
NSPS continuous emission monitoring requirements:	
Opacity	Continuous measurement of the attenuation of visible light by particulate matter in stack effluent with an averaging-time interval of 6 minutes
SO_2	Continuous measurement of SO_2 in stack effluent (30-day rolling average)
Oxygen (O_2) or carbon dioxide (CO_2)	Monitoring required to determine appropriate conversion factors for flue gas stream dilution

[a] Coal analysis:

Sulfur content, %	3.2	Ash content, %	16.0
Chlorine, %	0.06	Moisture, %	8.0
Carbon, %	60.0	Hydrogen, %	4.2
Oxygen, %	7.5	Nitrogen, %	1.0
Heating value, Btu/lb	11,000		

[b] Adapted from Smith et al. 1980.

[c] Assuming 80% of the coal ash leaves the boiler as fly ash.

[d] Assuming 100% of the coal sulfur leaves the boiler as SO_2.

Prevention of Significant Deterioration. New or modified powerplants are also subject to preconstruction review and to the requirements for Prevention of Significant Deterioration (PSD). These requirements are currently being implemented by the U.S. EPA while the States are incorporating them into their implementation plans.

Under the PSD program, clean areas of the nation [i.e., those whose pollutant levels are below the National Ambient Air Quality Standards (NAAQS)] are classified as Class I, II, or III, each class representing a specific amount or increment of allowable deterioration. Class I increments permit only minor air quality deterioration, Class II increments permit moderate deterioration consistent with normal growth, and Class III increments permit considerably more deterioration; in no case, however, can the deterioration reduce the area's air quality below that permitted by the NAAQS. Except for certain wilderness areas designated as Class I, the entire country is designated as Class II.

In addition to the increment concept and classification system, the PSD regulations require that each major new or modified source apply Best Available Control Technology (BACT). BACT is determined on a case-by-case basis for each pollutant; it must represent an emission limitation based on the maximum achievable degree of reduction, taking into account energy, environmental, and economic impacts. At a minimum, BACT must result in emissions not exceeding any applicable NSPS or National Emission Standards for Hazardous Air Pollutants (NESHAP). The PSD regulations also provide further protection for Class I areas in terms of "air quality related values," such as visibility.

Other Pertinent Regulations. In addition to the PSD requirements, EPA has recently published regulations to protect visibility in designated national parks and wilderness areas. Generally, these visibility regulations require the affected states to formulate provisions for installing Best Available Retrofit Technology (BART) at existing powerplants whose emissions impair visibility, to consider further controls beyond BACT for new sources, and to develop a long-term strategy to remedy and prevent visibility impairment. These regulations apply to 36 states containing 156 areas designated for protection as "mandatory Class I areas."

Local, state, and Federal regulations relating to water pollution or land use will also affect the overall system design. The RCRA regulations mentioned earlier provide guidelines for preventing the contamination of groundwaters. The Clean Water Act of 1977 further restricts the discharge of powerplant effluents into any natural water bodies. These regulations provide a strong impetus to development of closed-loop limestone FGD systems.

The major effect of closed-loop operation is the buildup of dissolved solids, especially chloride, in the circulating scrubber liquors. The chloride enters the scrubber in the makeup water and with the flue gas as HCl. It is converted to calcium chloride by reaction with the limestone. The dissolved chloride accumulates in the scrubber liquor because, unlike SO_2 and SO_3, it forms no insoluble calcium compounds and therefore can leave the system only with the liquid fraction of the sludge. The problem is exacerbated when low-sulfur, high-chloride coals are used in conjunction with tightly closed loops, forced oxidation, and high percentages of solids in the final sludges. Chloride removal techniques are described in Appendix B.

New limestone scrubbers designed to incorporate chloride removal techniques, with or without some of the available modifications in scrubber technology, could treat flue gases from combustion of high-sulfur and high-chloride coals without damage to the environment (Borgwardt 1978). In such cases, economic factors must be considered very seriously.

DESIGN BASIS

The primary factors in determining the size and configuration of the FGD system are the flow rate, composition, pressure, and temperature of the flue gas from the boiler. These parameters, together with the pollutant removal requirements, reagent stoichiometric ratio, and compositions of the limestone and makeup water provide the basis for FGD system design. Establishment of the overall system design is also based upon the available configuration options. Material and energy balances are performed to establish water and reagent requirements and to allow estimates of the amounts of sludge production and energy consumption. The following discussion deals with each of these major factors that establish the overall system design basis.

Flue Gas Flow and Composition

The flue gas entering the FGD system contains combustion products together with sulfur oxides, nitrogen oxides, HCl, and some fly ash. The volume of the flue gas and the design velocity determine the number of FGD scrubber modules required. The amount of SO_2 determines the liquid pumping rate and the volume of the effluent hold tank. A substantial part of the fly ash generated in combustion is removed upstream of the FGD system, most commonly by electrostatic precipitators, and to a lesser extent by fabric filters and wet scrubbers. Chlorides from the coal are absorbed by the FGD system, and attention must be given to the possible development of high concentrations of corrosive chlorides. Temperature and moisture content of the flue gas entering the FGD system determine the amount of water that evaporates when the gas is cooled (adiabatically) in the scrubber. Adiabatic saturation is discussed in Appendix E in relation to material balances.

Pollutant Removal Requirements

It is essential to know the SO_2 and particulate removal efficiencies required for compliance with NSPS regulations. The percentage of SO_2 removal varies with the type of coal fired, its sulfur content, and the heating value, as discussed earlier. The SO_2 removal requirement determines the amount of limestone required and the quantity of sludge produced. The required SO_2 removal may also be important when bypass reheat is considered (see Appendix E).

Reagent Stoichiometric Ratio

In its broadest sense, stoichiometry is a system of accounting applied to the materials participating in a process that involves physical or chemical change (Benenati 1969). As outlined briefly in Section 1, the stoichiometric ratio (SR) in a limestone scrubbing system is the number of moles of $CaCO_3$ fed to the limestone scrubber required to neutralize a mole of SO_2; the theoretical value is 1. In practice, however, the actual amount of limestone required to neutralize the SO_2 exceeds the theoretical value. The ratio of the actual to the theoretical amount is the SR. In most limestone-based FGD systems, the SR value, based on SO_2 removal, ranges from 1.02 to 1.30.

The SR depends upon the mass transfer capability of the scrubber, the quality of limestone, and the hold tank design. Some scrubbers provide good capability for mass transfer, and thus operate at relatively lower SR values. It should be noted that stoichiometry and pH are not independent. A higher SR may be needed when the inlet SO_2 loading is variable. The limestone requirement for an FGD system can be estimated on the basis of limestone quality and the requirements for both SO_2 and HCl removal.

Limestone Composition

Limestone consists primarily of calcium and magnesium carbonates ($CaCO_3$, $MgCO_3$) and inert material. On the basis of their chemical analyses, limestones are classified into three grades:

1. High-calcium limestone, containing at least 95 percent calcium carbonate.

2. Magnesia limestone, containing 5 to 15 percent magnesium carbonate and 80 to 90 percent calcium carbonate.

3. Dolomitic limestone, containing 15 to 45 percent magnesium carbonate and 50 to 80 percent calcium carbonate.

High-calcium limestone is the most widely used for SO_2 absorption. A typical composition of such a limestone is 95 percent calcium carbonate, 1.5 percent magnesium carbonate, and 3.5 percent inert materials. Small amounts of magnesium ions have a beneficial effect on scrubbing chemistry (Appendix A), and the use of magnesia limestone is gaining favor. Very often, the magnesium fraction of dolomitic limestone is unavailable to become magnesium ions for FGD chemistry.

The utility Project Manager, together with the A/E consultant and potential scrubber suppliers, must determine how much limestone of what size and grade is needed and the source of supply. Design of an efficient system will require that all project participants have a working knowledge of the chemical composition, reactivity, and grindability index of the selected limestone. If possible, pilot plant tests should be conducted with the candidate limestone to determine its performance characteristics. Recent experience in commercial FGD operation has demonstrated that a continuing supply of the specified grade of limestone is important to reliable system performance.

Makeup Water

Water is consumed in the scrubbing process during the quenching of the incoming flue gas and by chemical and physical binding of water in the sludge. Knowledge of the chemical composition of the makeup water is important in FGD system design. Sources of makeup water are raw water, fresh water, and cooling tower blowdown.

Raw water may come from a river, lake, or an untreated well; fresh water comes from a municipal water system. Filtered water is usually required for pump seals. Cooling tower blowdown can sometimes be taken from the powerplant water inventory for use as makeup water. This water should be closely monitored, however, because only limited use can be made of cooling tower blowdown that contains high concentrations of sodium, magnesium, or chloride ions, as discussed in Appendix A.

Table 2-6 shows typical makeup water analyses.

MATERIAL AND ENERGY BALANCES

A material balance and energy requirements should be calculated so that the specific configuration of the FGD system can be established. Appendix E presents detailed procedures for calculating material and energy balances, giving example calculations for a hypothetical 500-MW plant firing an eastern high-sulfur and a western low-sulfur coal.

Figure 2-3 depicts the overall inputs and outputs associated with the boiler-furnace system, the ESP, and the FGD system. Flue gas leaving the

TABLE 2-6. TYPICAL MAKEUP WATER ANALYSES[a]

Sources of makeup water	Normal fresh makeup water will be cooling tower blowdown and filtered well water. Additional makeup water may be obtained from the recycle basin which contains the following: ° Decanted bottom ash sluice water ° Plant equipment, floor, and roof drainage ° Coal pile runoff ° Area drains	
Typical chemical composition, mg/liter as $CaCO_3$ except as noted	Filtered well water	Cooling tower blowdown
Calcium	200	800
Magnesium	55	220
Sodium	30	120
Total alkalinity	225	200
Sulfate	25	800
Chloride	20	80
Silica, as SiO_2	15	60
Orthophosphate, as $PO_4^\equiv$		2
Total phosphate, as $PO_4^\equiv$		6
Total dissolved solids, as such	315	1375
Total suspended solids, as such	<1	7
pH	7.5	7.5-8.0
Conductivity, μmho/cm	505	2200

[a] Smith et al. 1980.

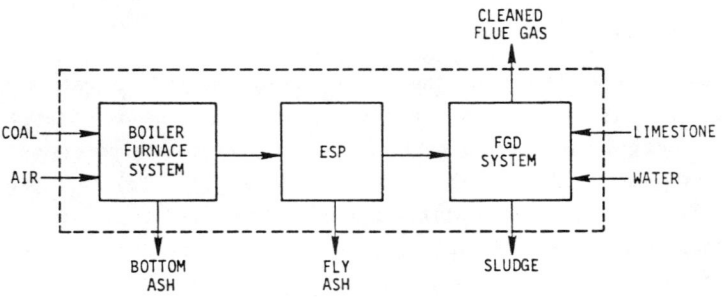

Figure 2-3. Overall inputs and outputs.

boiler-furnace system passes through the ESP, which removes enough particulate to bring the flue gas content below the maximum allowable particulate emission limit (0.03 lb/million Btu heat input for a new coal-fired boiler). Inputs to the FGD system include the SO_2-laden flue gas, limestone slurry, and makeup water. The outputs include cleaned flue gas (with residual SO_2, particulate, and evaporated moisture) and dewatered sludge.

Beginning with the known amount and composition of the flue gas entering the FGD system and the emission regulations, the material balance calculations are performed in five steps, as follows:

- SO_2 removal requirement
- Limestone requirement/slurry preparation
- Humidification of flue gas
- Recirculation loop and sludge production
- Makeup water requirement

Once the makeup water requirement is known, the overall water utilization is established on the basis of the mist eliminator (ME) wash procedure. The important interplay of the ME wash requirements and water balance in the limestone scrubbing system is discussed in Appendix E.

Within the FGD system, energy is consumed by the fans that drive the gas through the system, by recirculation and transfer pumps that drive the slurry and liquor, and by the reheater (if selected). Comparatively small amounts of energy are also consumed by the ball mill, thickener, dewatering devices, agitators, conveyors, bucket elevators, and similar components. The energy demands of the fans, slurry recirculation pumps, and reheater are calculated separately (as shown in Appendix E for the hypothetical 500 MW powerplant). Energy consumption by the other parts of the system is assumed to be 20 percent of that consumed by flue gas fans and slurry recirculation pumps.

SYSTEM CONFIGURATION OPTIONS

After determination of the major design basis and material/energy factors, several system configuration options must be considered. Most critical among these are the options for (1) particulate removal (separate versus integral), (2) location of the flue gas fan, (3) flue gas bypass (versus no bypass), (4) reheat (versus no reheat), and (5) sludge disposal. Redundancy and system flexibility, which are closely interrelated, are needed to some extent in every limestone FGD system. They are discussed here as options because decisions must be made as to the degree of redundancy and flexibility to be incorporated into the system design.

Particulate Removal

One of the major decisions required in preliminary planning is whether to assign any particulate removal function to the limestone scrubbers.

Slack (1977) outlines the following options:

1. Separate high-efficiency particulate removal.
2. Partial low-efficiency particulate removal.
3. Integral particulate removal.

Most of the current systems operate with a separate high-efficiency ESP upstream of the scrubber. Fabric filters have been used to a lesser extent in FGD systems but are becoming increasingly popular. In these configurations there is no need for additional particulate collection in the FGD system. The combination ESP/FGD system offers the advantages of simplicity, segregation of functions, and relatively low costs for operation and maintenance. The ESP's afford high reliability and can allow for emergency flue gas bypass around the scrubbing modules without load reduction or unit shutdown. The ESP/FGD system offers further benefits (Devitt, Laseke, and Kaplan 1980):

- Exotic construction materials can be used more selectively and in lesser amounts.
- Induced-draft and booster fans can precede rather than follow the FGD system, and thus are less subject to fouling and erosion.
- Sludge stabilization and fixation processes that use dry fly ash as an additive can make use of the precollected ash.
- Less total waste volume is produced by mixing dry fly ash and dewatered sludge.

Partial low-efficiency particulate removal can be achieved by use of a mechanical collector or low-efficiency precipitator (90 to 95% removal) upstream of the scrubber. Under this option the SO_2 scrubbers must be capable of removing the residual particulate. A low-efficiency particulate collector may be combined with the FGD system when there is no system bypass. This method has particular advantages when additional particulate collection is achieved by means of a wet ESP downstream of the scrubber, as now provided commercially (see Appendix D).

Integral particulate removal is done with a venturi or mobile-bed scrubber that accepts all of the particulate and SO_2. Some designs have two scrubbers in series. Low capital cost has been cited as an advantage of this option, but there are drawbacks (Slack 1977):

- The product solids cannot be dewatered to the degree attainable when dry ash is mixed with dewatered sludge.
- The system bleed must be increased because the effluent contains ash plus sludge with higher water content than the sludge alone. The higher blowdown reduces the steady-state chloride concentration and thereby minimizes corrosion. This advantage, however, is offset by a larger amount of effluent that is subject to leaching

at the final disposal site.

- Ash in the scrubber increases erosion and could increase the potential for corrosion caused by the accumulation of solids at wet/dry interfaces. Careful engineering can effectively eliminate these problems. If the fly ash has high alkalinity, it can be used for SO_2 absorption and can considerably reduce the limestone requirements.

- Full particulate removal in the FGD system eliminates the option for bypass of the system. Many utilities have a market for some or all of the ash, depending on ash properties and site-specific conditions. In these cases, full ash removal in the scrubber would not be feasible.

- The higher pressure drop required for particulate scrubbing (compared with SO_2 removal only) increases power consumption. With low-sulfur coal, the power consumption penalty is reduced because of the relatively high power requirement of the precipitator. The use of a wet ESP downstream of the scrubber may further reduce the power penalty.

When fly ash resistivity is very high or when outlet particulate emissions must be very low, a fabric filter may become a viable option for total particulate removal.

Fan Location

Fans are used to overcome the pressure drop associated with the scrubbing system by pushing or pulling the flue gas through the system. This pressure drop may be overcome by induced-draft (ID) fans in the main boiler or by booster fans in the FGD system. Booster fans generally supplement existing boiler ID fans and may be located upstream or downstream of the scrubbers. Fans located upstream of the scrubbers or downstream of the reheaters are considered dry fans; they do not handle saturated gas and are not sprayed with wash water. Fans located downstream of scrubbers where reheat is not provided are wet fans.

Wet booster fans offer a size advantage because the cool, wet gas has less volume than dry gas. Because of numerous operating problems, however, the trend is to the use of dry booster fans upstream of the scrubbers. In view of the poor performance record of wet fans, downstream fans should be specified only with a rationale for circumventing known problems of erosion, corrosion, and solids deposition.

Abrasion effects of particulate matter in a gas stream require a dry fan to be placed before a scrubber only when it is preceded by a particulate removal device.

The location and operational characteristics of boiler ID fans and/or scrubber booster fans will determine whether the FGD system operates at positive or negative pressures with respect to the atmosphere. Location of fans upstream of the scrubbers is currently favored in most installations.

It should be noted, however, that with upstream fans the scrubbers will be under positive pressure and any leaks in the system will allow emission of flue gas to the local environment. Any leakage problem would be aggravated where the FGD system is located in an enclosure. For this reason, special consideration should be given to ensuring a leak-proof design. Specific areas of concern include seal welding, access doors, dampers, and penetrations of the scrubber shell and ductwork.

Where fans are located downstream of the FGD system, the scrubbers and adjacent ducts operate at a negative pressure with respect to the atmosphere. If leaks occur, the result is inleakage of air to the FGD system, which has caused high natural oxidation (see Appendix A) but generally has no adverse impact on scrubber operation. The fans, however, are subject to potential corrosion and imbalance caused by contact with saturated flue gas and entrained slurry droplets. Although downstream fans have been used successfully, their use demands close attention to mist eliminator and reheat design.

Regardless of the type of fan selected, it is critical to maintain a balanced draft between the boiler and the FGD system. This stabilizes the boiler flow and thereby stabilizes the flame. A balanced draft also provides sufficient energy to move the flue gases through the FGD system in an efficient, stable manner.

Flue Gas Bypass (Versus No Bypass)

There are two basic types of flue gas bypass: emergency bypass, in which the total flue gas flow bypasses the FGD system; and continuous bypass, in which part of the gas bypasses the system. Emergency bypass of flue gas is used only when the FGD system is forced to shut down and the boiler must continue to operate. The current NSPS regulations require that a redundant module be provided for the FGD system when emergency bypass ductwork is installed. Thus emergency bypass would occur only upon failure of the total FGD system, including the redundant module.

The bypass system should be designed to prevent undesirable eddies or unintended reversal of flow. Safety in boiler operation is always a primary consideration. During a period of emergency bypass, the main concern is whether the bypass damper will open when the scrubber dampers are closed. Most current systems appear acceptable in this respect.

When bypassing of flue gas around the scrubbing system is not possible, the boiler operation is dependent on the performance of the scrubbing system; i.e., if the scrubbing system shuts down, so must the boiler. Hence the reliability of the boiler, an important consideration in utility operation, is directly linked with that of the scrubbing system.

Continuous bypass of portions of the flue gas is permissible only when the SO_2 emissions to the atmosphere and the SO_2 removal efficiency conform with NSPS regulations. If bypass is selected, the SO_2 removal efficiency of the treated portion of the gas must be significantly higher than that determined by simple material balance so as to compensate for the untreated gas (see Appendix E). Operation under the performance standard requiring 90 percent SO_2 removal efficiency, as when firing high-sulfur coal, essentially rules out continuous bypass. Continuous bypass of gas usually necessitates that a particulate control device (ESP) precede the scrubbing system. In a system with integral particulate-SO_2 removal, this device may be in the bypass duct. Bypassing hot gas around the scrubber provides an economical means of reheat. In addition to providing reheat energy, continuous bypass allows the use of smaller scrubbing modules and reduces the amount of water evaporated by the flue gas.

Reheat (Versus No Reheat)

At this time, a preference for stack gas reheat is evident among the systems now in service and some that are planned for the future. Incorporation of reheat capability into the system design does offer several significant advantages.

First, the greater buoyancy of reheated gas tends to reduce pollutant concentrations at ground level near the plant. Unheated flue gas returns to ground level more quickly than does reheated gas and thus produces potentially higher ground-level concentrations.

Reheat also helps to prevent condensation and the formation of a heavy steam plume. When the wet, cool gas from the scrubber is not reheated, condensation may take place, possibly in the exit duct or the stack. The effects of local stack fallout are under intensive investigation. In cold weather, operation without reheat can generate a heavy steam plume.

Finally, reheat can reduce downstream corrosion of scrubber components by preventing or minimizing condensation of the sulfurous acid produced from residual SO_2 in the gas. Under most FGD conditions, the inlet concentration of sulfur trioxide (SO_3) is 1 to 2 percent of the SO_2 concentration. Although the scrubber removes much of the SO_3, nearly half of it passes through as acid mist (Slack 1977), which can also condense in the outlet ductwork.

Some of the methods available for increasing the temperature of gas from a scrubber prior to discharge to the stack are indirect in-line reheat, direct combustion reheat, direct hot-air reheat, gas bypass reheat, exit gas recirculation reheat, and waste heat recovery reheat. The first four of these methods have been applied in commercial limestone FGD systems operating in the United States, and waste heat recovery is planned for two future installations. Among the systems that have operated or are currently in service, indirect in-line reheat has proved to be the most popular method, although not the most reliable.

The various reheat systems can be evaluated in terms of capital and operating costs, but it is very difficult to evaluate the intangible reliability factors. Bypassing flue gas to the degree feasible is the lowest-cost approach; a problem, however, is that designers often bring the bypass gas back into the system at a point near the stack, in which case much of the duct from the scrubbing modules to the stack is exposed to wet gas and is subject to corrosion. An in-line reheat system is the next lowest in cost, but tube corrosion and fouling have been major problems.

Recently, designers of some powerplants have selected a no-reheat system and have designed for condensation in the outlet ductwork, the ID fan, and the stack. An important feature of the no-reheat system is the low-velocity stack, in which the gas velocity is about 30 ft/s as compared with a conventional stack velocity of about 90 ft/s. At this low velocity, mist droplets can settle out and be collected in hoppers at the bottom of the stack. Such an installation requires larger, corrosion-resistant stacks, with resultant increases in costs and plume opacity. The plume opacity, which is due to liquid water droplets rather than solid particulate, could be a problem under some local regulations.

To limit corrosion in a no-reheat operation, the designer may either select materials that are inherently resistant to corrosion, such as high-alloy steels and acid-resistant brick mortar, or provide protective linings in the ductwork and the stack. Many installations with wet stack operation have problems with stack linings, which tend to blister and flake off. Once this happens, stack corrosion begins (PEDCo Environmental 1979b).

Sludge Disposal

The handling and disposal of the waste or sludge generated in limestone FGD systems has been of great concern. The sludge can be converted into gypsum for use in manufacture of wallboard or portland cement; however, the presence of fly ash in the sludge and the abundance of relatively pure natural gypsum have kept U.S. utilities from full-scale commitments to produce gypsum from FGD sludge. Several installations are undertaking gypsum production, but none are yet in commercial operation (see Appendix D). As a result, most of the sludge is discarded, usually in ponds or landfills (Jones 1977). There is, however, an increasing trend toward dewatering and/or stabilization of FGD wastes for disposal in managed landfills; correspondingly, the trend is away from pond disposal, which has served the functions of clarification, dewatering, and temporary or final storage of the sludge. The increasing emphasis on disposal in landfills and structural fills, in conjunction with the desire for closed-loop operation, has necessitated the use of thickeners, centrifuges, and vacuum filters for sludge treatment. In addition, several installations are using forced oxidation to enhance solids settling and filtration properties, to improve the sludge quality, and to reduce the land requirements for sludge disposal. Added benefits cited for forced oxidation are a general improvement in process chemistry and a reduction in scaling potential. Details of these and other aspects of FGD sludge disposal are presented by Knight et al. (1980).

There are four principal options for landfill disposal: landfill without treatment, treatment by blending with fly ash (stabilization), fixation with a chemical additive such as lime, and forced oxidation. The first three options do not require modification of process chemistry (Ansari and Oren 1980).

In landfill without treatment the sludge is thickened from about 15 percent solids to about 30 percent and dewatered in a vacuum filter to about 60 percent solids. The thickener overflow and the filtrate are returned to the FGD system. The filter cake is conveyed to a landfill with no additional treatment. Alternatively, the waste slurry or the thickened waste slurry can be pumped directly to a disposal pond.

In a stabilization process, the dewatered FGD sludge is mixed with fly ash, soil, or other similar material to induce only physical changes without chemical interaction between the additive and the sludge. Fixation is a type of stabilization involving the addition of reagents that cause chemical reactions with the sludge. The better known processes use lime and other alkaline materials such as blast furnace slag or alkaline flyash, which cause cementitious reactions in the sludge (Knight et al. 1980).

As an alternative to these landfill options, the limestone FGD system can be modified to include a step that forces the oxidation of calcium sulfite sludge to calcium sulfate (gypsum) by the introduction of air. Gypsum is a more desirable waste product than calcium sulfite because of its greater chemical stability and better settling properties, and because a lower volume of material is generated for disposal. Gypsum can be readily dewatered to greater than 80 percent solids, and the process reduces the thixotropic potential that is inherent in calcium sulfite sludge. In gypsum "stacking" (see Appendix D) the bleed slurry is merely pumped to the stack, where the gypsum settles to a dry state without use of either a thickener or filter (Ansari and Oren 1980).

If it is not to be marketed, the gypsum sludge can have a high fly ash content. With forced oxidation, the thickener can be much smaller, or may not be needed at all. Gypsum sludge can be dewatered satisfactorily by a centrifuge without previous thickening or by a hydrocyclone/vacuum filter system.

Overall System Design 45

Flexibility and Redundancy

Flexibility and redundancy are considered as options to the extent that decisions are needed concerning the degree to which they will be designed into the system. Flexibility and redundancy are interrelated, and both contribute to total system reliability. A new FGD system should be designed with the greatest possible degree of flexibility that is consistent with cost and with site-specific considerations. This can be done by providing a spare scrubbing module and spares of various components, especially pumps, and by providing surge capacity in tanks. Under the current NSPS, a spare scrubbing module must be provided in order to permit emergency bypass, so that no violation occurs in a full-demand power supply situation.

The design should provide for expected increases in the flue gas flow over the service life expectancy of the FGD system and the remaining life of the boiler. As the boiler ages and leaks occur, the volume of flue gas to the FGD system will increase as a result of increasing excess air requirements by the boiler or air leakage into the ductwork.

Much depends on the relative complexity of the FGD system. A complex system is more likely to need spares than a simple one. If the design is based upon the firing of an average coal, a spare scrubber module may be needed to accommodate the use of other coals with higher sulfur content or lower heat content.

System breakdowns are usually due to the failure of pumps, valves, piping, and other such ancillary components rather than to failure of the scrubber vessel itself. Hence proper redundancy of this ancillary equipment contributes to system reliability.

Other flexibility provisions could include the availability of a site for additional SO_2 scrubbing modules, or a built-in capability for additional SO_2 removal to accommodate a future boiler unit. The system should be flexible enough also to handle the use of additives that increase SO_2 removal efficiency, to allow modifications for byproduct utilization, and to permit the implementation of energy conservation measures.

COMPUTERIZED DESIGN GUIDES

Three major computer programs are available for use in the planning and operational stages of a limestone FGD project. These programs and their potential uses are described briefly in the following paragraphs; detailed information is given in Appendix C.

TVA Lime/Limestone Scrubbing Computer Model

The TVA and Bechtel National, Inc., have jointly developed a computer model for use in calculating major design parameters and costs of lime and limestone FGD systems. The model is based on results of tests at the Shawnee test facility, which was constructed by the TVA with Bechtel serving as test contractor. The test facility is owned by the U.S. EPA and operated by the TVA to determine the effects of various process parameters on system performance.

The model is structured to generate a complete conceptual design package for either a lime or a limestone scrubbing system. The package includes a detailed material balance, a detailed water balance, equipment specifications and quantities, and a breakdown of capital investment and annual cost requirements for the system. The model output is presented in the form of the following individual reports:

1. Input data.
2. Process parameters.
3. Pond size parameters and costs.
4. System equipment sizes and costs.
5. Capital costs.
6. First-year annual revenue requirements.
7. Lifetime annual revenue requirements.

Any or all of these reports can be obtained by entering appropriate values on the input cards. Also available are several system options such as the number of scrubber trains, number of spare scrubbing modules, and indirect cost percentages.

The input for the model is of two types:

1. Input needed to calculate process stream rates and sizes of the equipment items. This information consists of basic plant parameters such as system gas flow rates, fuel analysis, and SO_2 removal requirements.

2. Input parameters controlling the program options. This information is derived by multiple choice from among entries given in the users manual.

This model is intended for use by project managers and project engineers during the planning and decision-making stages. It is useful for comparison of various FGD scenarios. The model is equally useful for the manager and the engineer in that each can obtain the reports that meet his needs; for example, the manager might request summary reports, whereas the engineer might request process parameter reports for analysis of stream compositions and flow rates.

The model is developed for an IBM 370/165 computer system. The prospective user can obtain a copy of the program from the TVA or can supply TVA staff with the input needed for analysis of cases. The model is structured to accept input through punched cards for batch runs or through an interactive terminal.

PEDCo Flue Gas Desulfurization Information System (Experience Records)

For the past 6 years PEDCo Environmental under sponsorship of EPA has conducted an EPA Utility FGD Survey Program, whose purpose is to monitor and report on the nationwide status of utility FGD systems. The ongoing survey is conducted through telephone contacts with the system operators and suppliers, visits to the operating plants, written transmittals, and use of in-house data files. The goals are to maintain current design and performance information on the operating FGD systems and to obtain design and progress information on systems under construction or in various stages of planning. In addition to the design and performance data, capital and annual cost information is sought regarding both the operational and nonoperational systems.

These data are then reduced, verified, and loaded on a continued basis into the Flue Gas Desulfurization Information System (FGDIS), a collection of data base files stored at the National Computer Center (NCC) in North Carolina. The system is used to generate a quarterly EPA Utility FGD Survey Report, which summarizes current FGD developments. The report is distributed worldwide to recipients who are directly or indirectly involved in development of FGD technology.

In addition to EPA's printing of the report, the National Technical Information Service now makes the FGDIS available for on-line access by interested users. The more detailed design and performance data that cannot be conveniently included in the survey report are available for examination and analysis. Users thus have access to current and detailed information in the interim periods between quarterly reports.

Access to and manipulation of data within the FGDIS is accomplished through SYSTEM 2000, a general data-base management system. The system contains a complete set of user-oriented commands that offer flexible and extensive data retrieval capabilities. Both design and performance data can be accessed and tabulated in such a manner that virtually any information request can be satisfied.

Because of these flexible and comprehensive data retrieval capabilities and the extensive data base, the FGDIS is a valuable design tool. In preliminary design stages, the predominant FGD design schemes can be tabulated on the basis of current operating experience. Various design parameters can also be evaluated by analogy to those of operating systems. The designer can identify current problems as they are reported, e.g., problems of component failure or chemical imbalance, and can evaluate the effectiveness of the solutions. Additional means of using the FGDIS in technology transfer, together with details of the system, are given in Appendix C and in the system users manual (PEDCo Environmental, Inc. 1979a).

Bechtel-Modified Radian Equilibrium Computer Program

The Bechtel Corporation has fitted data to a computer program developed by Radian Corporation to predict the calcium sulfate (gypsum) saturation level of slurry at the scrubber outlet on the basis of measured concentrations of calcium, chloride, magnesium, and sulfate ions (Head 1977). This program for monitoring of gypsum saturation was developed because testing at EPA's Shawnee facility has shown that scaling usually occurs whenever the saturation level exceeds 130 percent.

Simplified equations for calculating the degree of gypsum saturation in scrubbing liquors at 25° and 50°C were fitted to predictions of the modified Radian program based on data from long-term reliability tests at Shawnee. Results obtained with the equations differed only slightly from the computer results of the Radian program. The equations are accurate for concentrations of total dissolved magnesium and chloride ions up to 15,000 ppm. Additionally, they have been expanded to include operation with adipic acid additive (Burbank and Wang 1980). They provide a convenient means for simple and accurate prediction of gypsum saturation levels by those not having access to the modified Radian program.

Details of this computer program and the simplified gypsum saturation equations are also presented in Appendix C.

REFERENCES FOR SECTION 2

Ansari, A., and J. Oren. 1980. Comparing Ash/FGD Waste Disposal Options. Pollution Engineering, 12(5):66-70.

Benenati, R. F. 1969. Industrial Stoichiometry. In: Kirk-Othmer Encyclopedia of Chemical Technology. 2d ed. Vol. 9. John Wiley & Sons, New York.

Borgwardt, R. H. 1974. EPA/RTP Pilot Studies Related to Unsaturated Operation of Lime and Limestone Scrubbers. Symposium on Flue Gas Desulfurization, Atlanta, Georgia, November 4-7, 1974. EPA-650/2-74-126a. NTIS No. PB-242 572.

Borgwardt, R. H. 1978. Effect of Forced Oxidation on Limestone/SO_x Scrubber Performance. In: Proceedings of the Symposium on Flue Gas Desulfurization, Hollywood, Florida, November 1977. Vol. I. EPA-600/7-78-058a. NTIS No. PB-282 090.

Bunicore, A. J. 1980. Air Pollution Control. Chemical Engineering, 87(13):92-94.

Burbank, D. A., and S. C. Wang. 1980. EPA Alkali Scrubbing Test Facility: Advanced Program Final Report (October 1974 to June 1978) EPA-600/7-80-115. NTIS No. PB 80-204 241.

Devitt, T. W., B. A. Laseke and N. Kaplan. 1980. Utility Flue Gas Desulfurization in the U.S. Chemical Engineering Progress, 76(5):45-57.

Galluzzo, N. G., and P. G. Davidson. 1979. Effects of Coal Properties on the Installed and Operating Costs of Power Plants. Presented at the 2d International Coal Utilization Conference in Houston, Texas, November 1979.

Head, H. N. 1977. EPA Alkali Scrubbing Test Facility; Advanced Program, Third Progress Report, EPA-600/7-77-105. NTIS No. PB-274 544.

Hudson, J. L. 1980. Sulfur Dioxide Oxidation in Scrubber Systems. EPA-600/7-80-083. NTIS No. PB-187 842.

Jones, J. W. 1977. Disposal of Flue-Gas-Cleaning Wastes. Chemical Engineering, 84(4):79-85.

Knight, R. G., E. H. Rothfuss, K. D. Yard, and D. M. Golden. 1980. FGD Sludge Disposal Manual. 2d ed. EPRI-CS 1515. Research Project 1685-1.

PEDCo Environmental, Inc. 1979a. Flue Gas Desulfurization Information System Data Base User's Manual. Cincinnati, Ohio.

PEDCo Environmental, Inc. 1979b. The Lime FGD Systems Data Book. Prepared for the Electric Power Research Institute under Research Project 982-1. EPRI FP-1030.

Slack, A. V. 1977. Design Considerations in Lime/Limestone Scrubbing. Proceedings: The 2d Pacific Chemical Engineering Congress (Pachec '77) Denver, Colorado. AIChE, New York.

Smith, E. O., W. E. Morgan, J. W. Noland, R. T. Quinlan, J. E. Stresewski, D. O. Swenson, and C. E. Dene. 1980. Lime FGD System and Sludge Disposal Case Study. EPRI CS-1631. Research Project 982-18.

Stephenson, C. D., and R. L. Torstrick. 1979. Shawnee Lime/Limestone Scrubbing Computerized Design/Cost-Estimate Model Users Manual. EPA-600/7-79-210. NTIS No. PB-80-123037.

Section 3

The FGD System

Major emphasis in this section is given to the equipment items that most strongly affect the operation and performance of a limestone FGD system: the scrubber, mist eliminator, reheater, and fans, together with equipment used in slurry preparation and sludge treatment. Each of these major equipment items is presented in a similar format: first, a description of the unit and its function, followed by discussion of the basic equipment types, including major design variations. Then follows a review of the principal design considerations, and finally a brief survey of actual applications in operational systems in the United States.

The remaining items of equipment that make up the scrubber system are treated in less detail. These items--pumps, piping, ductwork, and the like--receive less emphasis not because they are considered unimportant in scrubber operation but because they are common to many other types of major engineering systems and thus the functional operations and basic design options are better known. The discussions of these equipment items deal primarily with specific application to limestone scrubbing.

Additionally, this section gives an overview of process control and its application to the operation of a limestone scrubber system. Once again, the emphasis is on those aspects of the control loops and instrumentation that are peculiar to the scrubber system. An overall control philosophy is described, and examples of reliable operational process control systems are presented.

SCRUBBERS

Description and Function

The principal unit operation involved in a wet limestone FGD system is absorption of SO_2 from the flue gas stream. The scrubber is the principal component of the system because it provides the means of bringing the SO_2-laden flue gas into contact with the limestone slurry to undergo chemical reaction.

The scrubber promotes the transfer of energy and material between the flue gas and the liquid portion of the slurry. Within the scrubber vessel, the following transfers occur:

- Transfer of heat
- Transfer of water vapor
- Transfer of gas-entrained solids
- Transfer of water-soluble gases

Transfers of heat and water vapor occur as entering flue gas is quenched and cooled to approximately its adiabatic saturation temperature. In a well-designed scrubber, these transfers occur rapidly; the exit gas is saturated with water vapor, and its temperature is the same as the average temperature of the scrubbing slurry.

The transfer of gas-entrained solids (particulates) occurs as the gas and liquid become completely mixed. The particles become wetted and trapped in the liquid phase. This transfer occurs more slowly than the transfer of heat and water vapor. In a well-designed scrubber, the discharged gas stream contains virtually no unwetted particulates (fly ash); essentially all of the particulates that enter the scrubber become constituents of the scrubbing slurry. In a scrubber designed primarily for gas absorption, there is no net increase in particulate loading.

The transfer of water-soluble gases, the slowest transfer step, is the major concern in limestone scrubbing because it is the mechanism for removal of SO_2 from the flue gas. This discussion focuses on the design and performance aspects of scrubbers as they relate to SO_2 removal. Particulate removal is addressed only in relation to scrubber designs in which primary or secondary particulate removal is performed simultaneously with SO_2 removal.

Basic Scrubber Types (PEDCo Environmental, Inc. 1981; IGCI 1976; Calvert 1977; Saleem 1980)

Many scrubber designs have been developed for removal of particulate and gaseous pollutants from waste gas streams. A number have been adapted or developed exclusively for removing SO_2 (and particulates) from the flue gas of coal-fired boilers. Table 3-1 summarizes the various scrubber designs currently used or planned for use in commercial limestone FGD systems for coal-fired utility boilers. The generic types listed in this table represent a grouping of scrubber designs according to the basic collection mechanism. The specific types represent distinct design variations within each generic grouping. The trade names or common names have been assigned to special or proprietary scrubber designs. Five generic scrubber types encompassing 17 specific designs have been developed for commercial lime-

TABLE 3-1. BASIC SCRUBBER TYPES FOR COMMERCIAL LIMESTONE FGD SYSTEMS

Generic type	Specific type	Trade or common name
Venturi	Variable-throat/bottom-entry liquid distribution disk	Flooded-disk scubber
	Variable-throat/side-movable blades	
	Variable-throat/side-movable blocks	
	Variable-throat/vertically adjustable rod decks	Rod scrubber
	Variable-throat/adjustable drum	Radial flow venturi
Spray	Open/countercurrent spray	Vertical spray tower
	Open/crosscurrent spray	Horizontal spray chamber
Tray	Sieve tray	
Packed	Static bed	Marble-bed scrubber
	Mobile bed	Turbulent contact absorber (TCA)
	Rod deck	Ventri-sorber
	Grid	
Combination	Spray/packed	
	Venturi/spray	

stone FGD systems in service or planned for future service on coal-fired utility boilers.

Table 3-2 summarizes the numbers of units and the equivalent electrical generating capacities (MW) associated with each limestone scrubber design. The data indicate that spray towers are the predominant scrubber design for application to wet limestone FGD systems. It should be noted that all the venturi scrubbers identified in Table 3-2, except for one on a 235-MW unit, provide primary or secondary particulate control and remove only a portion of the inlet SO_2. These are two-stage scrubbing systems in which additional scrubbers provide primary SO_2 control.

The following paragraphs describe in more detail the various scrubber designs used or planned for use in commercial limestone FGD systems for coal-fired utility boilers. These detailed descriptions are presented in a manner consistent with the classifications in Table 3-1.

Venturi Scrubbers. The most prominent feature of the venturi design is a converging throat, which causes acceleration of the inlet flue gas to velocities between 150 to 400 ft/s. The scrubbing slurry, which is introduced at the inlet of the throat, is sheared into fine droplets of approximately 25 to 100 micron diameter by the high-velocity gas stream. A turbulent zone created immediately downstream of the throat promotes

TABLE 3-2. NUMBERS AND CAPACITIES[a] OF LIMESTONE SCRUBBER TYPES

Scrubber type generic/specific	Operational		Under construction		Planned		Total	
	No.	MW	No.	MW	No.	MW	No.	MW
Venturi[b]								
Variable-throat/bottom-entry liquid distribution disk	2	383						
Variable-throat/side-movable blades	1	550						
Variable-throat/side-movable blocks	2	1,155	1	235				
Variable-throat/vertically adjustable rod decks	4	2,025	1	575				
Subtotal	9	4,113	2	810			11	3,921
Spray								
Open/countercurrent	11	4,728	7	2,721	7	3,760		
Open/crosscurrent	2	700	2	1,380	3	1,280		
Subtotal	13	5,428	9	4,101	10	5,040	32	14,569
Tray								
Sieve	2	1,100						
Subtotal	2	1,100					2	1,100
Packed								
Static bed	2	1,480						
Mobile bed	3	1,176						
Rod deck	3	816						
Grid	1	550	2	980				
Subtotal	9	4,022	3	1,380			12	5,402
Combinations								
Spray/packed	9	3,927	3	1,475	1	166		
Venturi/spray			2	1,408				
Subtotal	9	3,927	5	2,883	1	166	15	6,976
Total[c]	42	18,590	19	9,174	11	5,206	72	31,968

[a] Gross unit generating capacity, MW.
[b] One unit (235 MW) uses venturi scrubbers for SO_2 control only. All of the others use venturis as part of a scrubbing train to provide primary or secondary particulate removal and remove only a portion of the inlet SO_2.
[c] Totals include venturi scrubbers. Except for the 235-MW unit, all venturi scrubbers are followed by another scrubber that provides primary SO_2 control.

thorough mixing of the gas and slurry droplets. Large differences in velocities of the slurry and gas occur in the turbulent zone and cause impaction. As the slurry droplets and gas decelerate in the divergent section of the venturi, the droplets agglomerate through collisions and a portion of the energy required to accelerate the gas through the throat is recovered. The solids (fly ash, limestone slurry, SO_2 reaction products) are separated from the gas stream by gravity as the stream moves to the next scrubber stage.

Venturis are considered high-energy devices because they normally operate in the pressure drop range of 10 to 20 in. H_2O (in limestone FGD systems). They can remove submicron particulate with increasing efficiency as a function of pressure drop. They also provide effective SO_2 absorption because the finely atomized slurry droplets present a large liquid surface area for contact with the gas stream. Absorption of SO_2, however, is much less efficient than in other scrubber designs because the contact time is short and the slurry is introduced into the gas stream cocurrently.

Variable-throat designs offer the option of changing the cross-sectional area of the throat to accommodate varying flue gas flow rates. With a variable-throat design a constant pressure drop can be maintained across the throat, and thus a relatively constant removal efficiency (particulate and SO_2) can be maintained even with widely fluctuating flue gas flow rates (turndown ratio of approximately 2:1). Designs currently used in limestone scrubbers to adjust the throat opening include liquid distribution disks, blades, blocks, rod decks, and adjustable drums. Figure 3-1 depicts a basic venturi scrubber configuration and the variable-throat design options.

Spray Tower Scrubber. A spray tower is an open gas absorption vessel in which scrubbing slurry is introduced into the gas stream from atomizing nozzles. The relative velocities of the gas stream and the slurry spray allow intimate gas/liquid contact for SO_2 absorption.

In typical spray towers, the pressure imparted to the scrubbing slurry discharged from the spray nozzles, together with the velocity of the incoming gas stream, produces fine liquid droplets as sites for SO_2 absorption. Nozzle pressures of 15 to 20 psig are normally utilized to produce droplets of 2500 to 4000 microns diameter. Droplets in this size range provide sufficient surface area for SO_2 absorption, and the entrainment problems normally associated with smaller droplets are minimized. Relatively low gas-phase pressure drops, approximately 1 to 4 in. H_2O, are normally encountered because the spray tower includes no internals other than spray headers.

Important design features of a limestone slurry spray tower are L/G ratio, gas distribution, and liquid distribution. High L/G ratios improve SO_2 removal by increasing the surface area for mass transfer and by reducing the high liquid film resistance to SO_2 absorption. Moreover, because calcium sulfite and sulfate tend to form highly supersaturated solutions by virtue of their low solubilities in water, operation at high L/G ratios prevents any instantaneous saturation beyond 30 percent of normal solubilities. This reduces the potential for scale formation in the scrubber.

Gas distribution is a critical feature with respect to spray tower performance. Uniform distribution of gas in the tower is achieved in vertical towers by action of the sprays on the rising gas; the sprays apparently impart enough energy to the gas stream to distribute it evenly.

The distributed liquid must completely cover the cross section of the tower. The tower must include enough spray nozzles to provide a spray zone of uniform density. Placement of nozzles so as to provide a considerable

The FGD System   55

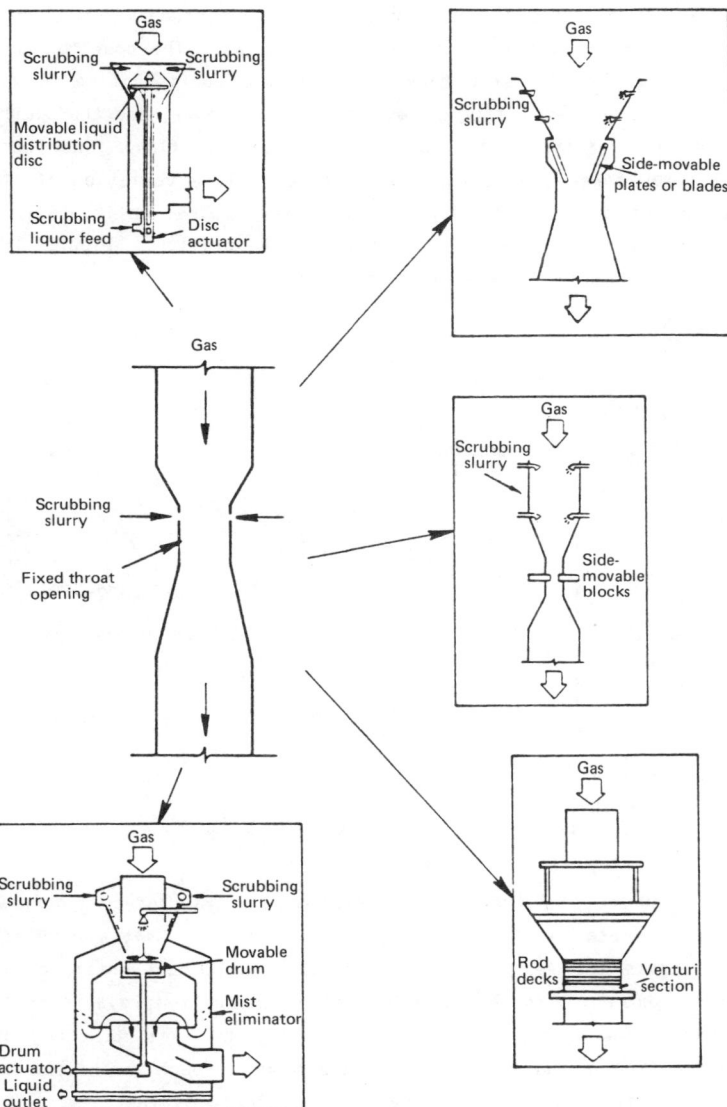

Figure 3-1. Basic venturi scrubber and design configurations.

overlap of slurry spray reduces the problems associated with nozzle failure, which could create a path of least resistance to gas flow. The number of spray headers (spray banks) through which slurry is fed to the nozzles varies with the amount of SO_2 loading and the required SO_2 removal efficiency. One to six spray headers are used in limestone spray towers in commercial FGD systems operating on coal-fired flue gas.

Spray towers currently used in commercial limestone FGD systems include open/countercurrent and open/crosscurrent designs. The open/countercurrent spray tower (vertical spray tower) is a simple configuration in which the gas stream passes vertically upward through the tower, and the slurry droplets fall by gravity countercurrently to the gas flow. The open/crosscurrent spray tower (horizontal spray chamber) is a variation of the open/countercurrent design, in which the gas stream passes horizontally through the vessel and the slurry droplets fall by gravity. Figure 3-2 shows simplified diagrams of these basic spray tower designs.

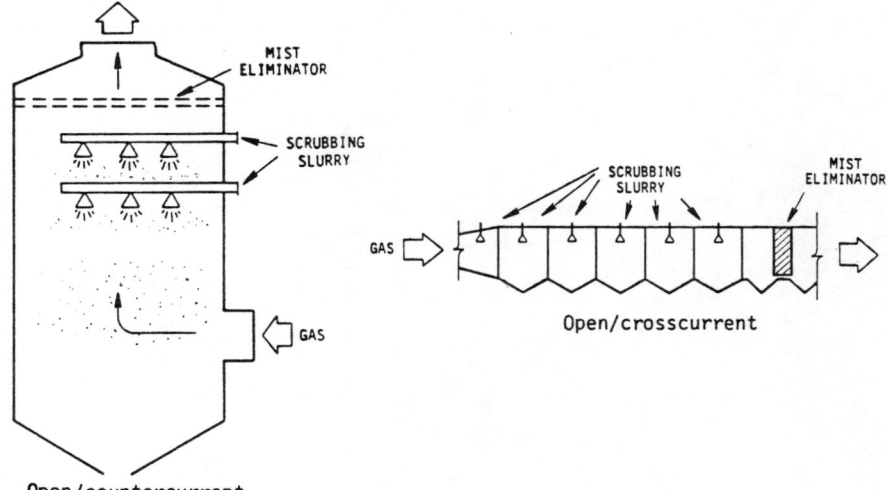

Figure 3-2. Basic spray tower designs.

Tray Scrubber. A tray tower incorporates a tray internal consisting of a horizontal metal surface perforated with holes or slots mounted transversely across the vessel. In this scrubber, flue gas enters at the base and passes upward through the holes while slurry is sprayed onto the tray from above. The slurry builds up on the tray until it has enough pressure to overcome the pressure of the gas passing up through the holes. An equilibrium is established between the gas and slurry on the tray. The slurry is vigorously agitated to a froth, which provides a large gas/liquid contact area. Absorption of SO_2 into the slurry occurs in drops suspended on froth.

The sieve tray is the simplest tray scrubber design. With the use of a conventional sieve tray, the gas velocities are such that gas passing up through the holes bubbles through the liquid on the tray and provides intimate gas/liquid contact. Overflow pipes or weirs divert a portion of the slurry through a "downcomer" to preceding stages in the scrubber or directly to the effluent hold tank.

A principal disadvantage of the sieve tray is its extremely limited turndown capability. A stable froth layer can be maintained only within a narrow range of gas flow rates. At rates below the lower limit, "weeping" may occur, whereas at rates above the upper limit the gas/slurry mixture is blown out the scrubber. Another disadvantage is the potential for plugging of the tray holes with accumulated reaction products (scale deposits), unreacted limestone, and/or fly ash. Such plugging can cause an increase in pressure drop across the scrubber, sometimes severe enough that the unit must be shut down for manual removal of the deposits.

Pressure drop across each tray varies with the open tray area, hole diameter, slurry recirculation rate, and overflow rate. Typical pressure drops are 1 to 4 in. H_2O. Typical hole diameter is 1-3/16 in. Typical superficial gas velocities through the holes are 15 to 20 ft/s.

A simplified diagram of a sieve tray scrubber is shown in Figure 3-3.

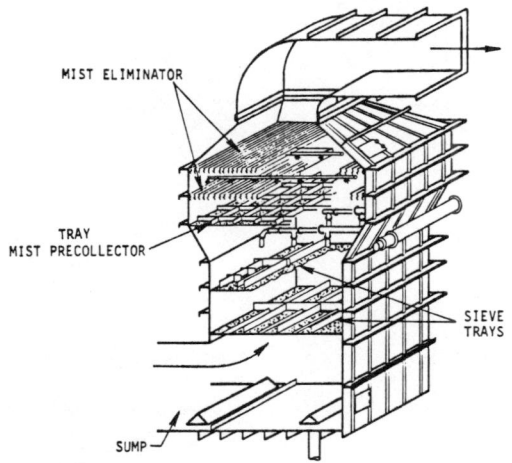

Figure 3-3. Sieve tray scrubber and detail of tray.

<u>Packed Tower Scrubber</u>. A packed tower incorporates a bed of stationary or mobile packing, which is mounted transversely across the vessel. In a packed scrubber, flue gas enters the base of the tower and flows up through the packing countercurrent to the slurry, which is introduced at the top of the scrubber by low-pressure distributor nozzles. The packing provides a large surface area for gas/slurry contact. The greater the contact area, the longer is the holdup time and the more effective is the mass transfer of SO_2 into the slurry.

Depending on packing type and configuration, packed towers usually operate at gas-side pressure drops of 3 to 10 in. H_2O, L/G ratios of 20 to 50, and gas velocities of 5 to 12 ft/s.

Three packed tower designs currently used in limestone FGD systems include static bed, mobile bed, and rod deck. A static bed is an essentially immobile bed of a packing material such as glass spheres. A mobile bed consists of a highly mobile bed of solid spheres, which are fluidized by the gas stream. A rod deck tower consists of a series of decks of closely spaced immobile rods placed on staggered centers. Figure 3-4 is a simplified diagram of a packed bed tower, showing the design variations.

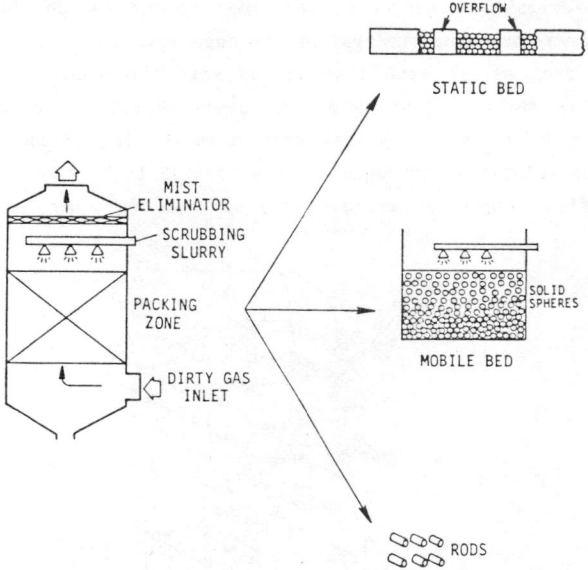

Figure 3-4. Packed tower and packing types.

Combination Scrubbers. Combination towers currently used or planned for use in limestone FGD systems include the spray/packed design and the venturi/spray design. In the spray/packed tower, the flue gas first contacts the limestone slurry in an open/countercurrent spray zone. The second stage consists of a fixed bed of stationary rigid packing of honeycombed material. Typical design parameters for this scrubber are gas-side pressure drops of 3 to 4 in. H_2O, L/G ratios of 40 to 80, and gas velocities of 10 ft/s in the spray zone and 15 ft/s in the packed zone. A slurry with low solids content (5 to 10 percent) and low pH (approximately 5) is recirculated in the spray zone and consists of spent slurry from the packed zone.

In the combination venturi/spray tower, the flue gas first contacts the limestone slurry in a variable-throat venturi. The second stage consists of an open/countercurrent spray zone. The venturi removes primarily particulate and a portion of the inlet SO_2, and spent slurry from the spray zone is used to improve limestone utilization while controlling sulfite oxidation. The spray tower removes primarily SO_2 and a portion of the residual particulate remaining from the venturi. Typical design parameters for this scrubber are a gas-side pressure drop of 10 in. H_2O, an L/G ratio of 80, and gas velocities of 150 ft/s in the venturi and 10 ft/s in the spray zone.

Design Considerations

Important mechnical design features to be considered when evaluating the various types of scrubbers for limestone FGD systems include L/G ratio, gas velocity and pressure drop, residence time, gas/liquid distribution, and turndown capability. A detailed discussion of the theoretical aspects of these design considerations is presented in Appendix B. Following is a brief review of these design considerations as they relate to the major aspects of limestone scrubber selection.

L/G Ratio. The scrubber L/G ratio represents the primary mechanical means of achieving the required SO_2 removal. The L/G ratio is one of the more variable parameters in scrubber design. For example, the L/G ratios of limestone scrubbers in commercial service or planned for commercial service on coal-fired utility boilers range from 20 to 80 gal/1000 acf. Pilot plant systems have been tested at L/G ratios of 5 to 120 gal/1000 acf.

With respect to mechanical design considerations, selecting the most appropriate L/G ratio for a specific application must take into account the gas/liquid contact area. In some scrubber designs, the gas/liquid contact area is directly proportional to the L/G ratio. In a venturi or spray tower design, for example, the only contact between the gas and liquid phases occurs as the liquid droplets pass through with the gas (venturi) or fall through the gas (spray tower). In these designs, the L/G ratio affects the quantity of SO_2 absorbed, typically in direct proportion to a change in L/G ratio. In other scrubber designs (tray, packed), the relationship between L/G and SO_2 is not so direct because of variations in other mechanical design factors that influence SO_2 absorption (e.g., the presence of internals that provide intimate mixing of liquid and gas).

Although precise L/G ratios cannot be specified to define optimal values for all scrubber designs, the following generalizations can be applied:

- If the scrubber design is one in which the gas/liquid contact area is directly proportional to L/G, any increase in L/G will increase total absorption of SO_2, independently of SO_2 concentrations.

- An increase in L/G will increase total absorption of SO_2, independently of scrubber design.

Table 3-3 summarizes the design L/G ratios of the limestone scrubbers currently operating or planned for service on coal-fired utility boilers.

Gas Velocity and Pressure Drop. The superficial gas velocity of the

TABLE 3-3. LIMESTONE SCRUBBER DESIGN L/G RATIOS

	L/G ratio, gal/1000 acf (inlet)					
	Operational			Planned		
Scrubber type	No.[a]	Range	Avg.	No.[a]	Range	Avg.
Venturi						
Variable-throat/bottom-entry liquid distribution disk	1	10	10			
Variable-throat/side-movable blades	1	10	10			
Variable-throat/side-movable blocks	2	12-14	13			
Variable-throat/vertically adjustable rod decks	4	16-18	17			
Spray						
Open/countercurrent	9	19-70	49	3	10-80	48
Open/crosscurrent				1	54	54
Tray						
Sieve	2	27-48	37			
Packed						
Static bed	2	10	10			
Mobile bed	3	41-60	50	1	45	45
Rod deck	3	50	50			
Grid	1	60	60			
Combination						
Spray/packed	1	20-60	40	1	60	60

[a] Represents number of systems for which design scrubber L/G is known.

flue gas in the scrubber is determined by volumetric flow rate through the vessel and by the cross-sectional open area of the vessel. Typically, a higher superficial gas velocity is preferred because a lower cross-sectional area is then required, and the scrubber vessel can be smaller. The upper limit on gas velocity is determined by the flooding potential (or pressure drop) of packed and tray towers and by the mist eliminator re-entrainment potential of venturi and spray towers.

The gas-side pressure drop through the scrubber depends on superficial gas velocity and scrubber design; it is increased, for example, by scrubber designs that include internal structures to impede the flow of gas and slurry.

Tables 3-4 and 3-5 summarize the design gas velocities and pressure drops of current and planned limestone FGD scrubbers.

TABLE 3-4. LIMESTONE SCRUBBER DESIGN GAS VELOCITIES

Scrubber type	Gas velocity, ft/s					
	Operational			Planned		
	No.[a]	Range	Avg.	No.[a]	Range	Avg.
Venturi						
Variable-throat/side-movable blocks	2	90-130	110			
Variable-throat/vertically adjustable rod decks	1	80	80			
Spray						
Open/countercurrent	2	10	10	3	8-10	9
Open/crosscurrent	1	22	22	2	22	22
Tray						
Sieve	2	10-15	13			
Packed						
Static bed	1	31	31			
Mobile bed	2	13-15	14			
Rod deck	1	13	13			
Grid	1	12	12			
Combination						
Spray/packed	2	7-10	8	1	10	10

[a] Represents the number of systems for which design scrubber L/G is known.

TABLE 3-5. LIMESTONE SCRUBBER DESIGN GAS-SIDE PRESSURE DROPS

Scrubber type	Gas-side pressure drop, in. H_2O					
	Operational			Planned		
	No.[a]	Range	Avg.	No.[a]	Range	Avg.
Venturi						
Variable-throat/bottom-entry liquid distribution disk	2	15	15			
Variable-throat/side-movable blades	1	5	5			
Variable-throat/side-movable blocks	2	3-7	5			
Variable-throat/vertically adjustable rod decks	4	9-13	11			

(continued)

TABLE 3-5. (continued)

Scrubber type	Gas-side pressure drop, in. H_2O					
	Operational			Planned		
	No.[a]	Range	Avg.	No.[a]	Range	Avg.
Spray						
Open/countercurrent	9	1-7	5	5	3-8	7
Open/crosscurrent	2	2	2	1	2	2
Tray						
Sieve	2	4-6	5			
Packed						
Static bed	2	2	2			
Mobile bed	3	6-12	8	1	14	14
Rod deck	3	8	8			
Grid	1	2	2			
Combination						
Spray/packed	8	1-6	4			

[a] Represents the number of systems for which design scrubber L/G is known.

Residence Time. Residence time is the amount of time that the slurry is in contact with the gas in the scrubber vessel. Generally, increasing the residence time increases SO_2 removal. Thus, a required SO_2 removal efficiency can be achieved by increasing the scrubber residence time, which is done by promoting slurry holdup through the use of internals and by reducing the slurry flow rate (L/G ratio). This alternative, however, increases the gas-side pressure drop, which increases the system energy consumption because of higher demand on the fans.

The relationship between residence time and SO_2 removal would be directly proportional if the SO_2 transfer rate were to remain constant. The relationship is not linear, however, because of changes in composition of the flue gas and the slurry. Moreover, since residence time is also a function of equipment size, a larger scrubber tends to capture more SO_2 than a smaller one.

Gas/Liquid Distribution. Gas/liquid distribution refers to the intermixing of gas and slurry in the scrubber vessel. Proper distribution of gas and liquid is essential for maintaining design SO_2 removal efficiencies. Improper distribution reduces both the effective residence time and the effective interfacial mass transfer area.

Uniformity of gas flow across the scrubber can be enhanced by the use of distribution vanes and by provision of sufficient freeboard distance between the gas inlet and first stage (tray, packing, spray header) and between the stages.

Uniformity of scrubbing liquor flow is achieved by atomizing the liquor into fine droplets to increase the interfacial area and by optimizing the angle of the spray from the nozzles. In venturi and spray tower designs, atomization of liquid is an important consideration; atomization is affected by pressure drop in the throat in a venturi scrubber, or by nozzle type, nozzle opening, and nozzle pressure drop in a spray tower scrubber. The fineness of droplets in these designs is limited by considerations of energy consumption and entrainment.

Turndown Ratio. Turndown capability is the ratio of maximum to minimum gas flow that a scrubber can handle without reducing SO_2 removal or causing unstable operation. This ratio is dependent on scrubber design. For example, although a reduction in gas flow rate should generally improve SO_2 removal efficiency, it also reduces the gas/liquid interfacial area by decreasing the gas dispersion in tray towers, the liquid agitation in packed towers, or the pressure drop in venturi towers. In spray towers, the gas/liquid interfacial area does not depend on gas flow rate or pressure drop.

Turndown capability is also a function of the mechanical design of the scrubber. For example, "weeping" may occur in tray towers at reduced gas flow rates because of the decrease in gas pressure. "Channeling" may occur in certain packed towers at low gas flow rates.

The range of turndown capability therefore depends on scrubber design. In general, the turndown capabilities of spray towers (4:1) are superior to those of tray towers (3:1), packed towers (2:1), and venturi towers (2:1).

Operational Systems

Table 3-6 summarizes data on selected utility limestone FGD systems according to scrubber design and operating parameters. The systems are representative of the major scrubber types described. One facility is listed for each major scrubber type; high-sulfur coal applications were selected preferentially because they represent the greatest severity of design and performance conditions.

MIST ELIMINATORS

Description and Function (Conkle et al. 1976)

A mist eliminator collects slurry droplets entrained in the scrubbed flue gas stream and returns them to the scrubbing liquor. In limestone scrubbing, small droplets of liquid are formed and carried out the scrubber with the gas. These mist droplets generally contain both suspended and dissolved solids. The suspended solids are derived from particulates (fly

TABLE 3-6. SCRUBBER DESIGN AND OPERATING CHARACTERISTICS
FOR OPERATIONAL LIMESTONE FGD SYSTEMS

Generic type	Specific type	Modules/unit	Dimensions, ft	Gas flow acfm	Gas temp., °F	L/G, gal 1000 acf	gas velocity ft/s	ΔP, in. H_2O	SO_2 removal, %
Spray tower[a]	Open counter-current spray	2	NR	296,000	129	74	NR	6.0	89
Tray tower[b]	Sieve tray	8	32x16x65	238,500	122	38	15	6.0	80
Packed tower[c]	Mobile bed packing	1	30x16x34	400,000	280	60	12	2.0	70
	Grid packing	3	30x16x34	400,000	280	60	NR	2.0	70
Combination[d]	Spray/packed	2	30x93 (dia.xheight)	387,500	280	NR	9	0.7	95

[a] Built by Babcock & Wilcox; operated on Marion 4, Southern Illinois Power Coop; boiler output 173 MW gross; coal sulfur content 3.75 percent.

[b] Built by Babcock & Wilcox; operated on La Cygne 1, Kansas City Power & Light; boiler output 820 MW gross; coal sulfur content 5.39 percent.

[c] Built by TVA; operated on Widows Creek 8, TVA; boiler output 550 MW gross; coal sulfur content 3.70 percent.

[d] Built by Research Cottrell; operated on Dallman 3, Springfield Water, Light & Power; boiler output 290 MW gross; coal sulfur content 3.30 percent.

NR - Not reported.

ash) collected in the scrubber, from limestone introduced into the scrubbing liquor, and from products of chemical reactions occurring within the scrubber. Similarly, the dissolved solids come from gaseous species absorbed from the flue gas, from limestone and reaction products, and from the system's makeup water (e.g., cooling tower blowdown, well water, river water).

Carryover of mist from the scrubber can affect both the FGD system and the ambient atmosphere. It can collect on the downstream equipment--the reheater ID fan, ductwork, dampers, and stack. Where an in-line reheater is used, the reheat energy requirement to effect a given temperature rise increases as mist carryover increases. Moreover, the mist carryover can collect on the heat exchange surfaces of the reheater and eventually cause plugging, which can reduce the reheater's heat transfer capability and contribute to corrosion.

Droplets can collect on the blades of an ID fan. The solids deposited from these droplets can cause vibration, possibly leading to failure of the blades, rotor, housing, and/or support structure.

Solids from the mist carryover can also be deposited in the ductwork, dampers, and stack, where they can accumulate and break off in chunks that are eventually blown through the system and out the stack. Moreover, if the equipment downstream of the mist eliminator is constructed of materials dependent on dry gas service, an increase in energy demand or a reduction in heat transfer capability may reduce the efficiency of the reheater and lead to corrosion attack on these components.

With respect to the ambient atmosphere, mist carryover can increase the particulate loading in a discharge gas stream to the extent that emissions do not comply with particulate and opacity regulations.

Basic Types (Conkle et al. 1976; PEDCo Environmental, Inc. 1981; IGCI 1975)

Two basic types of mist eliminator designs have been developed for use in limestone scrubbing systems: the primary collector and the precollector. A primary collector sees the heaviest duty with respect to mist loading and required removal efficiency. A precollector precedes the primary collector and is designed to remove most of the large entrained mist droplets from the gas stream before it passes through the primary collector. Most limestone FGD systems are equipped solely with primary collectors; only a few are equipped with precollectors.

Table 3-7 summarizes the various designs for primary collectors and precollectors used or planned for use in commercial-scale limestone FGD systems for coal-fired utility boilers. The generic classification desig-

TABLE 3-7. BASIC MIST ELIMINATOR TYPES

Generic type	Specific type	Trade or common name
Primary collectors		
Impingement	Mesh	Knitted wire mesh pad
	Tube bank	Vertical parallel tube bank
	Curved deflector plate	Gull wing
	Baffle	Open vane (slat) Closed vane
Electrostatic deposition	ESP	Wet ESP
Centrifugal separation	Radial vane	Spin vane Radial baffle
Cyclonic separation	Cyclonic separator	Cyclonic tower
Precollectors		
Bulk separation	Baffle slats	Bulk entrainment separator
	Perforated plate	Sieve tray
	Gas direction change	90° 180°
Knock-out collection	Wash tray	Valve (bubble cap) tray
	Trap-out tray	Irrigation tray

nates the basic collection mechanism by which the entrained mist droplets are removed from the gas stream. The specific classification includes the distinct design variations within each generic grouping. The common or trade name classification lists trade names or common names assigned to special or proprietary designs. This classification also includes variations of specific design types.

The four generic groupings established for primary collectors are based on collection by impingement, electrostatic, centrifugal, and cyclonic mechanisms, briefly defined as follows:

> Impingement (or inertial impaction) effects mist removal by collection on surfaces placed in the gas stream. The liquid droplets thus collected coalesce and fall by gravity (or drain) back into the scrubbing liquor.
>
> Electrostatic deposition effects mist removal through the use of an electrostatic field. The particles entrained in the gas stream are exposed to an electrostatic field, become charged, migrate to an oppositely charged surface, are collected, and then are returned to the scrubbing liquor.
>
> Centrifugal separation is based on the use of baffles that impart a centrifugal force to the gas. The mist droplets entrained in the gas are spun out to the walls of the separator chamber, where they collect and drain by gravity back to the scrubbing liquor.
>
> Cyclonic separation involves the use of tangential inlets which impart a swirl or cyclonic action to the gas as it passes through the separator chamber. The mist droplets entrained in the gas stream are spun out to the walls of the chamber, where they collect and drain to the scrubbing liquor.

Of these four generic types, impingement collectors have been used most extensively in limestone scrubbers. Virtually all of the impingement collectors used to date have relied on the baffle design, through the use of either open slats or continuous chevron vanes.

Of the two generic types of precollectors, bulk separation devices have been used most extensively in limestone scrubbers. These devices involve the use of slats (similar to the open-vane primary collector design), trays, or gas changes prior to passage through the primary collector. In the knock-out devices, trays are used to collect and recirculate wash water in a separate recirculation loop.

The balance of this discussion focuses on primary collectors. (Where precollectors are discussed, they are designated as such.) Moreover, because only the baffle type is in use or planned for use in commercial systems, this discussion is limited to the two design variations of baffle-type mist eliminators: open vane and chevron vane.

Open Vane. The open-vane, baffle-type mist eliminator consists of a series of disconnected slats. Two variations of this design are the zigzag baffle and the open louver. These design variations are illustrated in Figure 3-5.

The open-vane design has received recent interest in limestone applications because of ease of washing and the absence of "dead corners," which can become clogged. Recent model tests have shown that, during washing of the open louver unit, water actually circulates around the blades and thus cleans both sides.

<u>Chevron Vane</u>. The chevron-vane, baffle-type mist eliminator consists of a series of connected slats. Two variations of this design are the sharp-angle (Z-shape) chevron and the smooth-angle (S-shape) chevron. Figure 3-5 illustrates the difference between the Z-shaped and S-shaped bends. Basically, sharp-angle chevrons provide greater collection efficiency and construction stability. With smooth-angle chevrons, however,

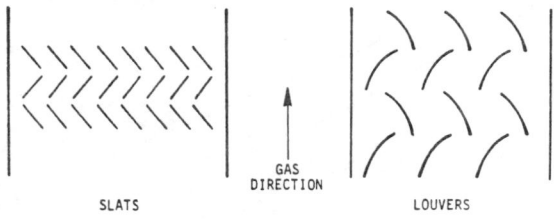

OPEN-VANE DESIGN

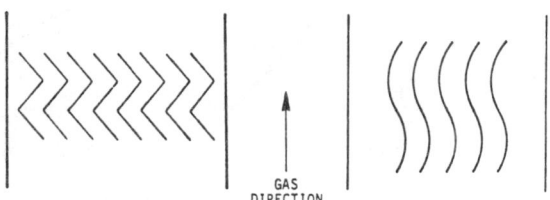

CLOSED-VANE DESIGN

Figure 3-5. Baffle-type mist eliminator designs.

there are fewer problems associated with plugging in "dead corners" and reentrainment.

Among all of the baffle-type mist eliminator designs, the sharp-angle chevron mist eliminator is the predominant type currently used or planned for use in limestone FGD systems.

<u>Design Considerations (Conkle et al. 1976; PEDCo Environmental, Inc. 1981)</u>

Calvert et al. (1974) developed theoretical equations for baffle-type mist eliminators. The equations pertaining to primary collection efficiency, pressure drop, and reentrainment potential are not rigorously applicable to the high-solids slurry environment typical of limestone scrubbers, but they do provide insight into the relative importance of specific design variables.

Figure 3-6 shows the calculated relationships between penetration and gas velocity with diameter of the mist droplets and baffle angle. Penetration is defined here as the ratio of the mass of drops at the outlet to the mass at the inlet of the mist eliminator; that is, it represents the quantity that escapes collection and thus is a measure of collection efficiency.

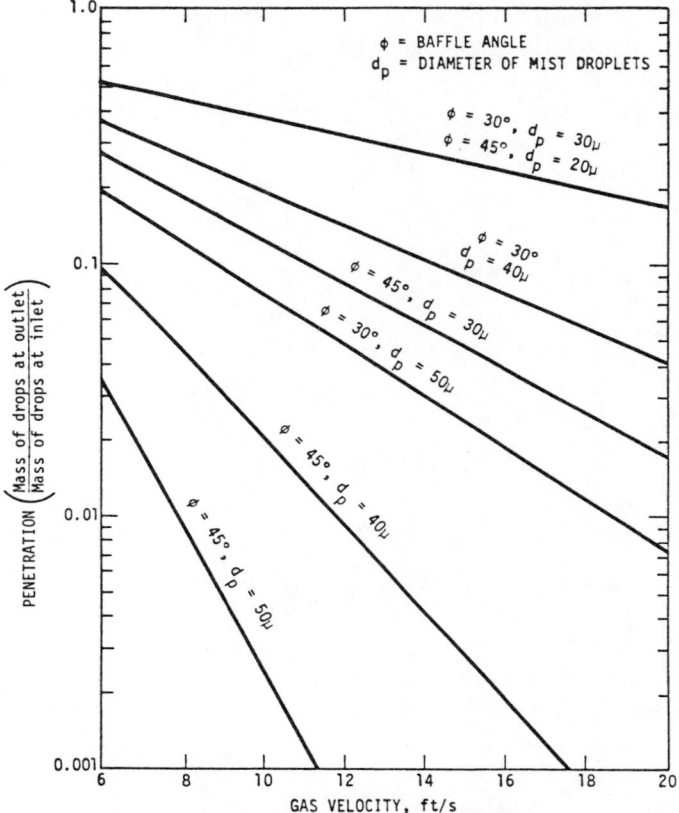

Figure 3-6. Relationship of penetration and gas velocity.

At moderate gas velocities, where collection is high and reentrainment low, this theoretical relationship agrees well with experimental data on overall collection efficiency. As velocities increase, reentrainment becomes significant and this relationship no longer adequately describes the observed collection efficiency. Since high gas velocity is desirable for efficient mist collection, designers must consider a trade-off between high gas velocity and reentrainment. Ostroff and Rahmlow (1976) have measured gas velocities in chevron mist eliminators and have suggested a method for determining the point of failure.

The pressure drop that can be tolerated before reentrainment occurs is determined experimentally. Then, the maximum gas flow rate can be calculated with known physical constants, gas properties, mist eliminator dimensions, a correction factor, and the experimentally determined pressure drop. Alternatively, if the flow rate is dictated by the application, the appropriate dimensions of the mist eliminator can be determined.

Factors affecting reentrainment include wash water rate and quantity, turndown ratio, gas velocity, and scrubber L/G ratio. Of these, the most important are gas velocity and L/G ratio. In relating these two factors, Figure 3-7 displays in the shaded area the region where reentrainment becomes observable. In that area, reentrainment constitutes 0.5 to 1 percent of the inlet loading to the mist eliminator. Thus, to maintain the scrubber in or below the desired low-reentrainment range, superficial gas velocity and L/G ratio cannot be set independently. Clearly, if the gas velocity is increased, the L/G ratio must be reduced and vice versa.

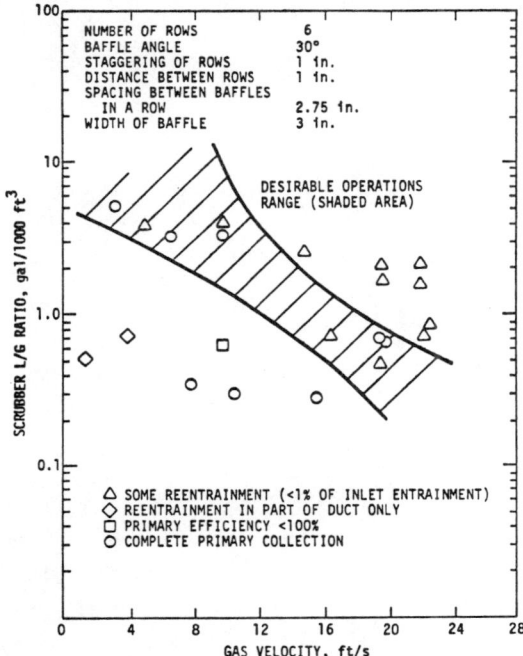

Figure 3-7. Effect of gas velocity and liquid loading on performance of baffle-type mist eliminators.

In the design of a mist elimination system, several conflicting objectives must be considered. The theoretical considerations of high collection efficiency and reduction of reentrainment must be weighed against such mechanical factors as washability and susceptibility of the unit to scaling and plugging. The balance of this discussion of design considerations deals with the following significant design factors: configuration, number of stages, number of passes, freeboard distance, distance between stages, distance between vanes, vane angle, and use of precollection devices.

Configuration. The two basic mist eliminator configurations are horizontal and vertical. In the horizontal configuration, the gas flows in a vertical direction and opposes the path of drainage. In the vertical configuration, the gas flows in a horizontal direction through vertically arranged vanes. Figure 3-8 illustrates these basic configuration arrangements.

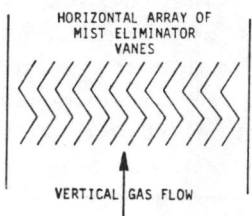

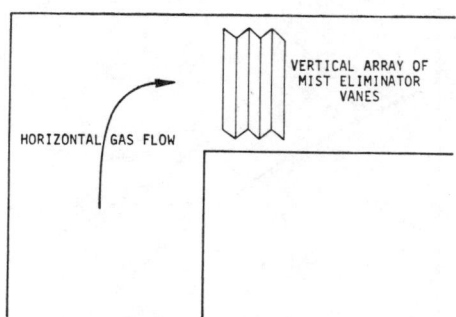

Figure 3-8. Horizontal and vertical mist eliminator configurations.

In the horizontal configuration, the mist droplet must overcome drag forces exerted by the gas stream before it falls from the mist eliminator blade. The balancing of drag and gravitational forces results in a longer residence time of droplets on the blade, which increases the chance of solids deposition and reentrainment. This is one of the disadvantages of the horizontal configuration. Another problem is that the direction of wash water is limited. Whereas the most effective washing is achieved when water flows longitudinally along the length of the vane, the horizontal configuration admits wash water only from the top or bottom of the mist eliminator.

In the vertical configuration, the mist eliminators are normally installed in a separate chamber after the scrubber. Sometimes they are mounted in the top section of the scrubber in a "suspended box" arrangement. Removal of droplets is efficient in the vertical configuration primarily because "piling up" of the collected liquor is avoided. Also, in the vertical configuration the unit can be operated at higher gas velocity without reentrainment. Even though it is more expensive, the vertical configuration reduces the load on the reheater because of its higher efficiency.

Number of Stages. Both single- and multiple-stage (two-stage is most common) mist eliminators are used in limestone scrubbers. The efficiency of a single-stage mist eliminator can be increased with the use of a precollector. Although the two-stage mist eliminator is more expensive and somewhat more complex, it offers several advantages. The first stage of a two-stage system can be washed rigorously from the front as well as from the back. Mist generated in the washing operation is collected in the normally unwashed second stage. A two-stage unit makes possible the use of a greater quantity of wash water at higher pressure for a longer period of washing; it also affords greater operating flexibility.

Number of Passes. The number of passes in a mist eliminator corresponds to the number of direction changes the gas stream must make before it exits. The greater the number of passes, the greater the collection efficiency and the gas-side pressure drop. Because of the high-solids environment of a limestone scrubber, however, the likelihood of plugging increases with the number of passes. Figure 3-9 shows two-pass, three-pass, and

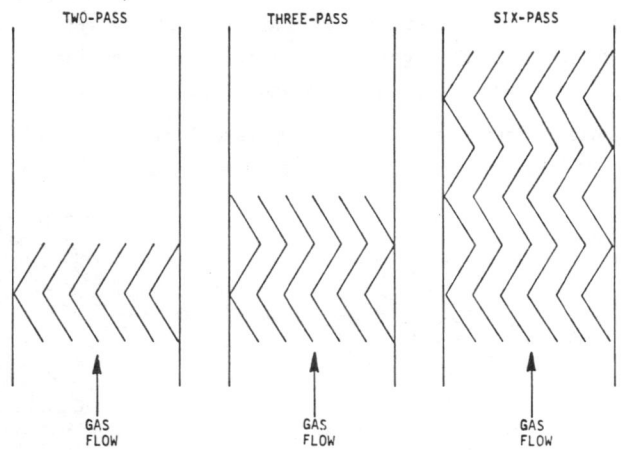

Figure 3-9. Schematic of one-stage mist eliminators with two-, three-, and six-pass arrangements.

multiple-pass chevron mist eliminator configurations. Three-pass collectors, the most commonly used in limestone FGD systems, provide good collection efficiency (>90 %) with adequate washability.

Freeboard Distance. Freeboard distance is the distance between the top of the scrubber section and the mist eliminator. It varies widely among installations, ranging from 4 ft to more than 20 ft. In the freeboard area, entrained particles can coalesce and return to the scrubber solution by gravity before encountering the mist eliminator. Most particles usually drop out in the first 8 to 10 ft; additional freeboard is not effective in removing the smaller entrained particles before they contact the mist eliminator.

Distance Between Stages. First-generation designs of multiple-stage mist eliminators typically provided less than 3 ft between stages. With such short distances, deposition of solids occurred frequently. Designers have subsequently achieved higher collection efficiency by allowing for better-washed first stages and including enough spacing between stages to allow entrained liquid droplets to settle out before they contact the second stage. The minimum requirement for distance between stages is 6 ft. This also allows sufficient space for personnel to walk between stages for cleaning and maintenance.

Distance Between Vanes. The spacing between individual baffle vanes is an important factor in mist eliminator design. The closer the spacing, the better the collection efficiency, but the greater the potential for solids deposition. In a single-stage mist eliminator, the spacing typically ranges from 1.5 to 3 in. If a second stage is used, the spacing is usually the same as that in the first stage. It can, however, be reduced to as low as 7/8 to 1 in. to provide higher efficiency for collection of the smaller droplets that pass through the first stage.

Vane Angle. Both sharp and smooth vane bends are used in limestone scrubbers, but sharp-angle vanes predominate. Figure 3-10 illustrates the

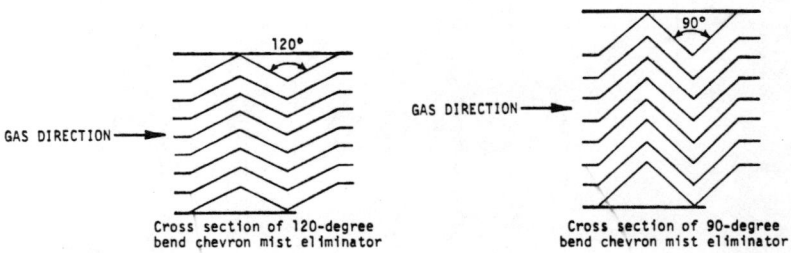

Figure 3-10. Mist eliminator vane angle configurations.

difference between the 120-degree and the 90-degree bends in a three-pass, continuous-chevron mist eliminator. Sharp-angle bends cause more sudden changes in direction of the gas and provide greater collection efficiency, but they also are conducive to reentrainment and provide convenient sites for the deposition of solids.

Precollectors. As indicated earlier, a precollector removes large mist droplets from the gas stream before it passes through the primary collector. Precollectors available for use in limestone scrubbers include bulk separation and knockout devices. The bulk separation devices are characterized by a low potential for solids deposition, a low gas-side pressure drop, and simplicity. The knockout devices are characterized by a higher potential for solids deposition, higher pressure drop (~3 in. H_2O), and greater complexity. Because of these characteristics, bulk separation devices are favored exclusively over knock-out devices in limestone scrubber applications.

Mist Eliminator Wash

Design of the mist eliminator wash system has advanced greatly since scaling and plugging problems first became evident. Factors important in the design and operation of a mist eliminator wash system include the wash water type, the direction, duration, flow rate, and pressure of the wash water, and the maintenance provisions.

Wash Water Type. Since the main purpose of mist eliminator washing is to remove accumulated solids, fresh water is preferred. In closed-loop operation, washing with 100 percent fresh water is not possible. The normal procedure in limestone scrubbing systems is to introduce all makeup fresh water through the mist eliminator wash system. Additional wash water, which is sometimes required, is obtained by recycling clear water from the solids dewatering system (thickener, vacuum filter, and/or sludge pond) and blending it with the fresh makeup water.

Wash Direction. The direction of wash water flow depends on the mist eliminator configuration and on the number of stages. With a horizontal configuration, washing is possible only from the bottom and the top of the column. With a vertical configuration, wash water can be directed horizontally from the front or back and vertically from the top. Experience has shown that washing in a direction countercurrent to the gas flow, or from the top in a vertical mist eliminator, generates large quantities of mist. A second-stage mist eliminator is therefore desirable when long-duration countercurrent washing is planned. When a single-stage mist eliminator is specified, countercurrent washing is normally limited to short duration at high pressure and high volume.

Wash Duration and Flow Rate. The mist eliminator wash system can be operated in a separate loop. This is possible with horizontal mist eliminators (vertical gas flow) when knockout devices are used. These devices can significantly increase the total quantity of wash water available, allowing the use, when necessary, of continuous, high-volume sprays.

Wash Water Pressure. Wash water pressure, important to design of the wash system, varies widely. Operational limestone scrubbing systems use an intermittent flush spray at pressures ranging from 20 to 100 psig. Incorporation of high-pressure washing procedures requires the use of stainless steel or reinforced plastic baffles to withstand the additional stress of high-pressure washing.

Shawnee Experience (Burbank and Wang May 1980). Much of the early effort at the Shawnee test facility was devoted to achieving reliable mist eliminator operation. This work concentrated on the resolution of problems with solids deposition that often arose in the first generation lime/limestone FGD systems. Methods initially investigated to control these problems, which were caused by scaling and plugging, included operation at reduced gas velocity using various combinations of mist eliminator washing techniques and hardware configurations.

The original mist eliminator used in the spray tower prototype FGD unit was a baffle-type, open-vane assembly consisting of one stage of vanes with three passes (see Figure 3-11). This same mist eliminator configuration was

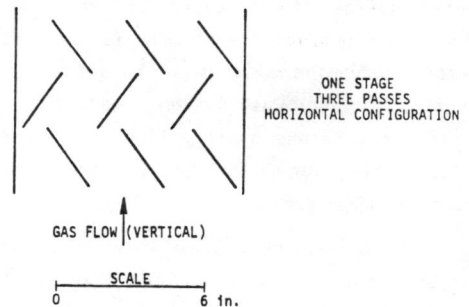

Figure 3-11. Original mist eliminator configuration at Shawnee test facility.

later used in the TCA prototype. Other mist eliminator configurations tested at Shawnee included the following: (1) a one-stage, six-pass, baffle-type, closed-vane mist eliminator, which was later equipped with an upstream wash tray; (2) a two-stage, three-pass unit, also of baffle-type, closed-vane design; (3) a cone-shaped, one-stage, four-pass unit, again of baffle-type, closed-vane configuration; and (4) a mesh pad unit placed atop the open-vane mist eliminator in the spray tower.

Because the experiences with these units were not completely favorable, a wide range of operating conditions was explored in an Advanced Test Program, in which open-vane mist eliminators were used in both the spray tower and TCA prototypes. The most significant finding involved the effect of alkali utilization on mist eliminator cleanliness. It was determined that the mist eliminator could be kept clean much more easily at high alkali utilization rates ($\geq$ 85 percent) than at lower utilization rates. The residual alkali in slurry solids deposited on the mist eliminator surfaces continues to react with the SO_2 (and O_2) in the exit gas, forming reaction products that are difficult to remove. In limestone scrubbing, which typically involves operation at low utilization values (< 85 percent), the mist eliminator was kept clean by operating at a lower pH level, which improved utilization to the required minimum level but was accompanied by a reduction in SO_2 removal efficiency.

Other significant findings included relationships between gypsum saturation level in the wash water and the incidence of gypsum scaling in the mist eliminator, between superficial gas velocity and loss of mist eliminator efficiency due to reentrainment, between solids levels in the scrubbing slurry and mist eliminator loading, and between the quality and quantity of wash water and cleanliness of the mist eliminator.

As a result of these findings, mist eliminator wash procedures were developed that maintained clean mist eliminators while minimizing any impacts on process chemistry, SO_2 removal, and closed-loop operation. These procedures, for bottom-side and top-side wash, are summarized in Table 3-8 as a function of high alkali utilization and low alkali utilization.

In operation at high alkali utilization rates, a periodic flush with makeup water was all that was required to keep the mist eliminator clean. The bottom side was flushed intermittently for 6 minutes every 4 hours with makeup water at a specific wash rate of 1.5 gpm per ft^2 of mist eliminator cross-sectional area. The top side was sequentially washed by one of six spray nozzles at a time for 4 minutes every 80 minutes at a specific wash rate of 0.5 gpm/ft^2. Total makeup water consumption with this scheme was equivalent to approximately 2.3 gpm on a continuous basis, well within the requirements for closed-loop operation.

In operation at lower utilization rates, the bottom side of the mist eliminator was washed continuously at a specific wash rate of 0.4 gpm/ft^2 with a blend of clarified scrubbing liquor and fresh water. At a blend of 80 percent clarified liquor and 20 percent fresh makeup water, all the makeup water available to maintain closed-loop operation was used. The top side was washed with each of the six spray nozzles activated in sequence for 3 minutes every 10 minutes at a specific wash rate of 0.5 gpm/ft^2. All

TABLE 3-8. MIST ELIMINATOR WASH PROTOCOLS AT SHAWNEE TEST FACILITY

Scrubber system	Alkali utilization >85 percent		Alkali utilization <85 percent	
Maximum flue gas rate, acfm at 300°F	Spray tower 35,000	TCA 30,000	Spray tower 35,000	TCA 30,000
Bottom-side wash				
Wash scheme	Low intermittent	Low intermittent	Continuous	Continuous
Number of nozzles	10	2	4	2
Nozzle location, in. below ME	10	31	20	31
Nozzle "On" time, min	6	6	Continuous	Continuous
Nozzle "Off" time, min	234	234		
Total on/off sequence, h	4	4		
Nozzle ΔP, psi	50	41	21	20
Flow rate per nozzle, gpm	7.5	37.5	5.0	10
Total flow rate, gpm	75	75	20	20
Specific wash rate, gpm/ft²	1.5	1.5	0.4	0.4
Makeup water (continuous basis), gpm	1.9	1.9	20	20
Top-side wash				
Wash scheme	Low intermittent	Low intermittent	High intermittent	High intermittent
Number of nozzles	6	6	6	6
Nozzle location, in. above ME	16	15	16	15
Nozzle "On" time, min	4	4	3	3
Nozzle "Off" time, min	76	76	7	7
Total on/off sequence, min	80	80	10	10
Nozzle ΔP, psi	13	13	13	13
Coverage area per nozzle, ft²	15	14.5	15	14.5
Flow rate per nozzle, gpm	8	8	8	8
Specific wash rate, gpm/ft²	0.53	0.55	0.53	0.55
Makeup water (continuous basis), gpm	0.4	0.4	2.4	2.4

available makeup water was used with 2.4 gpm as top-side wash and the remainder as clarified liquor diluent for bottom-side wash. With this wash scheme, some minor restriction of the mist eliminator was caused by the deposition of soft solids (plugging). The restriction, however, reached a steady-state level rarely above 10 percent and usually below 5 percent of the mist eliminator open area.

Operational Systems (PEDCo Environmental, Inc. 1981)

Table 3-9 lists selected limestone FGD systems currently in service according to mist eliminator design and operating parameters. The systems listed in this table are representative of the major designs.

REHEATERS (Choi et al. 1977)

Description and Function

In its most fundamental sense, stack gas reheat involves the addition of thermal energy to the gas stream discharged from the scrubber so as to raise its temperature. No equipment item, subsystem, or unit operation of the scrubbing process is more controversial than stack gas reheat. Questions are raised concerning the need for reheat and the effectiveness of the various strategies for achieving the desired level of reheat. The following reasons are generally advanced for using reheat:

- ○ To prevent condensation and subsequent corrosion in downstream equipment such as ducts, dampers, fans, and stack.
- ○ To prevent formation of a visible plume.
- ○ To enhance plume rise and pollutant dispersion.

The mechanisms by which a reheat system achieves these objectives are discussed in the following discussions of equipment types and design considerations.

Basic Types

A systematic grouping of the various stack gas reheat methods available for use in limestone FGD systems is provided in Table 3-10. The generic classification represents a grouping of reheat methods according to variations in configuration or energy source needed to raise the gas stream's temperature. The specific classification represents distinct design variations within each generic grouping, and the common classification indicates further design variations within each specific class.

<u>In-line Reheat</u>. In-line reheat involves the installation of a heat exchanger in the flue gas duct downstream of the mist eliminator. The heat exchanger is a set of tubes or tube bundles through which the heating medium

TABLE 3-9. MIST ELIMINATOR DESIGN AND OPERATING CHARACTERISTICS FOR OPERATIONAL LIMESTONE FGD SYSTEMS

Company name/unit name	Coal, % S	Generic design	Primary collector type generic/specific	Configuration	No. of stages	No. of passes/ stage	Freeboard distance, ft	Distance between stages/vanes, in.
Arizona Public Service Cholla 1	0.50	Spray tower	Impingement/ closed vane (1st stage) open vane (2nd stage)	Horizontal	Two	Two (1st stage) Four (2nd stage)	13.5	1.0/1.5 1.0/7.1
Central Illinois Light Duck Creek 1	3.66	Packed tower	Impingement/closed vane	Vertical	Two	Three	12.0	0.8/2.5
Colorado UTE Electric Assn. Craig 1	0.45	Spray tower	Impingement/closed vane	Horizontal	One	Three	5.0	NA
Kansas City Power & Light La Cygne 1	5.39	Tray tower	Impingement/closed vane	Horizontal	Two	Three	12.0	0.8/3.0
Kansas Power & Light Jeffrey 1	0.32	Spray tower	Impingement/closed vane	Horizontal	Two	Two	4.0	NR/2.0
Kansas Power & Light Lawrence 4	0.55	Spray tower	Impingement/closed vane	Horizontal	Two	Two	3.5	1.0/3.5
Northern States Power Sherburne 1	0.80	Packed tower	Impingement/closed vane	Horizontal	Two	Three	14.0	10.0/4.0
Salt River Project Coronado 1	1.0	Spray tower	Impingement/closed vane	Vertical	One	Three	29.0	0.7/1.2
South Mississippi Electric Pwr R.D. Morrow 1	1.30	Packed tower	Impingement/closed vane	Vertical	Three	One	NR	NR
Tennessee Valley Authority Widows Creek 8	3.70	Packed tower	Impingement/closed vane	Vertical	Two	Three	14.0	1.0/1.5

(continued)

TABLE 3-9 (continued)

Company name/unit name	ΔP, in. H_2O	Precollector type generic/specific	Wash system type	Wash system direction	Wash system duration	Wash water pressure, psig
Arizona Public Service Cholla 1	0.5	NA	Makeup water from well	Vertically downward (1st stage)	Intermittent	60
Central Illinois Light Duck Creek 1	1.0	NA	Fresh water and	Front spray/backspray pond overflow	Continuous	Low pressure
Colorado UTE Electric Assn. Craig 1	1.0	Sieve tray	Fresh water	Overspray	Intermittent	70
Kansas City Power & Light La Cygne 1	1.4	Bulk separation/ sieve tray	Blended pond and lake water (underspray); pond water (overspray)	Overspray/underspray	Continuous underspray/ intermittent overspray (every 8 hrs.)	80
Kansas Power & Light Jeffrey 1	NR	NA	Pond water	NR	NR	150
Kansas Power & Light Lawrence 4	1.0	Bulk separation/ bulk entrainment separator	Cooling tower blowdown, pond overflow, flash water	1st stage: vertically upward and downward 2nd stage: vertically upward	2 minutes every 8 hours	150
Northern States Power Sherburne 1	0.5	NA	Thickener overflow and cooling tower blowdown	1st stage: vertically upwards, 180° rototable; between stages, 360° rototable lance	2 minutes every 24 hours	120-150
Salt River Project Coronado 1	0.5	NA	Cooling tower blowdown	NR	NR	70
South Mississippi Electric Pwr. R.D. Morrow 1	NR	NA	Supernatant and fresh water	NR	Continuous	NR
Tennessee Valley Authority Widows Creek 8	1.0	NA	Fresh water	Vertically downward wash and horizontal front water	Continuous front; intermittent top	20

NA - not applicable.
NR - not reported.

TABLE 3-10. BASIC REHEATER TYPES

Generic type	Specific type	Common designs
In-line	Steam Hot water	Bare tube Fin tube Shell and tube
Direct combustion	In-line burner External combustion chamber	Natural gas or Oil
Indirect hot air	Boiler preheater External heat exchanger	Boiler combustion air Steam tube bundles
Waste heat recovery	Gas/gas Gas/fluid	Wheel type Tube bundles
Exit gas recirculation	Steam heat exchanger	
Bypass	Hot side Cold side	

is circulated. The gas passes over the tubes and picks up thermal energy from the surfaces of the heating tubes. These tubes are typically arranged in banks. Figure 3-12 is a simplified diagram of an in-line reheater.

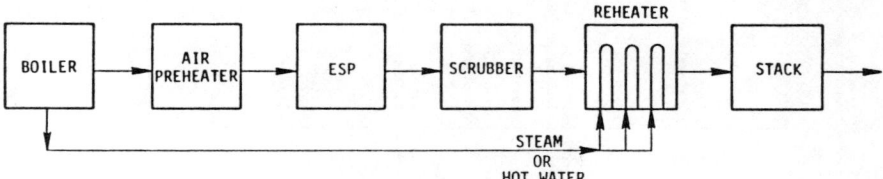

Figure 3-12. In-line reheat system.

In-line reheaters can be classified according to the heating medium: steam or hot water. When steam is used, the inlet steam temperatures and pressures range from 350° to 720°F and 115 to 200 psig, respectively. Saturated steam is preferred because the heat transfer coefficients of condensing steam are much higher than those of superheated steam. The configuration of a reheater using hot water is similar to that of the one using steam. Inlet temperature of the hot water ranges from 250° to 350°F, and the temperature drop over the heat exchanger is 70° to 80°F.

Fin-tubes are used in some applications because of their superior heat transfer capability. Soot blowers are generally used to periodically remove deposits that build up on the tube surfaces.

Because in-line reheaters are situated in the gas stream, they are prone to corrosion and plugging, the latter due to the reheater's dependence on proper operation of the upstream mist eliminator. Moreover, solids deposited on the tubes can reduce heat transfer considerably, increasing the potential for corrosion of downstream equipment.

Direct combustion--A direct combustion reheater eliminates the need for heat exchangers. As shown in Figure 3-13, gas or oil is burned and the

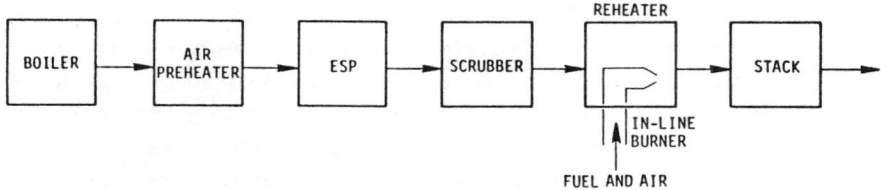

a. SYSTEM WITH AN IN-LINE BURNER

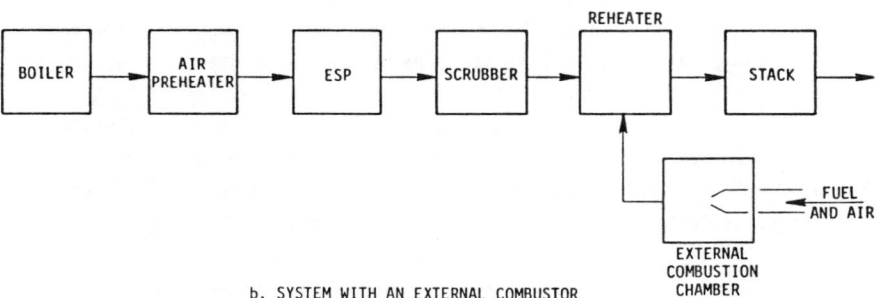

b. SYSTEM WITH AN EXTERNAL COMBUSTOR

Figure 3-13. Direct combustion reheat systems.

combustion product gas (at 1200° to 3000°F) is mixed with the flue gas to raise its temperature.

Direct firing requires some care in mixing the hot combustion gas with the cool scrubbing gas. If mixing is not effective, hot spots can develop downstream from the heater, causing damage to the duct lining. Also, maintenance of flame and flame stability are required for effective operation. The main problems with direct firing are the availability and cost of gas and oil.

Indirect hot air reheat--As shown in Figure 3-14, indirect hot air

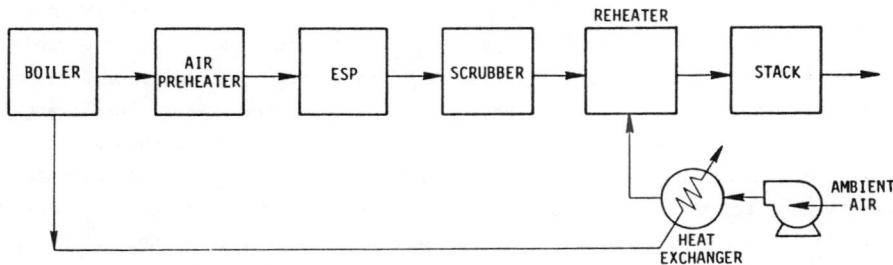

Figure 3-14. Indirect hot air reheat system.

reheat is achieved by heating ambient air with an external heat exchanger using steam at temperatures of 350° to 450°F. The heating tubes are arranged in two to three banks in the heat exchanger. Hot air and flue gas may be mixed by use of a device such as a set of nozzles or a manifold.

The advantage of indirect hot air reheat over in-line reheat is that the indirect system involves no corrosion or plugging because it is located outside of the scrubber gas discharge duct. The disadvantages include the need for an additional fan to convey hot air; the relatively large amount of space required for the reheat system (compared with other reheat methods); the increase in stack gas volume, which may be undesirable because of the limited capacities of ID fans and stacks; and the higher energy consumption, which is needed to heat air from the ambient temperature level. Another advantage, however, offsets the higher energy requirement to some extent: the dilution by the added air reduces the incidence of steam plume formation and gives better plume dispersion.

Another variation of the indirect hot air reheat method involves use of the air preheater associated with the boiler to provide hot air for stack gas reheat. In this case, the temperature of the combustion air entering the boiler would be lower than usual because part of the heat content of the flue gas is used to provide the hot air for stack gas reheat. Also, the preheater should be designed to heat a greater-than-normal amount of air to a lower-than-normal exit temperature.

<u>Waste heat recovery</u>--In the waste heat recovery method (see Figure 3-15), the sensible heat of unscrubbed flue gas, which would otherwise be

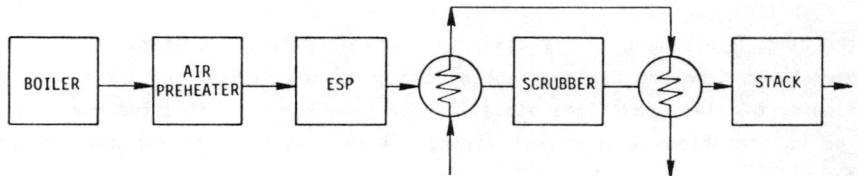

Figure 3-15. Waste heat recovery reheat system.

released in the scrubber, is recovered in an in-line heat exchanger. The heat is used to reheat the scrubbed stack gas. A direct gas/gas heat exchanger, especially the wheel type (e.g., Ljunstrom), can be used for this purpose. This type of reheat system requires a large heat-transfer area in the in-line heat exchangers because of the narrow temperature range involved. Another method involves gas/fluid heat exchange through a medium such as water or a fluid of high heat capacity (e.g., a glycol-water mixture), which is circulated through in-line heat exchangers (tubes) located upstream and downstream of the scrubber. With this method, there is a potential for corrosion due to acid condensation when the hot flue gas is cooled.

Exit gas recirculation--In reheat by exit gas recirculation, a portion of the heated stack gas is diverted, heated further to around 400°F by an external heat exchanger, and injected into the flue gases. An advantage of this type of reheat system over indirect hot air reheat is that it does not increase the total stack gas flow rate. Moreover, the reheat operation is less influenced by ambient air conditions.

A simplified diagram of recirculation reheat is presented in Figure 3-16. This type of reheat system can be used in combination with an in-line

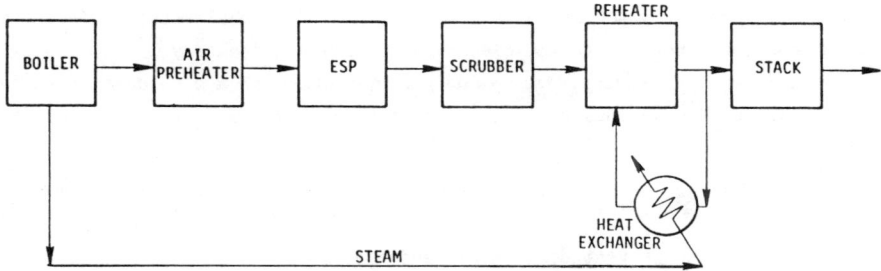

Figure 3-16. Exit gas recirculation reheat system.

reheat system to evaporate the mist droplets in the flue gas from wet scrubbers before they impinge on the heat exchanger tubes.

Bypass reheat--In the bypass reheat system, a portion of the hot flue gas from the boiler is allowed to bypass the scrubbing system and is mixed with flue gas that has been processed through the scrubber. Two variations of this method are "hot-side" bypass, in which the flue gas is taken upstream of the air preheater (Figure 3-17), and "cold-side" bypass, in

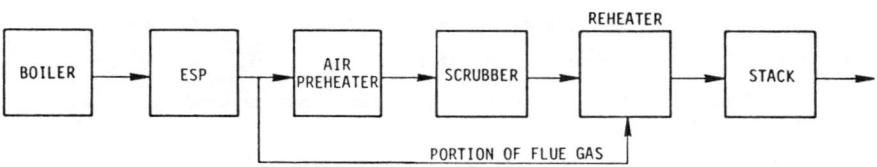

Figure 3-17. "Hot-side" flue gas bypass reheat.

which the flue gas is taken downstream of the air preheater (Figure 3-18). In the former, a separate particulate removal device (ESP or fabric filter)

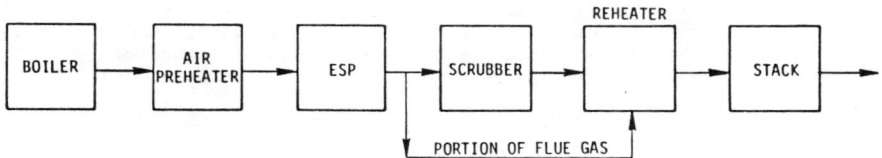

Figure 3-18. "Cold-side" flue gas.

is required for fly ash control if an upstream particulate collector is not used. The limiting conditions for application of this system are determined by: (1) the properties of the boiler fuel, such as heating value, sulfur content, and ash content; (2) particulate control (ESP) ahead of the scrubbing system; (3) the temperature of hot flue gas; (4) the efficiency of the scrubbing system; and (5) emission regulations.

Design Considerations

In new plants, and whenever possible in retrofit installations, the steam for reheat should be taken from a point in the boiler cycle where its withdrawal will not derate the electrical generating capacity.

Additionally, several important design factors should be considered in evaluation of the various stack gas reheat methods for a particular application:

1. The heat and energy requirements needed to effect normal operation with an external source of reheat energy, and at the same time
   ° to prevent downstream condensation
   ° to prevent formation of a visible plume
   ° to enhance plume rise and pollutant dispersion.
2. Constraints imposed by SO_2 standards when bypass reheat is selected.
3. The feasibility of using no reheat and the resultant impact on system design.

These considerations are discussed briefly in the following paragraphs with respect to reheat methods that are currently in use or likely to be used in commercial limestone FGD systems.

Definitions of pertinent terminology (from Perry 1973) are given here as an aid in understanding of the discussions.

Absolute humidity: The amount of water vapor carried by one unit mass of dry air.

Relative humidity: The partial pressure of water vapor in air divided by the vapor pressure of water at a given temperature.

Dew point or saturation temperature: The temperature at which a given mixture of water vapor and air is saturated. Dew point represents the temperature at which the relative humidity is 100 percent.

Wet-bulb temperature: The equilibrium temperature attained by a water surface when the rate of heat transfer to the liquid surface from the air-water vapor mixture equals the rate of energy carried away from the surface by the diffusing vapor.

Prevention of Downstream Condensation. The reheat requirement to prevent downstream condensation can be estimated by making a heat balance around the downstream system, including the stack. Condensation takes place when the vapor pressure of water in the stack gas exceeds the saturation value at a specific stack gas temperature and pressure. To prevent condensation (1) the gas temperature is raised above the dew point or (2) the gas is diluted with another gas so that the relative humidity of the mixture is less than that of the original stack gas. A combination of both measures can be used.

1. In-line reheat: A schematic of the heat balance around the downstream system, including an in-line reheater, is shown in Figure 3-19. The

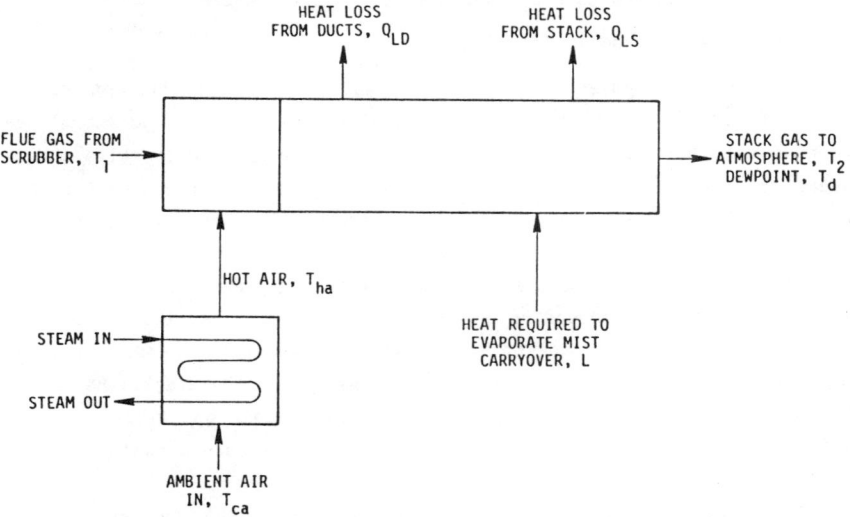

Figure 3-19. Schematic of heat balance around downstream system with indirect hot-air reheater.

gas temperature at the top of the stack should be at the dew point or higher to prevent condensation before emission. Because entrained moisture (mist) carried over from the scrubber is vaporized as the flue gas temperature is increased during reheat, the dew point at the top of the stack is slightly higher than the temperature of gas exiting the scrubber. The minimum heat requirement from steam or hot water is determined as follows:

| Heat input | = | Heat required to raise gas to its dew point | + | Heat loss from ducts | + | Heat loss from stack | + | Heat required to evaporate mist carryover | − | Heat gain due to fan |

or

$$Q = \frac{FC_p m}{k}(T_d - T_1) + Q_{LD} + Q_{LS} + L - Q_F \qquad \text{(Eq. 3-1)}$$

where:

Q = minimum heat requirement from steam or hot water, Btu/h

$C_p m$ = mean heat capacity of flue gas, Btu (lb-mol)°F

F = flue gas flow rate, scfh

T_d = dew point of stack gas at top of stack, °F

T_1 = temperature of flue gas exiting wet scrubber, °F

Q_{LD} = heat loss from ducts, Btu/h

Q_{LS} = heat loss from stack, Btu/h

k = 379 ft³/lb-mole

L = heat required to evaporate mist carry-over, Btu/h

Q_F = heat gain from ID fan, Btu/h

For estimation purposes, the overall heat-transfer coefficient can be assumed to be 10 Btu/ft² per °F per h for steam/gas (condensing steam) heat exchange and to be 6 Btu/ft² per °F per h for hot water/gas heat exchange. Because of the very small difference between T_d and T_1, the amount of heat required to raise gas to its dew point is very small. The method for calculating dew point is shown in Appendix E.

The values of Q_{LD}, Q_F, and Q_{LS} vary with the specific situation; Q_{LD} could be significant, depending on whether the duct is insulated, the length of duct, and the difference between stack gas temperature and ambient air temperature. The value Q_{LS} depends on the height of the stack, the materials of construction, and the difference between stack gas temperature and ambient air temperature. In stacks with an annular space between the stack and liner, the temperature drop should not be significant.

The heat required to evaporate mist carryover (L) depends upon the amount of carryover from the mist eliminator to the reheater.

2. Indirect hot air reheater: Figure 3-20 is a schematic of the heat

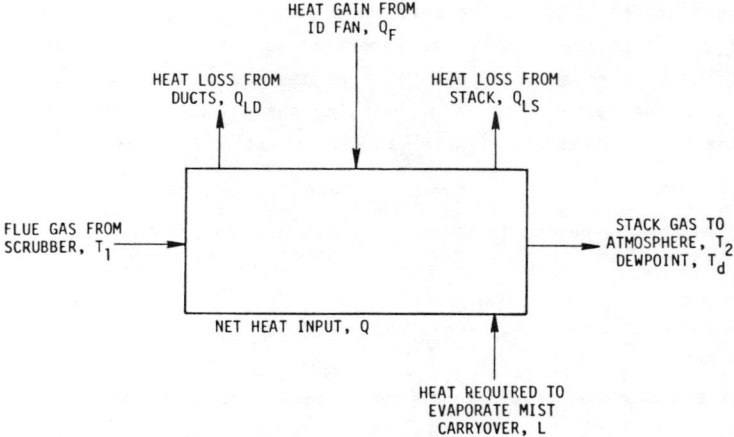

Figure 3-20. Schematic of heat balance around downstream system with in-line reheater.

balance around the downstream system, including an indirect hot air reheater. In this system, the flue gas from a wet scrubber is heated by addition of hot air. The minimum heat requirement from steam is determined as follows:

Heat input = Heat required to raise gas temperature to its dew temperature + Heat loss from ducts + Heat loss from stack + Heat required to evaporate mist carry-over = Heat gain from heated air

or

$$Q = \frac{F_a C_p m}{k}(T_d - T_1) + Q_{LD} + Q_{LS} + L = \frac{F_a C_p a}{k}(T_{ha} - T_d) \qquad \text{(Eq. 3-2)}$$

Net heat input = Heat required by amount of ambient air to reach temperature T_{ha}

or

$$Q_N = \frac{F_a C_p A}{k}(T_{ha} - T_{ca}) \qquad \text{(Eq. 3-3)}$$

where:

F_a = ambient air flow rate, scfh

$C_p a$ = specific heat of air, Btu/(lb/-mole)°F

T_{ha} = temperature of hot air, °F

T_{ca} = temperature of ambient air, °F

Q_N = net heat input, Btu/h

The total volume flow rate of stack gases is $F + F_a$, and the other symbols are as previously defined.

Equation 3-2 is used to determine the temperature of hot air (T_{ha}), then Equation 3-3 is used to determine the net heat input.

<u>Prevention of Visible Plume</u>. The following methods of determining heat requirement are currently applied in most reheat applications. For normal operation and prevention of visible plume, the temperature desired at the top of the stack is selected by the designer. It is usually between 125° and 220°F. The reheat requirement for normal operation and prevention of visible plume can be estimated by making heat balances as follows:

1. In-line reheat: The minimum heat requirement with in-line reheat can be calculated as follows:

Minimum heat required = Heat required to raise gas from temperature T_1 to T_2 + Heat loss from ducts − Heat gain due to fan + Heat loss from stack = Heat required to evaporate mist carryover or

$$Q = \frac{F_a C_p m}{k}(T_2 - T_1) + Q_{LD} - Q_F + Q_{LS} + L \qquad \text{(Eq. 3-4)}$$

where,

Q = minimum heat required, Btu/h

T_2 = stack gas temperature at the top of the stack, °F

All other symbols have been defined.

2. Indirect hot air: The minimum heat requirement with indirect hot air reheat can be calculated as follows:

| Minimum heat required | = | Heat required to raise temperature from T_1 to T_2 | + | Heat required to evaporate mist carryover | = | Heat required by amount of ambient air to reach temperature T_{ha} from T_1 | − | Heat loss from ducts | + | Heat gain due to fan | − | Heat loss from stack |

or

$$Q = \frac{F_a C_p m}{k}(T_2 - T_1) + L = \frac{F_a C_p a}{k}(T_{ha} - T_1) - Q_{LD} + Q_F - Q_{LS} \qquad \text{(Eq. 3-5)}$$

− Net heat input = Heat required by F_a amount of ambient air to reach temperature T_{ha}

or

$$Q = \frac{F_a C_p m}{k}(T_{ha} - T_{ca}) \qquad \text{(Eq. 3-6)}$$

Equation 5 is used to determine the temperature of hot air (T_{ha}); then equation 6 is used to determine the net heat input.

<u>Enhancement of Plume Rise and Dispersion of Pollutants</u>. Using a limestone scrubber without reheat could result in poor plume rise and dispersion characteristics, which in turn could cause undesirably high ground-level concentrations of pollutants (residual particulate, SO_x, and NO_x). The computation of reheat requirements to enhance plume rise and dispersion requires a quantitative analysis of plume behavior in the atmosphere, which is a function of such variables as meteorological conditions, stack size, and characteristics of the stack gas at the emission point.

A family of curves was developed using mathematical models that predict maximum ambient concentrations under certain given conditions. As shown in Figure 3-21, each curve represents a given percentage of SO_2 removal from the gas. The reduction in ambient SO_2 concentration is equal to the SO_2 removal when the stack gas is reheated to the scrubber inlet temperature, because ground-level concentration is directly proportional to the amount of SO_2 emitted from the stack. Because the temperature of the scrubber gas is lower (unless reheated to scrubber inlet temperature), the ambient concentration will be a higher value.

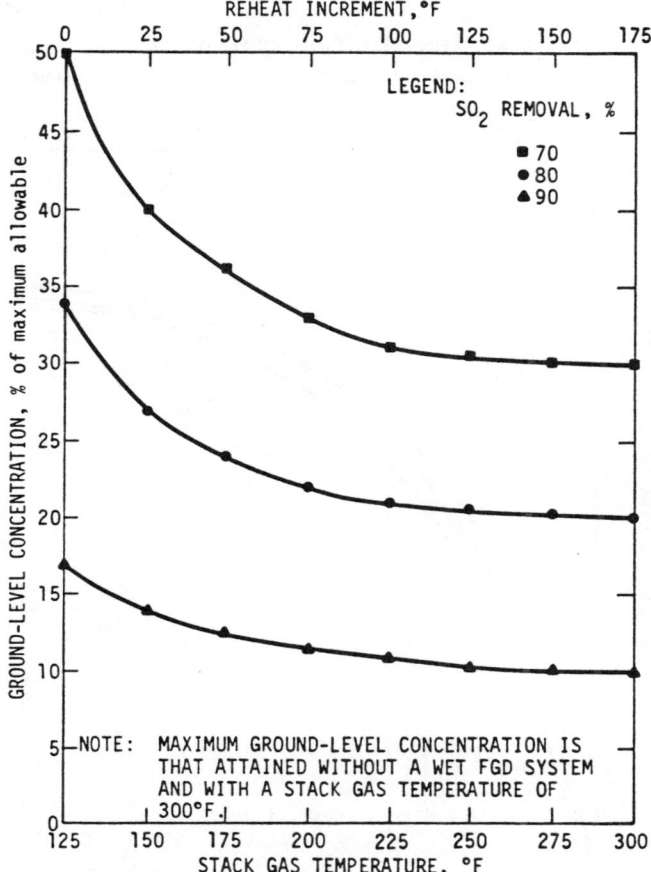

Figure 3-21. Effect of reheat increment on ground-level SO_2 concentration.

The curves show that at any SO_2 removal efficiency, the amount of reheat has a pronounced effect on ground-level SO_2 concentration. At 70 percent SO_2 removal and a stack gas temperature of 175°F, for example, the ground-level concentration can be reduced from 50 percent with no reheat to 36 percent of the maximum values--a 28 percent improvement. With the same amount of reheat at 90 percent SO_2 removal, the ground-level concentration of SO_2 is reduced from 17 percent to 12.5 percent of the maximum values--a 26.5 percent improvement. Thus it can be concluded that even with increasing SO_2 removal efficiency, reheat definitely offers a substantial benefit. A much greater benefit can be obtained, however, by increasing the SO_2 removal efficiency of the scrubber.

The relative improvement in the ground-level air quality remains almost identical for various SO_2 removal efficiencies over a wide range of reheat increments, as indicated in Figure 3-22. At a reheat increment of 150° to

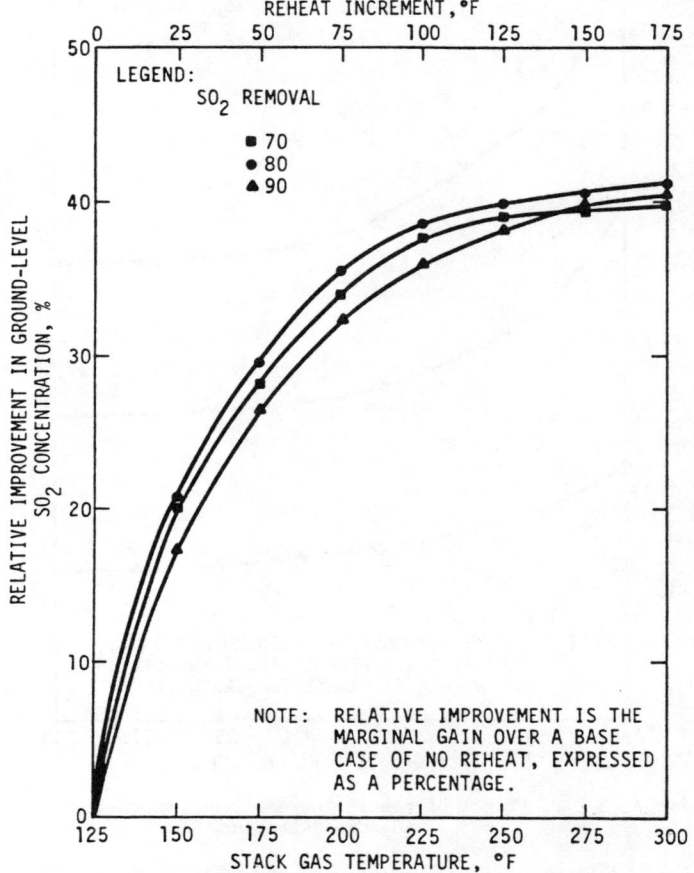

Figure 3-22. Relative improvement in ground-level SO_2 concentration as a function of degree of reheat.

175°F (stack gas temperatures of 275° to 300°F), the curves approach a value of about 40 percent relative improvement asymptotically. This means that theoretically a maximum relative improvement of 40 percent can be attained. The decision on amount of reheat to be used, however, must depend on the economic considerations as well as plume rise and pollutant dispersion.

It should be noted that the higher water vapor content (15 percent versus 6 percent) in the gas offsets to some extent the adverse effects of gas cooling. Since the water vapor is of lower density than other constituents of the gas, it makes the plume more buoyant. The effect is small, however, and has been omitted in developing the curves.

Constraints on Bypass Reheat. Bypass reheat offers the advantages of low capital investment and simple operation. The maximum amount of reheat that can be obtained, however, is limited by the constraints of pollutant emission standards. A regulation requiring 90 percent SO_2 removal efficiency could limit or completely rule out the bypass reheat option. The limitation on use of bypass reheat to meet the emission standard for sulfur dioxide can be written as:

$$\text{(Fraction of bypass flue gas stream)} = 1 - \frac{1}{\text{fractional sulfur removal efficiency}} + \frac{A}{(2)\,(\text{fuel required})\,(\text{sulfur content in fuel})\,(\text{fractional sulfur removal efficiency})}$$

or

$$X = 1 - \frac{1}{E} + \frac{A}{2WSE} \qquad \text{(Eq. 3-7)}$$

where:

X = fraction of bypass flue gas stream

E = fractional sulfur removal efficiency of the wet scrubbing system

A = maximum allowable SO_2 emission, $lb/10^6$ Btu

W = amount of fuel required to generate 1 million Btu, lb

S = weight fraction of sulfur in the fuel

Details of the heat balance around a bypass reheat system are given by Choi et al. (1977).

The No-Reheat Option. As indicated earlier, the use of stack gas reheat in wet scrubbers is not universally accepted in the industry. Results of a recent survey on reheat practices indicated that most designers and suppliers who responded to the survey do not think reheat is necessary and generally do not recommend the use of reheat (Muela et al. 1979). Therefore, alternative design strategies have been developed to handle such consequences as condensation and subsequent corrosion in downstream equipment. Basically, these strategies require the use of fans upstream of the scrubber (unit or booster fans) and the use of a suitable construction material to protect the downstream ductwork and stack from acid corrosion attack.

Most systems with wet stacks have reported problems with the stack linings, which tend to blister and eventually flake off. Once this happens, stack corrosion begins. To limit corrosion in no-reheat operation, one may either select materials that are inherently resistant to corrosion, such as high-grade alloys, or use organic or inorganic linings to cover corrosible base metals. These options are addressed in the Materials of Construction subsection.

Operational Systems

Table 3-11 summarizes selected limestone FGD systems currently in service according to reheater design and operating parameters. These FGD systems are representative of the stack gas reheat methods discussed in the foregoing.

TABLE 3-11. EXAMPLES OF OPERATIONAL STACK GAS REHEATERS

Power plant	Type	Heat source	ΔT, °F	Tube type	No. of tube banks	No. of tubes per bank	Soot blowing
Cholla 1 Arizona Public Service	In-line	Steam	40	Circular	2	NR	Steam, once per shift
Cholla 2 Arizona Public Service	In-line	Steam	40	Circular	2	NR	Steam, once per shift
La Cygne 1 Kansas City Power and Light	In-line	Steam	65	Circular	4	8	Steam, every 4 hours
Lawrence 4 Kansas Power and Light	In-line	Hot water	50	Finned	1	66	Two to three times per shift
Lawrence 5 Kansas Power and Light	In-line	Hot water	50	Finned	1	66	Two to three times per shift
Sherburne 1 Northern States Power	In-line	Hot water	40	Finned	3	15	Once per shift
Sherburne 2 Northern States Power	In-line	Hot water	40	Finned	3	15	Once per shift
Widows Creek 8 TVA	Indirect	Steam	50	Finned	NR	80	None
Petersburg 3 Indianapolis Power and Light	Indirect	Steam	30	Finned	NR	65	None

NR - Not reported.

FANS

The fans in a limestone scrubbing system are used to move the flue gas through the system. They are designed to generate a static head great enough to overcome resistance to flow of the flue gas within the system. The criteria for specification and selection of fans for this service are outlined in this section. Because the fan applications in a scrubber system are very similar to those in a boiler system, most power plant engineers are familiar with the operation and the basic mechanical design of these fans. This discussion therefore emphasizes features that are specific to FGD system applications.

Description and Function

The fans in an FGD system are required to handle gas flow rates typically ranging from 200,000 to 800,000 acfm. Gas temperatures, which depend on fan location in the system, range from 200° to 320°F after the boiler preheater and from 120° to 180°F after the scrubber exhaust reheater. Along with moisture, SO_2, and acid mist, the gases may contain particulates that are usually abrasive and tend to accumulate in the form of scale on fan blades.

The fans used in a scrubber system may be either axial or centrifugal types. Axial fans have been used in FGD systems at only two plants, whereas centrifugal fans are used in many FGD installations. The major reason for this predominance is that the centrifugal fan with radial blades is better suited to operation with gas streams containing particulate matter.

Because the fans are seldom able to operate continuously at constant pressure and volume, some convenient means of controlling the volume of flow through the fans is needed to meet the demands of variable scrubber and boiler loads. Control is commonly achieved with variable-inlet vanes or dampers, and with variable-speed controls on the fans.

The centrifugal action of a fan imparts static pressure to the gas. The diverging shape of the scrolls (the curved portion of the fan housing) also converts a portion of the velocity head into static pressure. Although the normal static pressure requirement is approximately 20 in. H_2O, scrubber system designers commonly add 15 to 25 percent to the net static pressure requirement as a margin of safety to allow for buildup of deposits in ductwork and for inherent inaccuracies of calculation.

Design Parameters

Location. In this manual fans are called either boiler induced-draft (ID) fans or scrubber booster fans, depending on the location of the fans with respect to the boiler. Location will determine whether the FGD system operates at positive or negative pressure with respect to the atmosphere and whether the fans operate in a wet or dry gas environment.

The main boiler ID fan must produce enough pressure to overcome flow resistance downstream of the boiler and maintain adequate gas flow until the point at which the natural stack draft provides the energy to exhaust the gas. When the ID fan is located ahead of the scrubbing system, it is considered a dry fan and it provides a positive pressure that pushes the flue gas through the scrubbing system. Any leaks that develop are easily detectable by the escaping flue gas. The leaking gas can be hazardous if the scrubber is in an enclosure.

An ID fan located after the scrubbing system and the reheater is also considered to be a dry fan because it does not handle saturated gas. This fan operates at negative pressure relative to the atmosphere because it pulls the flue gas through the FGD system. If leaks occur, the result is inleakage of air into the FGD system; such inleakage has caused high natural oxidation but essentially no adverse impact on scrubber operation. Inleakage of air can increase the volume of gas that must be handled, and detection of inleakage is very difficult. Figure 3-23 depicts typical ID fan

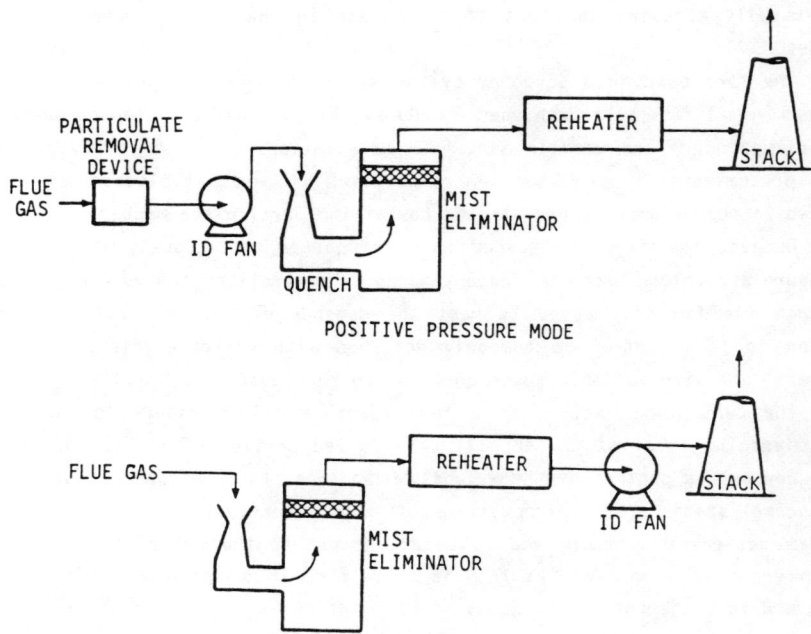

Figure 3-23. Typical dry ID fan applications.

applications for limestone scrubbing FGD systems.

Scrubber booster fans are ID fans that supplement the existing boiler fan and their location and functions are the same as the boiler ID fans previously described.

When the ID fan is located immediately after a scrubber, it operates in a wet environment because the gas is saturated with condensing water and contains extra moisture caused by entrainment. A typical wet ID fan application is shown in Figure 3-24. The gas from the prescrubber is saturated with water and contains acid mists and abrasive carryover materials. Even

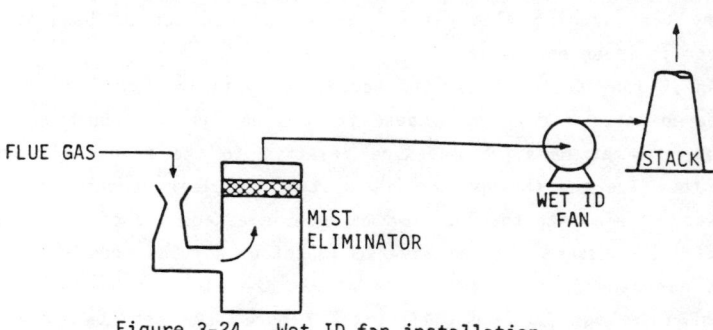

Figure 3-24. Wet ID fan installation.

though water sprays are installed to clean the fan internals, corrosion, erosion, and scaling can occur. Buildup of solids of the fan blades can cause severe corrosion and rotor imbalance, which in turn lead to high noise levels, excessive vibration, and fan failure. Wet ID fans offer a size advantage because the cool, wet gas has less volume than dry gas. Because of numerous operating problems, wet fans have been in disfavor and the trend is to use dry ID fans upstream of the scrubbers. Wet fans should be specified only with a rationale for circumventing these known problems.

Temperature Increase. The adiabatic gas compression caused by the use of a fan increases the gas temperature through the system, typically by about 0.5°F per inch of water pressure increase. The temperature increase due to the fan operation is an advantage when the fan is located after the scrubber in that it supplements the stack gas reheat (Green Fuel Economizer Co. 1977).

Materials of Construction. Fans operating on hot flue gases can be constructed of carbon steel because they are not subjected to corrosive conditions. They may be subjected to erosion by particulate, however, if particulate removal is inefficient or nonexistent. Carbide wear plates are used to protect against such erosion.

Where fans are subjected to a wet gas environment, with high levels of acidity and chlorides, corrosion-resistant alloys, polymer coatings, or rubber linings should be used for the fan components. The major problem with rubber-lined housings has been damage by impingement of loosened scale deposits from ductwork.

Cleaning and Inspection. All fans should have adequate cleanout doors. This is especially important on dry ID fans. Inspection ports are also useful for checking possible accumulation of deposits.

Operational Systems

Following is a discussion of fans used at some limestone scrubbing facilities, and their performance histories.

Cholla. The two booster fans provided for Cholla 1 maintain constant gas stream pressure to the scrubber and eliminate the need for extra loading on the ID boiler fans. The mild steel booster fans are rated at 240,000 acfm of flue gas at 276°F. Operation of the FGD system began in 1973. During the period December 1973 to April 1974, vibration occurred in the Module B booster fan as a result of uneven buildup of scale on the fan blades caused by intermittent operation. The blades were sandblasted, cleaned, and rebalanced to eliminate vibration. No further problems have been reported (Laseke 1978a).

Petersburg. Petersburg 3 of Indianapolis Power and Light Company has four scrubber modules. Four booster fans (one per module), each with a pressure drop of 18 in. H_2O, force the gas from the ESP's through the FGD system. Each fan handles a gas flow of 475,000 acfm at 279°F. Louver dampers permit isolation of one or more FD fans and crossover ducts. The FGD system was put into operation in October 1977, and through July 1978 no fan problems were reported (Laseke 1978b; Laseke et al. 1978).

La Cygne. La Cygne 1 of Kansas City Power and Light Company is fitted with eight scrubbing modules for control of particulate and sulfur dioxide emissions. Flue gases at 285°F from the boiler are pushed through a common plenum by three fans. The gases then enter the eight scrubbing modules. After passing through the FGD system, including a reheat section, the gases enter a common plenum at a temperature of 172°F and are fed to the stack by six ID fans through six ducts. The system cannot be bypassed. Each of the three FD fans has a design capacity of 765,000 acfm of gas flow at 105°F; the six ID fans are each rated at a gas flow of 445,200 acfm at 172°F with a 7000-hp motor drive (Laseke 1978c).

Problems have occurred with the ID fans since they were first balanced in September 1972. Initially the fans were prone to imbalance, and operation at close to the critical speed caused severe vibration in the fan housing, resulting in cracks in the inlet cones. Additional stiffeners were installed to strengthen the housing.

The temperature was finally controlled by cutting oil grooves in the thrust collar and installing forced-lubrication systems on all the fans. These modifications reduced the thrust collar temperatures to a range of 140° to 160°F.

Problems with the ID fans began with the initial firing of the boiler. Fly ash and slurry, carried over from the scrubber and deposited on the blades of the impeller, aggravated the tendency of the fans to become unbalanced and promoted fan blade erosion. Examination of all the blades by magnaflux testing revealed several cracks, indicating a need for reinforcement. By June 1974 all ID fan rotors were replaced with units of heavier design. Shaft diameter was increased, and thickness of the radial tip blades and side plates of the wheels was increased. The thick center plate was scalloped to reduce its weight, and the fan blades were modified to reduce the tendency to vibrate. The leading edge of each blade was covered with a stainless steel clip to deter erosion. Although fly ash carryover still necessitates intermittent washing of the fans, the cleaning frequency has been reduced.

Lawrence. Lawrence 4 of Kansas Power and Light Company operates with a retrofitted limestone FGD system that began operation in 1977.* The FGD system consists of two 50 percent capacity scrubber modules, to which the flue gas is conveyed through new ductwork. Flue gas from each module exits through a mist eliminator and reheater and is conveyed by two ID fans (one per module) to the stack. A bypass duct is provided for each module. Isolation dampers are located at the inlet and outlet of each module and in the bypass ducts. Each fan handles a gas flow of 181,500 acfm at 144°F (Green and Martin 1978).

A new FGD system has been installed on Lawrence 5, identical to that on Lawrence 4 except in flue gas handling capacity. It is equipped with two ID fans (one per module), each with a design capacity of 600,000 scfm. No qualitative evaluation of fan operation on these units is yet available.

Sherburne. Sherburne 1 and 2 of Northern States Power Company have limestone FGD systems containing 12 scrubbing modules each. Flue gas from the boiler passes through the scrubber, the mist eliminator, and the reheater before reaching the ID fan. Only 11 of the 12 scrubber modules are required for full-load operation. Reheat provides protection to the downstream ductwork and ID fans and controls potential plume formation before final discharge of flue gas to the atmosphere. The scrubbing modules cannot be bypassed. Each of the boiler units is equipped with six fans: two FD, secondary-combustion-air fans upstream, and four ID fans downstream of the scrubbing system. The ID fans are 30 percent capacity, axial-flow, variable-drive fans. The flow capacity of each ID fan is 702,000 acfm at 171°F (Laseke 1979).

Since startup of the FGD system in 1976, no problems with the ID fans have been reported (Melia et al. 1980).

Winyah. The FGD system on Winyah 2 of South Carolina Public Service treats only 50 percent of the flue gas. The remainder bypasses the FGD system and is recombined with the scrubbed portion for reheat before discharge to the stack. After passage of the flue gas through an ESP, a booster fan drives 50 percent through the scrubber. The booster fan handles 407,000 acfm of flue gas at 270°F with a design pressure drop of 13.5 in. H_2O. The FGD system was placed in service in 1977. No problems with the scrubber FD fan have been reported (Melia et al. 1980).

Southwest. The FGD system at Southwest 1 of Springfield City Utilities consists of two scrubber modules of 50 percent capacity each. The boiler flue gas first passes through an ESP and is then driven through the two modules by booster fans. The cleaned flue gas then passes through a set of

---

* The modified limestone venturi/spray tower FGD system was placed in service in 1977. The original limestone injection marble-bed scrubbers were in service from 1968 to 1976.

mist eliminators before it is discharged to the main stack. One or both modules can be bypassed during emergency or malfunction periods by the use of seal-air gas dampers.

Design capacity of the booster fans, manufactured by the Green Fuel Economizer Co., is 454,200 acfm of flue gas at 335°F, with 4390-hp, 800-rpm motors. No fan problems have been reported (Melia et al. 1980).

Widows Creek. The limestone FGD system on Widows Creek 8 consists of four equal capacity scrubbing trains. Flue gas from ESP's enters the FGD system through four fans, each with a flue gas capacity of 400,000 acfm at 280°F. The fans, manufactured by the Green Fuel Economizer Co., are powered by 4000-hp, 890-rpm motors.

Extensive problems were encountered after startup of the FGD system in August 1977. These problems included erosion of fan blades and trouble with the drive units. Guillotine dampers became inoperable because of jammed gear boxes. The seals around the dampers corroded, allowing leakage of flue gas and particulate matter. Serious erosion of the FD fan rotors continued. All of these fans have been rebuilt.

The problems at Widow's Creek stemmed from the operation of old existing ESP's, which collected only 50 percent of the inlet fly ash. As a result, large pieces of fly ash literally sand-blasted the fan blades, resulting in erosion and failure (Wells et al. 1980).

THICKENERS AND MECHANICAL DEWATERING EQUIPMENT

This section describes the types of equipment used to dewater the slurry generated by limestone scrubbing. Dewatering is classified as either primary or secondary. The primary dewatering process takes the slurry bleed stream from the scrubber at 5 to 15 weight percent solids content and performs liquid/solid separation as an initial step, typically increasing the solids content to 20 to 40 weight percent. The secondary dewatering process can dewater sludge solids fed from a primary dewatering device to a solids content of 60 to 85 weight percent. Thickening and dewatering can be accomplished by any one or a combination of the following mechanisms:

- Settling ponds
- Thickeners
- Vacuum filters
- Centrifuges
- Hydrocyclones

Settling ponds are not considered here because the trend is toward the use of mechanical dewatering equipment. Additionally, plants that burn high-sulfur coal must use thickeners in combination with secondary dewatering equipment. These plants do not generate enough fly ash to mix with only primary dewatered sludge to make a stabilized material for dry landfill. Because of their simplicity, thickeners are almost always justified and hence are now used in more than half of the limestone scrubbing systems. Although more expensive than a settling pond, a thickener is usually preferred because it removes solids more easily, can be located more flexibly in the plant environs, and requires less space.

A recent EPRI publication (Knight et al. 1980) provides additional information on the details of thickeners and dewatering equipment as applied to limestone scrubbers. Also, the EPA has published an excellent manual on the treatment and disposal of sludge from wastewaters that includes a review of mechanical dewatering equipment (EPA 1979).

The main purposes of dewatering are:

To reduce the volume of sludge for transport and disposal, thereby reducing transport costs, land area requirements, and final disposal cost.

To recover the liquid for further use.

To reduce the moisture content so the waste can be handled and disposed of dry.

To minimize leaching and pollution of ground water.

To recover sludge for utilization.

Depending on the sludge characteristics, the thickening/dewatering process may also be beneficial to sludge blending, equalization of sludge flow, and sludge storage, as well as reducing the chemical requirements for sludge conditioning. Figure 3-25 shows the effect of dewatering on the total sludge volume. As the water is taken out and the solids content increased, the volume for disposal is substantially reduced. It is evident, therefore, that transport and disposal costs can be reduced by reducing the volume of water carried to the disposal site.

On the other hand, the size and cost of dewatering equipment increases as the degree of dewatering increases. As a result, there is an economic trade-off between disposal cost and dewatering cost. Similar trade-offs occur in other aspects of sludge thickening and dewatering. For any given situation, an optimum system can be determined by evaluation of the alternatives.

Dewatering facilities should be placed as close as possible to the scrubber facilities. This will minimize the length of large-diameter piping required to convey scrubber bleed to the dewatering facility and to return the portion of the liquid fraction that is removed. Placing a settling pond

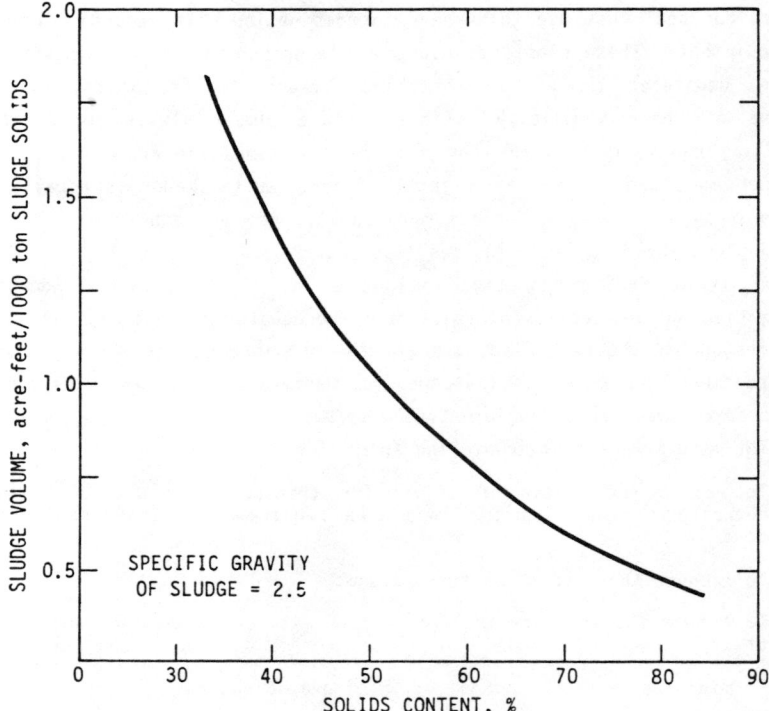

Figure 3-25. Relationship between sludge volume and solids content.

close to the scrubber might not be possible because of the relatively large amount of land required for a pond.

Dewatering facilities may be installed singly, in parallel, or in series, depending upon the degree of treatment required. Mechanical equipment should be sized to allow sufficient storage, or sufficient surge capacity should be provided to permit flexible operation. Enough equipment should be provided to ensure uninterrupted operation of the scrubber in the event of failure in the sludge processing system. Usually this is done by providing redundant or oversized equipment or by installing a bypass with emergency storage facilities.

Description and Function

Thickener. A thickener is a sedimentation device that concentrates a slurry by gravity so that the settled solids may be disposed of and the clarified liquid recycled. Thickeners (also known as clarifiers) have been used in many industries, from which conventional design has been applied successfully to limestone scrubbing.

A typical thickener (Figure 3-26) consists of a large circular holding

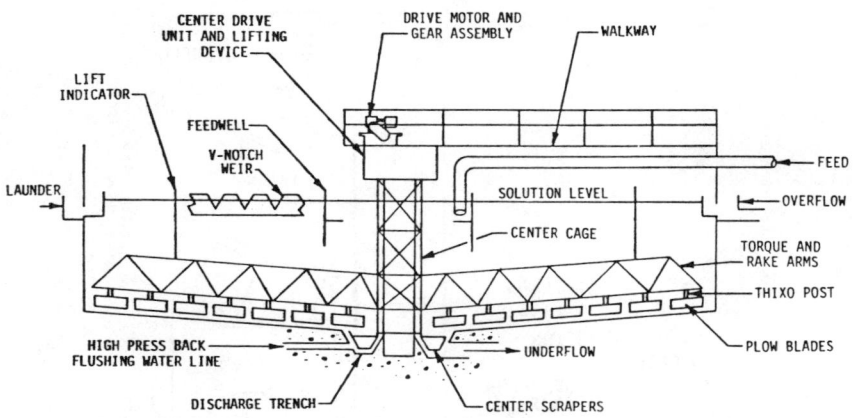

Figure 3-26. Cross-section of a conventional gravity thickener.

tank with a central vertical shaft that is supported either by internal structural design, by a center column, or by a bridge. Two long, radial rake arms extend from the lower end of the vertical shaft; two short arms may be added when necessary to rake the inner area. Plow blades are mounted on the arms at an oblique angle (with the trailing edge toward the center shaft), with a clearance of 1.5 to 3 in. from the bottom of the tank. They can be arranged identically on each arm or in an offset pattern so that the bottom is swept either once or twice during each revolution. The bottom of the tank is usually graded at a slope of 1:12 to 1.75:12 from the center. The settled sludge forms a blanket on the bottom of the thickener tank and is pushed gently toward the central discharge outlet. Center scrapers clear the discharge trench and move the solid deposits toward the underflow discharge point.

The scrubber slurry is fed through a feedwell into the thickener at a concentration of 5 to 15 percent solids, and the underflow is discharged at a concentration of 30 to 40 percent solids.

The use of inorganic polymer flocculant can reduce solids concentration in the overflow from a range of 50 to 100 ppm to a range of 10 to 70 ppm. Additionally, the size of a thickener in which a flocculant is used may be only about half that of one without flocculation because the sludge settles much more rapidly. Flocculants in powder form are dissolved to make solutions of 0.5 to 1 percent concentration for addition to thickener feed slurry.

When space is limited, a high-capacity thickener may be used. Such a unit is the Lamella® plate-type thickener that has been used successfully in the phosphate industry. The Lamella® thickener (Figure 3-27) consists of two tanks: an upper rectangular unit contains a series of parallel, slotted

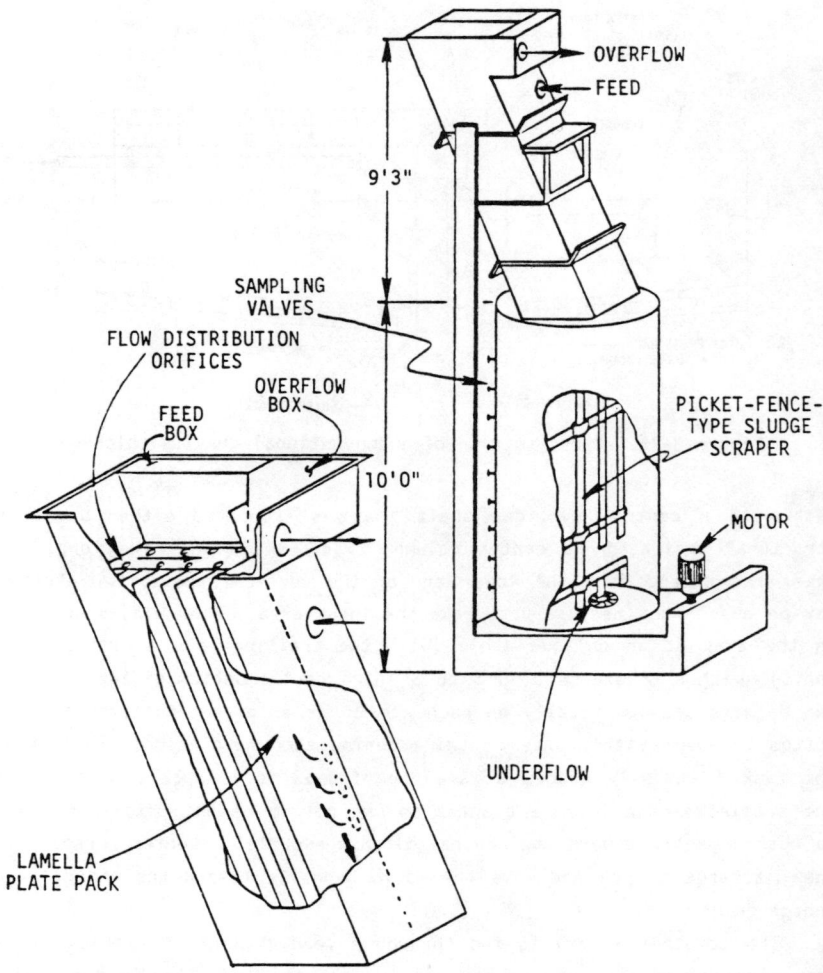

Figure 3-27. The Lamella® gravity settler thickener.

plates inclined at a 55-degree angle and serves as a high-rate gravity settler; a lower circular tank functions as a thickener with a picket-fence-type sludge rake and houses the underflow outlet. The arrangement of plates in the rectangular tank increases the settling capacity of the thickener and substantially reduces space requirements.

A plate-type thickener has been tested at the Shawnee facility. Results indicate that the Lamella® thickener can be used at installations where a dilute or rapidly settling slurry such as gypsum requires clarification.

Continuous Vacuum Filters. Vacuum filters are economical for continuous service and are widely used because they can be operated successfully at relatively high turndown ratios over a broad range of feed solids concentrations. A vacuum filter also provides more operating flexibility than other types of dewatering devices, and a drier product. Because a vacuum filter will not yield an acceptable filter cake if the feed solids content is too low, it is frequently preceded by a thickener or a hydrocyclone.

Four types of vacuum filters are applicable to limestone-generated sludge systems: drum, disk, horizontal belt, and pan. Each has different characteristics and applicability. The drum type, which is the most widely applied, is discussed here.

A rotary-drum vacuum filter is depicted in Figure 3-28. The drum is divided into sections, each connected through ports in the trunnion to the discharge head. The slurry is fed to a tank in which the solids are held uniformly in suspension by an agitator. As the drum rotates, the faces of the sections pass successively through the slurry. The vacuum in the sections draws filtrate through the filter medium, depositing the suspended solids on the filter drum as cake. As the cake leaves the slurry, it is completely saturated with filtrate and undergoes dewatering by the simultaneous flow of air and filtrate in the cake drying zone. Drying is

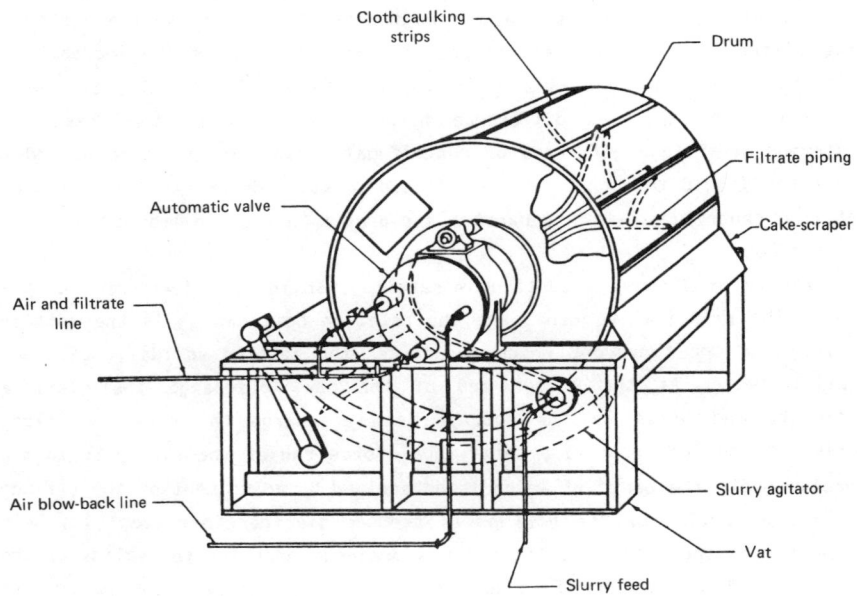

Figure 3-28. Cutaway view of a rotary drum vacuum filter.

negligible when the air is at room temperature. Finally, the cake is removed in the discharge zone either by a scraper or a string discharge.

A vacuum filter produces filter cake of 45 to 75 percent solids from feed slurries containing 20 to 35 percent solids. The filtrate containing 0.6 to 1.5 percent solids is recycled to the thickener.

The filtration rate in limestone scrubbing applications in which some calcium sulfate solids are present ranges between 150 and 250 lb/h per ft^2 (Heden and Wilhelm 1975).

Normally, the filters are installed at an elevated location so that the cake solids discharging from the filter can drop into a chute leading to a storage hopper for easy loading into a truck. If an elevated position is undesirable, a belt conveyor may be used to collect the solids discharged from the filter and carry them to a raised storage hopper, again to permit easy loading.

Centrifuge. Centrifuges are widely used for separating solids from liquids. They effectively create high centrifugal forces, about 4000 times gravity. The equipment is relatively small and can separate bulk solids rapidly with a short residence time. A centrifuge is a reliable and efficient machine, yielding products that are consistent, uniform, and easily handled.

Centrifugal separators are of two types: those that settle and those that filter. A settling centrifuge uses centrifugal force to increase the settling rate over that obtainable by gravity settling; this is done by increasing the apparent difference between densities of the phases. A filtering centrifuge generates by centrifugal action the pressure needed to force the liquid through a septum. This discussion describes the continuous settling centrifuge, which separates a slurry into a clear liquid and a very thick sludge.

Figure 3-29 shows a continuous settling, solid-bowl dewatering centrifuge. The principal elements are the rotating bowl, which is the settling vessel, and the conveyor, which discharges the settled solids. Adjustable overflow weirs at the larger end of the bowl discharge the clarified effluent, and ports on the opposite end discharge the dewatered sludge cakes. As the bowl rotates, centrifugal force causes the slurry to form an annular pool, the depth of which is determined by adjustment of the effluent weirs. A portion of the bowl is of reduced diameter to prevent its being submerged in the pool; thus it forms a drainage deck for the solids as they are conveyed across it. Feed enters through a stationary supply pipe and passes through the conveyor hub into the bowl. As the solids settle to the outer edges of the bowl, they are picked up by the conveyor scroll and continuously overflow the effluent weirs. In a recent study with feed of 16

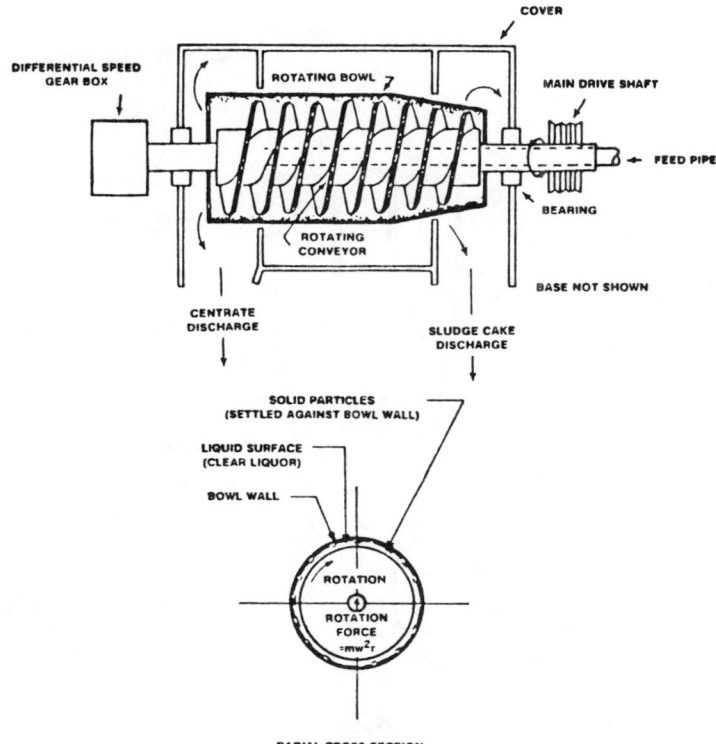

Figure 3-29. Cross section of solid-bowl centrifuge.

to 20 weight percent solids, the resultant solids concentrations ranged from 0.7 to 4 weight percent in the effluent and from 60 to 70 percent in the cake. The concentrations were dependent on feed rate and bowl rotation speed (3300 to 5400 rpm) (Wilhelm and Kobler 1977).

Hydrocyclone. A hydrocyclone is a small, simply constructed device used extensively for the classification and dewatering of slurries. It can rapidly separate bulk solids from process streams. Although it is an excellent low-cost dewatering device, it is not very effective in producing clarified overflow. The product of a hydrocyclone, therefore, is often fed to another unit, such as a centrifuge or filter, for additional dewatering. The hydrocyclone also requires frequent maintenance because of high rates of wear, erosion, and corrosion by fly ash and slurries from the SO_2 scrubbing system. A hydrocyclone also can be installed in a slurry recycle line for a venturi scrubber to prevent plugging of the nozzles by removing large particles.

A typical hydrocyclone (Figure 3-30) consists of a vertical cylinder with a conical bottom, a tangential inlet near the top, and an outlet for

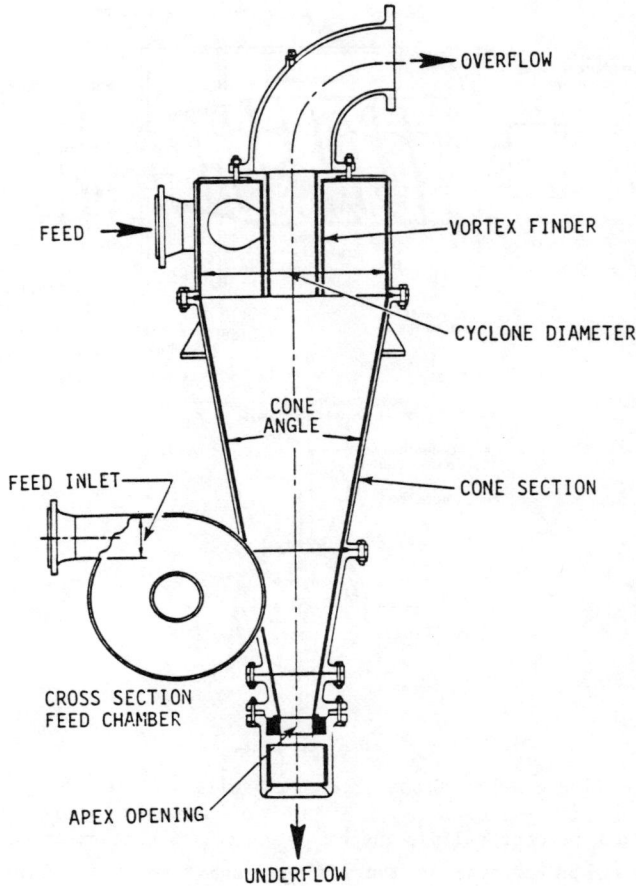

Figure 3-30. Hydrocyclone cross section (Dorrclone®).
Courtesy: Dorr-Oliver, Inc.

solids at the bottom of the cone. The fluid overflow pipe is usually extended into the cylinder to prevent short-circuiting of fluid from inlet to overflow.

The inlet piping imparts a rotating motion to the incoming fluid. The path of the fluid follows a downward vortex, or spiral, adjacent to the wall and reaching to the bottom of the cone. The fluid stream moves upward in a tighter spiral, concentric with the first, and leaves, still whirling, through the outlet pipe. Both spirals rotate in the same direction. The particles settle to the side walls and slide down the inclined wall to the apex of the cone. They are then discharged through an underflow orifice.

Basically, a hydrocyclone is a settling device in which a strong centrifugal force is used for separation. The centrifugal force in a hydrocyclone ranges from 5 times gravity in large, low-velocity units to 2500 times gravity in small, high-pressure units.

Design Considerations

Thickener. The usual criteria for sizing of a clarifier/thickener unit are solids loading and hydraulic surface loading. For a thickener and clarifier/thickener, the solids loading factor is more important. Solids loading without polymer flocculant may range from 12 to 30 ft^2/ton per day to achieve 25 to 35 percent solids in the underflow. To achieve higher underflow solids concentration, in the range of 35 to 45 percent, the solids loading without polymer ranges from 8 to 12 ft^2/ton per day. With addition of polymers, the solids loadings are reduced to ranges of 3.5 to 11 and 1.7 to 4.5 ft^2/ton per day, respectively, for the two ranges of underflow solid concentrations.

Recommended surface loadings range from 300 to 4000 gal/ft^2 per day of granular solids; 800 to 2000 gal/ft^2 per day of slow-settling solids; and 1000 to 2000 gal/ft^2 per day of flocculated particles. The latter range is normal for the thickeners used in dewatering FGD scrubber sludge.

In limestone scrubbing systems, the thickener diameters typically range from 50 to 100 ft; sidewall heights typically range from 8 to 14 ft. The rake arms are driven by a 2- to 5-hp motor through a worm gear connection at a speed of 10 to 20 min/rev. The thickener is usually a lined or painted (mild) steel tank with a steel or concrete bottom. Because the pH level of slurry in the thickener tank varies and chloride content may be high, most submerged parts are protected from corrosion by an epoxy or rubber coating.

Rotary Vacuum Filter. A number of variables affect the operation of the filter system:

Feed solids concentration

Filter cycle time

Drum submergence

Agitation

Filter vacuum requirements

Filter media

The size of a filter for a given application is inversely proportional to the solids concentration of the slurry feed. Thus, if a thickener is installed upstream, it is important to determine the minimum solids concentration in the underflow that occurs in average operation of the unit. When the filter is sized to accommodate this minimum solids concentration, it will have adequate capability to dewater the solids output of the plant.

If large variations in solids handling capability are expected, it is often more desirable to install two smaller filters than one large filter. Then when the amount of solids is reduced substantially over a long period, one of the units may be shut down; the other unit may then operate at the proper submergence level and thereby optimize performance. In addition, vendors recommend installation of a spare to ensure uninterrupted plant operation.

A standard rotary drum vacuum filter can be purchased from many suppliers. Correct sizing, i.e., determination of the correct filter area, is important economically because size usually accounts for an appreciable portion of both capital and operating costs. Enough filter area must be provided to maintain a rate of solids removal that will prevent excessive accumulation of solids in the plant.

To prevent corrosion in limestone slurry applications, the piping and support members in a vacuum filter should be made of stainless steel. Drum heads in the filter can be made of carbon steel if corrosion allowance is provided. Polypropylene is the customary filter medium.

<u>Centrifuge</u>. The following design parameters must be considered in design of a centrifuge:

Bowl diameter: A centrifuge with large bowl diameter and low speed (rpm) may produce the same particle settling velocity as a high-speed machine with a small bowl diameter.

Bowl length: Increasing the length of a bowl will result in a proportional increase in retention time.

Bowl speed: Bowl speed controls the centrifugal force that will be produced in a machine with a fixed bowl diameter. The general trend is to select machines that operate at low speeds to reduce operating costs and increase machine life. Recent application of special wear-resistant materials such as tungsten carbide and ceramic tile with long-life bearings has reduced the maintenance needs significantly.

Conveyor speed: Increasing the conveyor speed allows more rapid movement of solids through the machine. With most sludges, higher conveyor speeds will produce a dryer cake by causing the wetter solids to be left behind.

Conveyor pitch: In fluidlike and sulfite-rich FGD sludge, increasing the scroll pitch will produce results similar to those produced by increasing the conveyor speed.

Because scrubber sludge is erosive and sometimes corrosive, all liquid contact materials in the centrifuge should be made of 316L stainless steel. The tips of the conveyor should be made of tungsten carbide to reduce abrasive wear.

<u>Hydrocyclone</u>. The chief factors that influence hydrocyclone design include hydrocyclone diameter, cone angle, size of orifices, length of

cylindrical section, feed pressure, feed concentration, and particle size. Because of the number of significant variables, a hydrocyclone manufacturer is usually consulted regarding particular applications.

The most important factor influencing the application and efficiency of a hydroclone is diameter. The smaller the particles to be removed, the smaller must be the diameter of the hydrocyclone. For any specific application, the hydrocyclone manufacturer will determine the appropriate diameter, height, and cone angle.

When size and cone angle are established, sizing of the flow orifices for feed, overflow, and underflow is the next most important factor. Plant operating personnel can change the orifices as needed to suit process requirements.

By virtue of the high velocities within the device, the portions of the hydrocyclone exposed to the high velocities must be fabricated of erosion-resistant metal alloys. The bulk of the device can be constructed of type 316L stainless steel to resist chloride corrosion.

Operational Systems

Thickener. More than half of the operational limestone scrubbing FGD systems in the United States use thickeners. Some operational systems are described below.

Lawrence--

At Lawrence 4, the dewatering system was designed with separate slurry hold tanks for the combination venturi rod and spray tower scrubber. Staging of the liquid side of the system enables the addition of fresh limestone to the scrubber (spray tower) effluent hold tank (EHT). The solids content in the EHT is controlled at 5 percent. Slurry is bled from the EHT to the collection tank. The solids content of the collection tank is controlled at 8 to 10 weight percent by varying the effluent bleed pump flow. This pump delivers the effluent bleed to the system thickener (40 ft diameter), where the slurry is concentrated to about 50 percent solids before being pumped to a disposal pond. The pH in the thickener is 6.5 to 7.0, and the pH of the slurry leaving the thickener is about 8.5. The inlet temperature is 115° to 120°F, and the outlet temperature is 80° to 90°F. Water from the pond is returned to the makeup water surge tank, where it is combined with the thickener overflow water to provide makeup water for the scrubber system.

Sherburne--

Units 1 and 2 of Northern States Power Company's Sherburne County generating plant are basically the same. For control of the solids content in the EHT, an effluent bleed flow of 160 gpm is drawn from the spray water pump discharge and sent to the thickener through the slurry transfer tank. The flow is controlled by a nuclear density meter, which maintains the solids content at 10 percent. Of this 10 percent solids, 2 to 3 percent consists of calcium sulfate seed crystals. The continually forming calcium sulfate adheres to the seed crystal. As the crystals start to grow, they precipitate out of the solution.

Air is introduced into the EHT to provide enough oxygen to complete the oxidation of calcium sulfite to calcium sulfate. The air enters near two horizontal-entry mixers, which ensure that the air is mixed with all the available sulfite. In the thickener, the slurry is concentrated and sent to a fly ash pond. Clarified water from both the thickener and settling pond is recycled as makeup water.

Shawnee--

The EPA facility at Shawnee tested a Lamella® Gravity Settler Thickener, 20 ft high, consisting of two tanks constructed of epoxy-painted carbon steel. Solids content of the underflow is 40 percent and of the clarified liquid, 0.5 percent. The tests were done with oxidized limestone with fly ash.

Vacuum Filter. Both rotary drum and horizontal belt filters have been used successfully as secondary dewatering devices for limestone sludge. The following plants are using or plan to use rotary drum vacuum filters:

San Miguel 1, San Miguel Electric Coop

Paradise 1 and 2, TVA

Widows Creek 7, TVA

The following plants are using horizontal belt vacuum filters:

R. D. Morrow 1 and 2, Southern Mississippi Electric Power Association

Southwest 1, Springfield City Utilities

Much of the experience with vacuum filters has been with lime wet scrubbing systems or with the dual alkali process. The Lime FGD Systems Data Book (Morasky 1979) gives details on use of vacuum filters with lime scrubbers.

Rotary drum vacuum filters have been used as the secondary dewatering device by International Utilities Conversion System (IUCS) as the principal means of drying sludge (Morasky 1978) for mixing lime with fly ash to fixate FGD sludge. To date, IUCS has a total of 16 FGD systems either in operation or under contract for treating waste sludge by use of their rotary drum vacuum filter process (Knight et al. 1980).

Centrifuge. The use of a centrifuge for secondary dewatering is becoming more prevalent.

Shawnee--

The Shawnee test facility operates a continuous centrifuge as one of several processes used to dewater scrubber waste sludge and to recover the dissolved scrubbing additives. Normal operating conditions with an unoxidized slurry are a feed stream flow of 15 gpm at 30 to 40 weight percent solids, a centrate of 0.1 to 3.0 weight percent solids, and a cake of 55 to 65 weight percent solids. About 30 percent of the total solids is fly ash; the remainder is predominantly calcium sulfate and sulfite.

The machine is a Bird 18- by 28-in. solid-bowl continuous centrifuge, which operates at 2050 rpm and requires 30 horsepower. It is made of 316L stainless steel with stellite (Colmanoy) hardfacing on the feed ports, conveyor tips, and solids discharge ports. The bowl head plows and case plows are replaceable. The pool depth is set at 1-1/2 in. No cake washing is performed in this machine.

The centrifuge was inspected in June 1978 after 6460 hours of operation. The inspection was prompted by a gradual and continuing increase of suspended solids in the centrate to a level of approximately 3 weight percent. The machine was judged to be in generally fair condition, but certain components were in need of factory repair. Serious wear was observed at the conveyor tips on the discharge end at the junction of the cylinder and a 10-degree section of conveyor. The casing head plows and solids discharge head near the discharge ports were also worn. The bowl and effluent head were in good condition (Rabb 1978). The machine has required little maintenance.

Mohave--

An early experimental investigation of limestone scrubbing was conducted at Southern California Edison's Mohave Station, and the scrubbing units have since been dismantled. The centrifuge at Mohave was of the solid-bowl type, having a bowl 2 ft in diameter by 5 ft long. The auger/bowl speed ratio of 40/39 was established with an eddy current clutch device, and the centrifuge drive was designed to allow evaluation of dewatering characteristics over a range of rotating speeds. The device was designed to handle 120 lb/min of solids (dry basis), but was tested at 160 lb/min and could have gone still higher.

The centrifuge was controlled with relative ease. With inlet solids concentrations of 10 to 15 percent and 30 to 35 percent, solids content of the centrifuge cake ranged from 68 to 72 percent at approximately 2000 rpm and various flow rates. The only operating problems were high torque during the feeding of 5 percent slurry solids in one test, and cementing of the auger to the bowl with dried solids during outages. Normal centrifuge operation with solids between 25 and 35 percent in the feed resulted in average solids content of 71 percent in the cake and average suspended solids of about 0.15 percent in the centrate. Suspended solids in the centrate decreased when the feed solids content was reduced (Robbins 1979).

Centrifuges are also in use or planned for use at the following plants:

>Marion 4, Southern Illinois Power
>
>Martin Lake 1, 2, 3, and 4, Texas Utilities
>
>Thomas Hill 3, Associated Electric Coop
>
>Laramie River 1 and 2, Basin Electric Power Coop
>
>Craig 1 and 2, Colorado Ute Electric Association

Hydrocyclone. Both Peabody Process Systems and Research Cottrell Corporation use hydrocyclones to separate solids from slurry. A hydrocyclone system supplied by Research Cottrell is being used in conjunction with forced oxidation at the Dallman 3 unit, Springfield Water Light and Power.

SLUDGE TREATMENT

Treatment Processes

The method selected for sludge handling and disposal must be economical and must result in a properly placed, structurally stable material that is environmentally acceptable and complies with provisions of the Resource Conservation and Recovery Act (RCRA).

Detailed guidance that will direct a Project Manager through the sludge treatment decisions is not within the scope of this book. A recent EPRI publication (Knight et al. 1980) provides such guidance details. The brief introduction presented here is adapted from that publication; it is intended to indicate the basic considerations involved in the sludge treatment decision. This decision should be made early, because delay will reduce the number of options available.

Since it is very likely that today's waste may be a natural resource of the future, strong consideration should be given to reclaiming FGD scrubber wastes as well as the fly ash generated in burning the coal. Therefore, the first decision is whether to co-dispose the fly ash with scrubber sludge or to stockpile each separately to facilitate future reclamation.

Decision paths lead the Project Manager through the steps shown in Figure 3-31, which are required for all systems, wet or dry.

The first step is selection of the processing steps, which include the options of whether to (1) change the sludge characteristics by forced oxida-

The FGD System 113

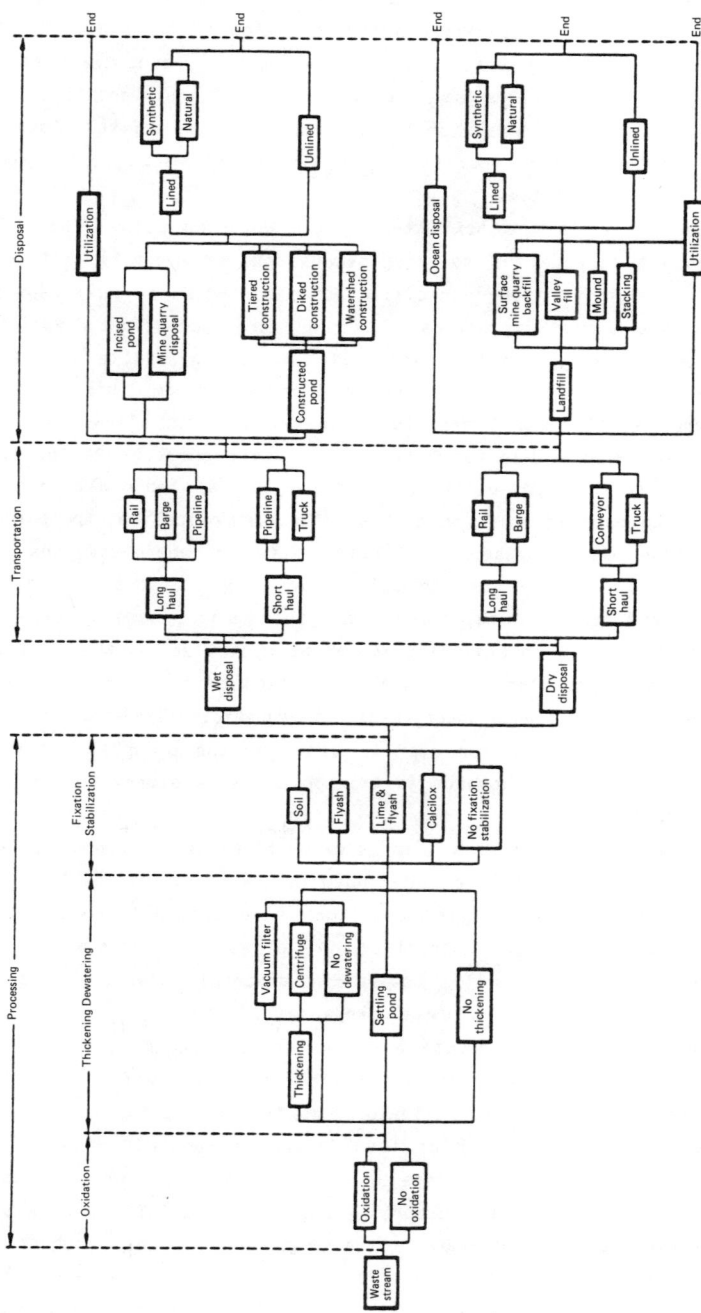

Figure 3-31. FGD sludge disposal alternatives (Knight et al. 1980).

tion to the calcium sulfate or gypsum form, (2) dewater by gravity or mechanical means, or (3) stabilize by the addition of fly ash or fixate by the addition of a fixating agent with or without fly ash. Each of these options has disadvantages, ranging from added cost of equipment or operation to increased solubility of calcium sulfate over calcium sulfite. Advantages may include better water separation, less volume of waste to handle, and better structural properties.

Transportation is the next step in the decision path. There is no choice when wet disposal is selected because the only practical transport mode is by pipeline, although specially equipped trucks or rail cars are a possibility but appear impractical. For dry transportation a variety of motorized earth-moving equipment and rail cars are available and in use.

Disposal/utilization is the final step in the system; this factor must be considered early in the decision process so as to allow the primary choice of wet or dry disposal. Cost is a very important factor in the decision. To this point, utilization is not yet a viable alternative to disposal. With more interest in and use of forced oxidation, the potential for utilization has increased. Calcium sulfate or "abatement gypsum" has found limited experimental use in agricultural applications, in wallboard manufacture, and as a set-retarding agent in cement. Some systems under construction are based on utilizing abatement gypsum as a saleable byproduct. A new method for storage/disposal of abatement gypsum has been used in a prototype scrubber test under EPRI sponsorship. The method, called "stacking," has been used for over 20 years by the phosphate fertilizer industry in Florida. The gypsum is transported as a slurry to the stack, for dewatering by gravity.

All sludge disposal systems can be categorized as wet (ponding) or dry (landfilling). Wet disposal has been used for years for disposal of bottom ash and fly ash. Many utility operators have naturally selected this familiar method for sludge disposal because of ease and economy of operation. Practice is now tending toward dry disposal, however, because of lower capital costs and environmental pressures.

Sludge processing to produce a dry material is done by stabilization and fixation. Stabilization improves the physical properties of sludge by blending the material with dry solids such as fly ash, bottom ash, or earth. The resultant moisture content of the mixture is reduced, resulting in a stable material for landfill disposal. Fixation involves chemical reactions between the fixative and the scrubber sludge. Figure 3-32 shows a sludge treatment flow diagram that depicts the processing of the scrubber bleed through the primary dewatering and/or the secondary dewatering equipment in conjunction with mixing of fly ash and/or fixatives to produce a dry material for landfill disposal.

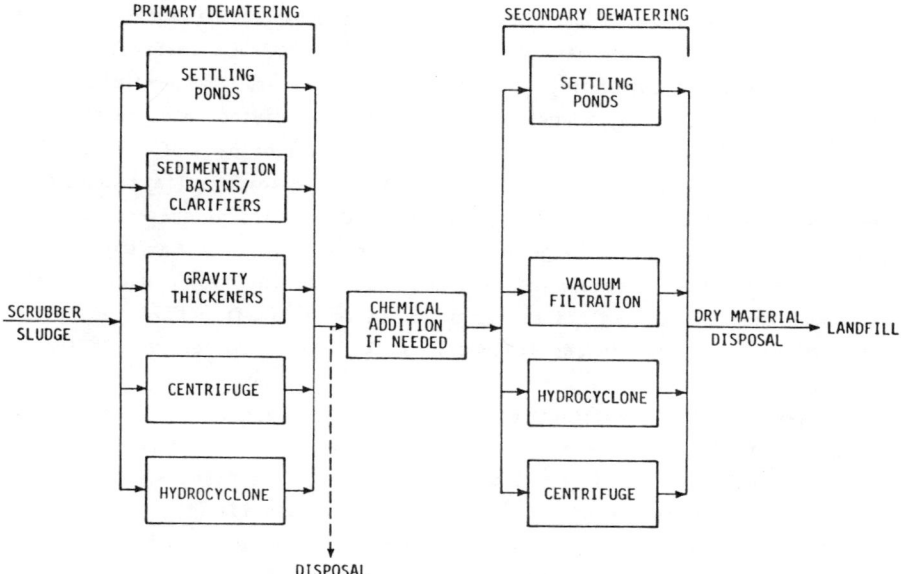

Figure 3-32. Sludge treatment flow diagram.

The terms "stabilization" and "fixation" are used interchangeably in the industry, and the terminology is somewhat confusing. Stabilization is generally understood to mean addition of dry fly ash, soil, or other dry additive to reduce the moisture content and improve handling characteristics without a chemical interaction between the sludge components and the additive. The term is used in this context throughout this manual.

In fixation processes, the chemical reactions bind the sludge particles together, thus increasing shear and compressive strength and reducing permeability. The structural stability and environmental characteristics of the waste product are thereby improved. The most widely practiced method of fixation involves the blending of lime, fly ash, and vacuum filtered sludge in a pug mill under controlled conditions. The resultant mixture is stacked for 2 to 3 days, during which time pozzolonic (cementitious) reactions responsible for improving the physical and chemical properties of the mixture occur. The fixated product is ultimately disposed of in a landfill. Landfill produced by such a process is chemically and physically stable, and generally characterized by low permeability. This approach is currently offered by several scrubber vendors as well as companies such as IUCS, which specialize in FGD sludge treatment. The process has the advantage of being nonproprietary and utilizes a generic fixative (lime). Alternative approaches such as the Dravo process, in which thickener underflow is mixed with a proprietary fixative (Calcilox) are also available.

An important preliminary step in any sludge disposal system is consideration of the chemical/physical characteristics of the material. The predominant solid components are calcium sulfite, calcium sulfate, and fly ash. The quantities of fly ash can vary considerably, depending on the presence and efficiency of upstream fly ash removal devices and, in the case of alkaline fly ash, the use of the fly ash as all or part of the scrubber reagent. The proportions of calcium sulfite and calcium sulfate depend on many process factors. Estimates of relative quantities of these components can be made from a knowledge of the boiler system, upstream particulate removal equipment, fuel and reagent analyses, and the liquid cycle of the scrubber/disposal system. These data can be obtained from pilot or prototype operation, from similar full-scale systems, or from equipment/material specifications.

Likewise, the quantities of sludge to be processed, transported, and disposed of can be estimated by a series of assumptions and calculations.

With this type of information and other pertinent data, a decision can probably be made regarding wet or dry disposal. Consideration should then be given to disposal site and design, which will involve a thorough exploration of environmental requirements.

Regulating Considerations

Because FGD sludge is a relatively new material, many regulatory agencies are not familiar with its properties and characteristics, nor are they familiar with disposal technologies. As a result, few regulations pertaining specifically to the disposal of FGD sludges have been issued. In the absence of specific Federal guidance, state and local agencies often rely on standards and regulations developed around the disposal of other types of waste, many of which are inappropriate or irrelevant in relation to FGD sludges. This situation, however, is rapidly changing.

The first two major pieces of Federal legislation addressing solid wastes were the Solid Waste Disposal Act of 1965 and the Resource Recovery Act of 1970, both aimed primarily at municipal solid waste, with little direct applicability to FGD sludges. Furthermore, there was little regulatory authority centered in these laws. The Federal government issued general guidelines and provided technical assistance, but had little power to regulate operations. This situation has now changed.

The Resource Conservation and Recovery Act of 1976 (RCRA) greatly expanded the role of the Federal government in regulating solid waste handling and disposal. Subtitle C of RCRA addresses hazardous wastes, and Subtitle D focuses on nonhazardous wastes. As required by the Act, the EPA is developing a detailed program for the regulation of hazardous wastes. Wastes defined as hazardous are subject to "cradle-to-grave" considerations.

In EPA's regulations of May 19, 1980, for identifying hazardous wastes, large-volume utility wastes, including FGD sludge, were specifically excluded from regulation under Subtitle C. With this exclusion, FGD sludge will be primarily regulated as a nonhazardous waste under Subtitle D. The nonhazardous waste regulations, however, are quite general and not specifically oriented toward FGD wastes. This generalized approach to the regulation of utility wastes may change in the near future as the EPA develops standards that are specifically applicable to large-volume utility wastes. The regulations evolving from this program will probably not be promulgated for several years. Therefore, while nonhazardous waste regulations will govern current FGD disposal activities, compliance with more detailed standards will likely be required in the future.

Subtitle D of RCRA prohibits open dumping and requires that environmentally acceptable practices be utilized. As provided for in the Act, the EPA has developed criteria for classifying solid waste disposal sites. Facilities that cannot meet the criteria will be considered "open dumps," which are illegal. The EPA has also proposed guidelines for landfill disposal of solid waste, describing waste disposal practices recommended for attaining compliance with the aforementioned criteria. Although the EPA has issued criteria and guidelines for nonhazardous waste management, the primary authority for implementing nonhazardous waste regulatory programs will continue to be with State agencies.

Current state regulations applicable to FGD sludge vary widely across the country. Most states have sufficiently broad legislative authority to regulate FGD sludge disposal to some degree. Some state laws allow direct control of solid wastes. Other states achieve indirect control through implementation of regulations designed to protect the ground water or limit activities in floodplains. Most states proceed cautiously with FGD wastes, issuing permits on a case-by-case basis.

The primary concern with pond disposal of FGD wastes has been the integrity of dikes and impoundments. Regulations pertaining to such structures are concerned with physical damage to persons or property and effects on ground or surface waters. Primary concerns with landfilled wastes are those of structural stability and leachate/runoff effects on ground and surface water.

As a result of RCRA, many state programs are being revised. With EPA's encouragement and financial support, numerous state agencies are developing new regulations and plans to implement them. Most state programs are being modeled after the Federal program so as to minimize complications in obtaining authorization from the EPA. Therefore, delays in finalizing EPA regulations are, in turn, postponing the promulgation of state regulations.

In carrying out this new Federal program, the state agencies will continue to play an important role in issuing permits for FGD sludge disposal sites. In spite of the increased Federal role, state requirements and procedures may differ substantially from state to state. For this reason, operators of scrubbing systems are advised to contact the appropriate state agencies to determine the prevailing permit requirements.

LIMESTONE SLURRY PREPARATION

The bulk material handling properties of limestone are very similar to those of coal, and the conventional powerplant procedures for coal handling and storage can be easily adapted to limestone. Because of the close similarity, this section emphasizes those aspects of limestone slurry preparation that apply specifically to limestone scrubbing.

All operating scrubbing milling facilities use wet grinding equipment, specifically some version of the wet ball mill.

In closed-circuit grinding, as shown in Figure 3-33, all oversized

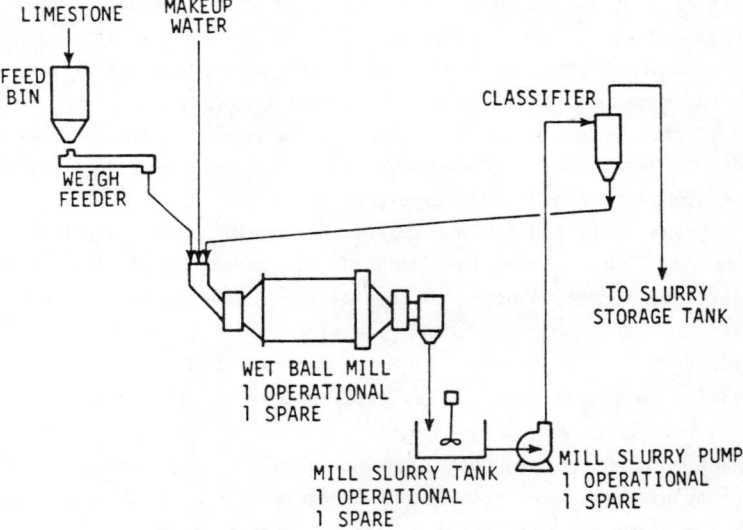

Figure 3-33. Typical closed circuit grinding system.

particles are recycled until they are ground away. Therefore, feed rock should contain the minimum possible level of siliceous impurities. It is not necessary to include facilities for slurry dilution in a closed-circuit installation; the 15 to 30 percent slurry from the classifier is suitable for direct feed to the scrubber.

The classification device used most often in wet closed circuits is the hydraulic cyclone (hydrocyclone). This is a static apparatus that uses energy from a circulating pump to develop a centrifugal force that concentrates particles of higher mass into a fraction of the flowing stream. The underflow stream from the hydrocyclone is a thick slurry containing most of the larger (heavier) particles; this stream is recycled to the ball mill. The overflow stream, containing most of the water and smaller particles, is sent to a storage tank. A classifying hydrocyclone can make a fairly sharp separation if slurry composition and flow rate and pressure at the hydrocyclone inlet remain reasonably constant. In plants where feed rock composition is expected to be inconsistent, automatic instrumentation to control density and flow rate should be included to ensure proper hydrocyclone performance.

Slurry Storage

Finely ground limestone slurry does not settle quickly and can be stored for long periods in properly designed, agitated storage vessels. Since limestone does not react chemically either with air or with constituents in recycled water, no scale is formed and there are no special considerations other than the application of proven engineering design principles for the storage of viscous slurries.

Each storage tank must be equipped with a constantly operating agitator to prevent settling of the solids. Agitators should be vertical-shaft, top-mounted units located axially within the vessel. Agitators with internal bearings should be avoided because they present maintenance and repair problems. Turbine impellers are best for slurry agitation, usually with motors connected to the shafts through speed-reduction gears. High-speed agitation is not needed.

Use of a pump to move slurry from the bottom of the tank to the top also improves vertical mixing and thereby maintains an even slurry concentration at all levels within the storage tank.

Slurry Feed

Limestone slurry is pumped from the storage tank to the scrubber vessel at varying rates to maintain proper pH in the scrubber. This operation is accomplished by a combination of pumps, piping, and controls designed to minimize erosion of the transfer equipment and solids deposition during transfer. Four basic arrangements, shown in Figure 3-34, have been developed to realize these design objectives.

Arrangement 1 utilizes a variable-speed pump drive to deliver the required volume of slurry. In this arrangement, there are no control valves and no piping restrictions. Operating pressure is therefore minimal, and the system consumes the least amount of pumping energy. Separate pumps and pipelines are needed, however, for each point of limestone application, and the system is consequently expensive for plants that have multiple scrub-

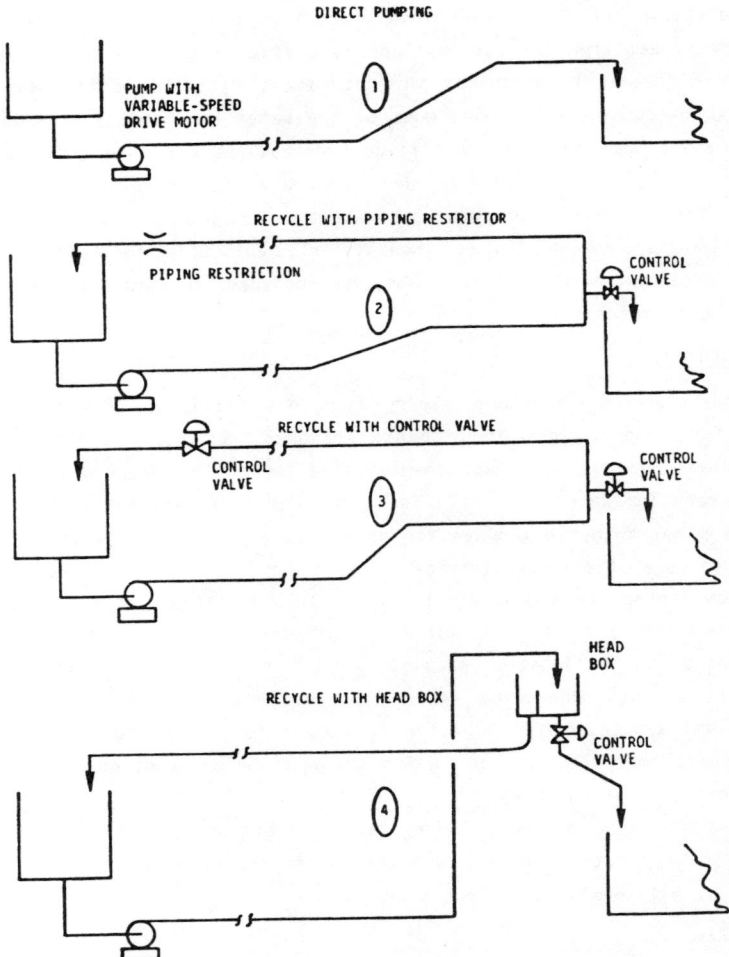

Figure 3-34. Limestone slurry feed arrangements.

bers. The system is not applicable to plants that have significant turndowns in scrubber operating rate.

Arrangement 2 is used in other industries that handle heavy slurries. A pump delivers slurry through a piping loop that is routed to pass close to each point of application, and the slurry returns through a piping restriction to the storage tank. At each point of application, valves are used to control the slurry flow. Success with this arrangement depends upon accurate hydraulic calculations and proper pump selection. The pump must be a "flat head" design, in which the discharge pressure remains relatively constant regardless of pumping rate. The piping restriction is usually a "flow choke," a block of hardened metal drilled through to a calculated diameter. Control valves, often angle-type with hardened trim, are close-coupled to the piping loop and are located so as to discharge vertically downward to the application point.

Arrangement 3 is the same as Arrangement 2, except that a control valve is substituted for the static flow restrictor. This system provides greater rangeability in the slurry loop and is used where the piping hydraulics cannot be calculated accurately. The best applications of this system incorporate instrument-activated control devices to throttle the return valve and thereby maintain a calculated minimum flow rate in all sections of the piping loop, regardless of rate of slurry usage. The instrumentation may be complex.

Arrangement 4 is used in plants where there is sufficient difference in elevation of the slurry storage and point of feed and no great distance separates them. In this arrangement, slurry is pumped continuously to an elevated head box. Control valves below the box allow flow of slurry as needed, and excess slurry overflows a weir in the box and returns by gravity to the storage tank. Control valves in this system are quite large but are long-lasting because they operate with only a slight pressure drop. Any type of pump can be used, providing it delivers slightly more slurry to the head box than the maximum total rate of usage.

Velocity of flow in all arrangements must be high enough to provide turbulence to keep solids from settling. A frequent error is to provide too much velocity; the amount needed is less than many piping designers assume. Shawnee experience indicates that a slurry velocity of 8 to 10 fps should be maintained in piping.

<u>Design Considerations</u>

Following are two basic considerations for selection and handling of dry limestone for scrubbers:

Procure the grade of rock that is most inexpensively ground to very small size.

Design materials-handling equipment to permit the handling of unclassified or waste limestone.

At least two feed bins or "day tanks" are needed to provide flexibility in the filling schedule and to allow partial drying of excessively wet feed rock before use. This improves the feeding characteristics of the material. Limestone feeders should also be duplicated, since they are usually high-maintenance items and spares may be needed to ensure continuity of production.

The size of a ball mill is determined by four factors:

1. The mill capacity, usually based on required tonnage of feed rock per 24-h day and hours per day of mill operation.

2. The size analysis of the feed rock, including both maximum particle size and size distribution.

3. The required size of product from the mill.

4. The Bond Work Index (grindability) of the rock, which is a measure of its resistance to grinding.

The size of a ball mill and to a large extent its initial cost are established by the degree of size reduction necessary. The larger the feed size and the smaller the required particle size, the larger a mill must be. For a given ball mill, the throughput capacity is reduced by 20 percent if the feed size is increased from 0.25 to 1 in. Capacity is reduced by 65 percent if product requirements are changed from 200 to 325 mesh.

The most fundamental and difficult procedure in selection of limestone milling facilities is the development of specifications for the finished slurry. These specifications significantly affect the initial and operating costs of the scrubber installation. There is, however, very little basic information that defines the most economical degree of grinding. The trend is apparently toward finer grinding: one major equipment manufacturer reports that specifications now range from 70 percent passing through a 200-mesh screen to 95 percent passing through a 325-mesh screen; most are in the range of 60 to 80 percent passing through 325 mesh.

A large volume of slurry storage, divided into at least two tanks, should be provided to permit maintenance of equipment without interrupting the scrubbing operation. Storage for 72 hours of operation may be necessary to permit major maintenance, such as replacement of the ball mill liner. Because agitation equipment is costly and consumes power continuously, the best compromise may be to provide a portion of the storage with only compressed air agitation (sparging), which is used only when needed.

The arrangement of pumps and piping to deliver slurry to the scrubbing system should be based on engineering comparisons of all possible arrangements.

The main experience factor gained from current limestone milling operations has been the effect of silica and other impurities contained in the supply limestone on the ball mill and its associated classification system. The quality of delivered limestone must be continuously checked to ensure that it meets the scrubber design specifications. Table 3-12 presents design specifications for several representative FGD installations. For details, the reader is referred to EPA's quarterly utility survey report and to Appendix F, which lists the U.S. plants with operating limestone scrubbing systems.

PUMPS

A coal-burning power plant that installs a limestone FGD system must

TABLE 3-12. LIMESTONE SLURRY PREPARATION AT OPERATIONAL FACILITIES

	Duck Creek 1	LaCygne 1	Lawrence 4 & 5	Sherburne 1 & 2	Southwest 1	Widow's Creek 8
Ball mills						
Number	2	2	2	2	2	1
Capacity, ton/h	40	110	6	24	8	40
Limestone slurry concentration, %	65	66	NR	60	40	40
Size (diameter x length), ft	NR	NR	6x12.5	7x30	NR	NR
Motor, hp	NR	2,000	150	500	NR	NR
Product size, % through required mesh	90 (200 mesh)	95 (200 mesh)	NR	80 (200 mesh)	NR	90 (200 mesh)
Type of circuit	Closed	Closed	Closed	1-closed 2-open	Closed	Closed
Slurry handling						
Number of slurry tanks	1	2	1	2	1	1
Capacity of storage tanks, gal	79,500	3,000	3,000	270,000	30,000	200,000
Agitator motor, hp	NR	12	NR	1	NR	NR
Number of dilution tanks	NA	2	1	2	NR	NR
Dilution tank capacity, gal	NA	186,000	1,000	4,200	NR	NR
Storage						
Mode	Open stockpile	Open stockpile	Open stockpile	Open stockpile	Silo	Open stockpile and silo
Number of silos	NA	NA	NA	NA	1	1
Storage capacity, No. of days	NA	30	NR	100	NR	NR
Conveying						
No. of feed bins	1	2	1	2	1	NR
Capacity of feed bins, (ton) each	NA	1,200	30	600	NR	NR
Type of feeders	Weigh	Belt conveyor	Belt conveyor	Belt conveyor	Belt conveyor	Weigh feed
No. of feeders	1	2	1	2	1	1

NR - Not reported.
NA - Not applicable.

provide for handling of the slurries generated in removal of SO_2 from the flue gas. Because experience with slurry pumping is the basis for achieving successful performance, the utility should obtain performance data from A/E firms, consultants, and pump manufacturers. Even with these helpful records, however, the project manager must be aware that his slurry pumping application is unique and must weigh all potential operating factors when he assembles the fluid handling system.

Operating records of the early limestone FGD units show numerous failures of pumps, valves, piping, and instrumentation (O'Keefe 1980). These failures can be traced to the overconfidence of designers, who envisioned the FGD slurry conditions as being fairly easy compared with those of mining, metallurgical, and many chemical processes. Additionally, the operating experience with pumps was not well delineated because the early FGD systems performed at low availability and reliability. Shutdowns for various reasons allowed ample time for maintenance of slurry-fluid handling systems. As the overall system performance improved, deficiencies in pumps and piping became obvious. These deficiencies are now being corrected through cooperation among operators, A/E firms, and equipment manufacturers.

The following discussion gives design information on pumps for the handling of scrubber recirculation slurry, limestone slurry feed, thickener overflow, thickener underflow, and sludge, and also on pumps that service clear liquid streams.

Description and Function

Recirculation pumps. The recirculating pumps are the largest pumps in the limestone slurry system. They receive the slurry directly from the bottom of the scrubber effluent hold tank. The discharge slurry is continuously recirculated through the scrubber. Normally, a portion of the recirculation stream is bled to the solids disposal system.

Many features of the recirculation pump distinguish it from the typical centrifugal pump used for clear liquids. Wall thicknesses of wetted-end parts (casing, impeller, etc.) are greater than those used in conventional centrifugal pumps. The cutwater, or volute tongue (the point on the casing at which the discharge nozzle diverges from the casing), is less pronounced so as to minimize the effects of abrasion. Flow passages through both the casing and impeller are large enough to permit solids to pass without clogging the pump. Because the gap between the impeller face and suction liner will increase as wear occurs, the rotating assembly of the slurry pump must be capable of axial adjustments to maintain the manufacturer's recommended clearance. This is critical to the maintaining of design heads, capacities, and efficiencies. Other special features include extra-large stuffing boxes, replaceable shaft sleeves, and impeller back-vanes that act to keep solids away from the stuffing box. Although the impeller back-vanes also reduce axial thrust by lowering the stuffing-box pressure, these vanes can wear considerably in abrasive service. Hence, both the radial and the axial-thrust bearings on the slurry pump are heavier than those on standard centrifugal pumps (Dalstad 1977).

Slurry pumps are available in a variety of materials of construction to meet the requirements of various applications for withstanding abrasion and corrosion. A comprehensive service description must be developed as a guide to selection of recirculation pumps. This necessitates detailed analysis of slurry characteristics, including composition, pH, specific gravity, viscosity, and other factors, all of which are discussed briefly below.

Composition of the slurry is critical to pump selection. The erosive circulating fluid contains many solid and dissolved species. The major solids are undissolved limestone, fly ash, calcium sulfite ($CaSO_3$), and calcium sulfate ($CaSO_4$). Solids content ranges from 10 to 20 percent by weight. The dissolved species are calcium, magnesium, sodium, sulfite, sulfate, chloride, and carbonate ions. Chemical analysis of the slurry is particularly important with closed-loop operation, because some species such as chloride ions can build up to high levels. Information about concentrations and particle size of the solids will help determine abrasion-corrosion resistance and the mechanical strength required of the pump.

The pH of the slurry in the scrubber is maintained between 5.0 and 5.8. The pH of a given solution varies with temperature, especially in the range from 50° to 150°F.

The specific gravity of the recirculating slurry is usually between 1.05 and 1.14. The system incorporates automatic solids control to keep the specific gravity of the slurry constant.

Knowing details of the rheology of the slurry makes it possible to evaluate the reduction in pump performance due to the viscosity of the mixture and the added slip between the fluid and the solid particles as the mixture accelerates through the pump impeller. This slip is greater in mixtures with high settling velocities.

The slurry may contain significant amounts of fly ash, depending on the coal and on whether an ESP mechanical collector precedes the scrubber. The amount and composition of the fly ash must be determined before equipment is specified.

Gas entrainment in the recirculating slurry could cause fluctuation of the slurry in the pump from all liquid to essentially all gas. These variable conditions can cause deflection of pump shafts, which may lead to bearing failure and abnormal packing wear. Gas entrainment also reduces the liquid flow and thus reduces SO_2 absorption.

<u>Limestone Slurry Feed Pumps</u>. Concentrated slurry feed is usually handled by rubber-lined centrifugal pumps. Positive displacement pumps with variable-speed drives are also applicable. Because the basic design of the centrifugal limestone slurry feed pumps is the same as that of the recirculation pumps, only the salient features of the positive displacement (screw) pumps are discussed. Cast iron, erosion-resistant alloy, and rubber-lined pumps are common in limestone scrubber systems. Some utilities prefer to use rubber-lined pumps from a single manufacturer for uniformity throughout the plant.

The limestone as received contains tramp materials such as rocks, metal, and wood. With proper design and operation, the screening process before the ball mills should remove these impurities so that they do not enter the limestone slurry preparation system.

The limestone slurry is alkaline. Following are typical limestone slurry conditions:

pH	8
Solids, wt. percent	20 to 60
Solids	$CaCO_3$
Temperature, °F	50 to 104
Specific gravity	1.1 to 1.3

Although centrifugal pumps are widely used, the screw pump that handles limestone slurry feed is a special type of rotary positive displacement pump in which the flow through the pumping elements is truly axial. The liquid is carried between screw threads on one or more rotors and is displaced axially as the screws rotate and mesh. In all other rotary pumps the liquid is forced to travel circumferentially; thus the screw pump, with its unique axial flow pattern and low internal velocities, offers a number of advantages in those few applications where centrifugal pumps cannot be used. The screw pump can handle liquids with viscosities ranging from that of molasses to that of gasoline. Because of the relatively low inertia of their rotating parts, screw pumps can operate at higher speeds than other rotary or reciprocating pumps of comparable displacement. Screw pumps, like other rotary positive displacement pumps, are self-priming and have a delivery flow characteristic that is essentially independent of pressure.

As with any rotary pump, the arrangement for sealing the shafts is very important and often is critical. All types of screw pumps require at least one rotary seal on the drive shaft. For drive shafts, rotary mechanical seals as well as stuffing boxes or packings are used. Double back-to-back arrangements with a flushing liquid are sometimes used for very viscous or corrosive substances.

The mechanical seal is gaining wider use with the advent of new elastomers such as Viton, Butyl, and Nordel. The rotary components are made of carbon, bronze, cast iron, Ni-Resist, carbides, or ceramics. The mechanical seal can be designed to be completely or partly independent of the fluid pressure to which it is exposed, and it also can operate at subatmospheric pressure without drawing in air.

Although it is the maximum viscosity and the expected suction lift that determine the size of the pump and set the speed, it is the minimum viscosity that determines the capacity. Screw pumps must always be selected to give the specified capacity when handling fluids of the expected minimum viscosity since this is the point at which maximum slip, hence minimum capacity, occurs.

Viscosity and speed being closely linked, it is impossible to consider one without the other. The basic speed that the manufacturer must consider is the internal axial velocity of the liquid going through the rotors. This is a function of pump type, design, and size.

Rotative speed should be reduced for handling of liquids of high viscosity. The reasons for this are not only the difficulty of filling the pumping elements, but also the mechanical losses that result from the shearing action of the rotors on the substance handled. Reducing these losses is often more important than obtaining relatively high speeds, even though the latter might be possible given positive pressure inlet conditions (Karassik 1976).

The actual delivered capacity of any screw pump is theoretical capacity less internal leakage or slip when handling vapor-free liquids. For a particular speed, the actual delivered capacity of any specific rotary pump is reduced by a decrease in viscosity and an increase in differential pressure. The actual speed must always be known; often it differs from the rated or nameplate specification. Actual speed is the first item to be checked and verified in analyzing pump performance.

Thickener Overflow Pumps. Thickeners are normally arranged for overflow by gravity to a collection tank, from which one or more pumps return the clarified overflow to the system. Thickener overflow may be used for preparing the limestone slurry, as prequencher water, as part of a blend with fresh water for mist eliminator wash, as level makeup to the effluent hold tank, or as makeup for an ash disposal system. The FGD system supplier should select the pump to satisfy the capacity and head requirements for the intended use. If some other use of the thickener overflow is contemplated, the utility must determine the end use before the pump specification is prepared. The pump head will vary considerably, depending upon where the thickener overflow is used.

Although largely free of suspended solids, the thickener overflow may still contain some suspended solids. Since corrosive ions such as chlorides may build up in the liquid in a closed-loop operation, the main design consideration for wetted parts of the pumps is corrosion resistance.

Following are typical properties of thickener overflow liquid:

pH	7 to 8 (normal)
Solids	500 ppm
Temperature, °F	100
Specific gravity	1.0

The thickener overflow pumps are usually centrifugal, with direct drive.

Thickener Underflow Pumps. Underflow from the thickener is pumped either to a dewatering system or to a sludge pond, which is often located thousands of feet and sometimes miles from the thickener. The pump may also recirculate solids to the thickener to maintain solids concentration. Depending on the sludge destination or whether it must feed directly to the downstream equipment, pumping head requirements may vary significantly for a slurry having uniform solids content. If solids are recirculated to maintain solids concentration in the thickener, an oversized pump must be used.

The thickener underflow is a thick slurry containing all the solids species present in the scrubber (e.g., calcium sulfate, calcium sulfite, fly ash, excess limestone).

The concentration of solids in the thickener underflow is limited to about 40 percent maximum because a centrifugal pump could not handle higher concentrations without causing nonuniform flow. Because the slurry contains high concentrations of abrasive solids, the main design consideration is erosion.

Following are typical thickener underflow characteristics:

pH	7 to 8
Solids, wt. percent	40
Solids	Fly ash, calcium sulfate, calcium sulfite, excess limestone
Temperature, °F	100

Thickener underflow pumps are either centrifugal or positive displacement types, with belt drive. The positive displacement pumps may have neoprene stators and high-chrome-alloy rotors. Discharge from the positive displacement pumps is nonpulsating, uniform, and reversible. Rubber-lined centrifugal pumps provide extra protection against corrosion during upset conditions. The underflow pump must be specified to handle high solids concentrations with abrasive components and under corrosive conditions during upsets.

Sludge Disposal Pumps. The distance to the disposal site determines the type of pump selected. The sludge may have to be pumped as far as several miles to remote areas for disposal or processing. If this is the case, a high head (700 psi) is required and reciprocating pumps are necessary. If the sludge disposal is close to the utility site, centrifugal pumps similar to thickener underflow pumps can be used.

Where the sludge has a high solids content and contains abrasive fly ash, pumps must be rubber-lined or made of erosion-resistant alloy.

Pond Water Return Pump. In systems where the sludge does not undergo secondary dewatering treatment prior to landfill, large sludge ponds are used to separate suspended solids. Clarified water from the sludge pond is recycled by pump to the scrubbing system to balance the overall water usage.

The pond water return pump should be specified to take a pH range of 6.5 to 8. Since chloride levels can become high in closed-loop systems, the pump should be chloride-corrosion resistant. Chloride corrosion is best deterred by using rubber-lined pumps.

With good pond design the pond return water should contain few solids; however, the pond water pumps return should be designed to accommodate some suspended solids.

Filtrate or Centrate Pump. Clarified filtrate or centrate from a vacuum filter or centrifuge is pumped back into the scrubbing system, usually to the thickener. In either case, the solids content of the filtrate or centrate should be low and the pH should be 7 to 8 or greater. This pump sees the mildest duty of any process pump within the scrubbing system.

Operating upsets in the vacuum filter or centrifuge may allow solids to contaminate the filtrate or centrate. Although these occurrences are rare, the filtrate or centrate pump must be designed to protect against possible corrosion and erosion. Erosion-resistant alloys and rubber-lined pumps are suitable.

Fresh Water Pump. In a limestone FGD system, the most likely points at which fresh water would enter are the ball mills, pump seals, and the mist eliminator wash system. A fresh water pump for this service can be a standard centrifugal pump. The important items to be specified are properties of the service water, available net positive suction head (NPSH), materials of construction, type of drive, and type and size of motor.

Design Considerations

The following discussion of pump design parameters is based on the service functions of the slurry recirculation pumps. Many of these considerations are applicable also to the other types of pump service in the FGD system.

Flow/Head. Low-speed operation is one of the most important wear-reducing features of a recirculation slurry pump; abrasive wear increases proportionally to the third power of rpm. The impeller tip speed of rubber-lined pumps is limited to 3500 to 4500 ft/min. This limitation restricts the rpm of rubber-lined pumps to 400 to 600 rpm, which corresponds to a maximum discharge head of about 100 ft. Additionally the discharge slurry velocity must be maintained at 7 to 11 feet per second (fps) as a design compromise between abrasion and settling in the piping.

Total liquid flow rate required for the scrubber is determined by the design L/G ratio. The normal recirculation flow range, corresponding to the rpm and head limitations, is 6000 to 10,000 gpm. Capacities, however, can go to 20,000 gpm at present, and 30,000 gpm is seen as a possible peak in the near future. Larger pumps are needed because larger scrubbers are being installed to reduce the number of modules required and thus improve overall system reliability (O'Keefe 1980). The number of pumps required per scrubber, other than spares, is determined by the need for redundancy to achieve system reliability.

Net Positive Suction Head (NPSH). The available NPSH must be determined accurately. More pump troubles result from incorrect determination of available NPSH than from any other single cause. As the available NPSH for a given pump decreases, its capacity and efficiency decrease, and a low-suction pressure is developed at the pump inlet. The pressure decreases until a vacuum is created and the liquid flashes to vapor (if the pressure is lower than the liquid vapor pressure). This condition, which can lead to cavitation damage, can be avoided by ensuring that the available NPSH is greater than the required NPSH.

Pump Efficiency and Energy Requirements. The pump manufacturer should be given data regarding energy costs, service conditions, and flow and head requirements. For example, an energy cost of 3¢/kWh and a penalty of $750 per additional kW can be specified as the basis for preliminary comparisons with the most efficient pump (Dublin 1977). Variations in pump size and efficiency result from each manufacturer's effort to choose, from his standard line of pumps, the one that most closely meets the specified requirements. Hence, specifications should not be so restrictive as to exclude high-efficiency pumps. Finally, when the efficiency penalty is less than 10 percent, the pump with lower speed should be selected because the longer pump life will compensate for the slightly higher operating costs (Reynolds 1976).

Drives. Slurry pumps operate at relatively low speeds, ranging from 400 to 600 rpm. Since the motors operate at either 1200 or 1800 rpm, some type of speed reducer must be used. The most common way of driving a limestone slurry pump is by using a V-belt drive with a fixed ratio; this method has the advantages of flexibility and low cost. The V-belt drive can be overhead-mounted or can be side-mounted on horizontal pumps. Because it is difficult to determine friction values of certain slurries for which data are not readily available, it is advisable to use V-belt drives with variable-pitch diameters. Without greatly increasing the initial purchase cost, these drives simplify balancing of the system at startup and enable the pump to accommodate future changes in flow rate and head. They also reduce deterioration of pump performance due to wear, and allow correction to initial system design for pumping of a particular slurry.

Seals. In horizontal centrifugal slurry pumps the shaft that passes through the pump casing must be sealed to prevent leakage. Mechanical seals, which are used for handling clear liquids, should not be used with slurries. Packed stuffing boxes are customarily used to seal the shafts because they cost less, allow faster repair, and usually last longer in abrasive service.

At an intermediate position in the packing, a continuous flow of clear water should be introduced into a lantern ring. This flush water prevents abrasive solids from entering the critical stuffing box and shaft sleeve area and greatly extends the life of the packing and the sleeve. Because abrasive solids may enter the packing during a shutdown or upset, the pump should be designed with a shaft sleeve of hardened alloy. Even in the best operations, abrasive slurry may enter the packing.

The flush water flows past the packing into the process or out the stuffing box. The volume of flush water that mixes with the recirculating slurry does affect the scrubber system water balance. In closed-loop systems, however, the sump water is reused within the system, depending on its quality. As a housekeeping measure, a large-diameter drain line should be provided to carry the leakage from the stuffing box to the sump.

The flush water supply system must be external to the scrubber system and reliable enough to deliver a minimum quantity and pressure. Process water can be used if the suspended particles are less than 40 microns in diameter and the maximum particle concentration is 1000 ppm by weight (Wilhelm 1977). The required clarity may be achieved with addition of a filter in the water line. Slurry pump design requires that the pump impeller have back-vanes, which remove slurry from the stuffing box region. Because this makes the pressure in the stuffing box assembly essentially the same as the suction pressure to the pump, the sealing fluid must be supplied at a pressure at least 5 to 10 psi above the suction pressure. When a pump lacks such back-vanes or is excessively worn, the pressure in the stuffing box region may rise to the discharge pressure. Then it is necessary to supply the sealing fluid at 5 to 10 psi above the discharge pressure. As a precautionary measure, an alarm may be installed on the seal water feed line to indicate low pressure or low flow of seal water.

Wear Rings. A wear ring provides an economically renewable leakage joint between the impeller and casing. The purpose of the wear ring is to minimize leakage from the discharge to the suction of the impeller by maintaining a close running clearance between the vanes and the wear face. Running clearance has a profound effect on the head capacity and efficiency of the pump. To reduce the rate of wear of the wear rings and thereby extend the life of the pump, the designer must consider the corrosion and wear characteristics of the ring material.

Operational Systems

Table 3-13 lists specifications of recirculation pumps installed at operational limestone FGD facilities. The following is a brief summary of performance histories of limestone slurry recirculation pumps.

Cholla 1. Rubber linings in the recirculation pumps at Cholla have been damaged many times, primarily because of frequent plugging of the strainer at the suction end of the pump (Laseke 1978a). As plugging stops the flow or drastically reduces it, the pump cavitates and the liner is sucked into the path of the impeller and shredded. Initially the suction strainers were located inside the recirculation tank and could be cleaned

TABLE 3-13. SPECIFICATIONS OF RECIRCULATION PUMPS IN OPERATIONAL LIMESTONE FGD SYSTEMS

	Individual pump ratings				No. of pumps	
	Flow, gal/min	Head, ft	Speed, rpm	Motor, hp	Total	Spare
Cholla No. 1 Arizona Public Service	2,670 9,300	90 100	NR NR	NR NR	2 2	1 1
Petersburg No. 3 Indianapolis P & L	3,000	NR	NR	NR	12	4
La Cygne No. 1 Kansas City P & L	5,000 9,000	98 114	500 550	350 400	8 8	1 1
Lawrence No. 4 Kansas P & L	5,300 3,600	83 NR	550 NR	200 200	2 2	1 1
Lawrence No. 5 Kansas P & L	NR NR	NR NR	480 480	500 500	2 2	1 0
Sherburne No. 1 Northern States Power	5,500 8,000	95 NR	500 500	200 NR	12 12	1 1
Winyah No. 2 South Carolina Public Service	6,600 13,700	130 84	500	NR NR	1 1	0 1
Southwest No. 1 Springfield City Utilities	13,800	100	500	NR	4	2
Widow's Creek No. 8 Tennessee Valley Authority	10,400 3,380	100 85	500 500	NR NR	10 6	4 1
Martin Lake No. 1 Texas Utilities	3,150	125	500	500	18	NR

NR - Not reported.

only by draining the tank. Pump problems were solved, however, by installation of a hydrocyclone in the recirculation line of each module, an installation made for other purposes as well. The hydrocyclone separates the larger particles of scale from the main slurry stream by centrifugal action. The strainer therefore is no longer needed, and its removal eliminates plugging problems.

Another preventive measure at Cholla is rubber lining of the process piping network to protect the carbon steel base from abrasive slurry. This has been generally successful, although a few incidents of wear in the area of the spent slurry valves have been reported. The wear is attributed to the throttling action of the valve to modulate the flow of slurry. The problem was solved by operating the valve only in a completely open or completely closed position.

La Cygne 1. Problems at La Cygne have been more in the piping system than in the recirculation pumps. Sediment built up several times in dead spaces in the pipelines and valves of idle pumps and also in process lines. This buildup occurred when slurry velocities in the pipe were low (during periods of reduced operating rate). The problem was resolved by redesigning some pipes to eliminate potential dead pockets. To prevent valve freezing due to sediment buildup, some valves were repositioned and flushout lines were installed.

Some pipe liners eroded (e.g., in the scrubber tower pump inlet piping). Some of the erosion was caused by unsatisfactory liner materials and some by high flow velocities through pipes and fittings. The rubber lining in some pipes cracked, primarily because of defects in fabrication. Piping modifications helped to reduce the erosion (Laseke 1978b).

Lawrence 4 and 5. The original FGD system on the Lawrence No. 4 unit entailed limestone injection into the boiler along with a marble bed scrubber. Limestone injection has been discontinued, and the marble bed scrubber was replaced by a rod scrubber/spray tower system.

The original system was plagued with erosion, corrosion, and plugging. The modifications involved replacement of all steel pipes with FRP and rubber-lined pipes, and installation of new slurry pumps and strainers (Teeter 1978). In the new system, the recirculation pumps at first required packing replacement every 2 or 3 days. The gland seal arrangement was redesigned, and the seal water flow was increased from 7 gpm to 15 gpm.

Pump problems at Lawrence 5, installed in 1971, were similar to those at Lawrence 4. This unit was also modified to a rod scrubber/ spray tower design.

Sherburne County 1 and 2. The most significant problem affecting availability of the overall scrubber system has been plugging of spray nozzles, originally caused by failures in the Zurn duplex strainers on the discharge of the spray water pumps. The strainers were designed to remove solid particles greater than 1/4 in. to prevent plugging of the nozzles. Frequent plugging of the strainers, mechanical failures, and bypassing of solids large enough to plug the spray nozzles led to reduced availability of the scrubber unit and high maintenance requirements. Maintaining the duplex operation of the strainers proved to be impractical; subsequent operation with a single strainer basket required shutdown of the module for cleaning. Extensive efforts to correct the problem were not successful (Kruger 1978a).

When the duplex strainers were abandoned, Combustion Engineering installed new in-tank strainers consisting of a large, perforated, semicircular plate installed over the spray water pump suction with an oscillating and retracting wash lance for periodic backwashing. After installation of one strainer in September 1976, plugging of the spray nozzles was significantly reduced. Because of this successful operation, all modules on Sherburne 1 and 2 were converted by March 1977. Plugging of nozzles has continued, however. The apparent cause is formation of scale inside the perforated plate and piping headers; the scale then breaks off and plugs the nozzles (Kruger 1978b).

When plugging occurred, the blower was used to clean the strainer perforated plate. Because the cleaning was not complete, the plate was plugged again and the blower used again. As this continued, the slurry level in the reaction tank became too high and the solids content too low. The problem is believed to lie in the supply pressure and capacity to the blowers, and a new system for the supply has been designed.

Failures of the LaFavorite rubber lining downstream of orifices have occurred in the main recirculation and effluent bleedoff piping. When these failures occur, the rubber breaks off in chunks and plugs the downstream nozzles or headers. Such failures were minor on Unit 1 but major on Unit 2. The rubber lining has been removed to prevent further plugging. A test program was undertaken to evaluate the performance of reinforced fiberglass, stainless steel, and rubber-lined spool pieces.

Operation of the recirculation pumps indicates that the Ni-Hard impeller must be replaced about every 6000 hours and the Ni-Hard suction side wear plate about every 4000 hours. Pump internals of 28 percent chrome-iron and rubber-lined internals have been installed on selected modules for evaluation.

Operation of Slurry Feed Pumps. Following is a brief summary of reported experience with operating limestone slurry feed pumps.

The positive displacement pumps at Sherburne 1 and 2 (Kruger 1978b) were underdesigned, and capacities were upgraded from 1 to 12 gpm to 2 to 20 gpm. The positive displacement pumps at Lawrence 4 and 5 require replacement of the stator and the rotor every 10 to 15 weeks (Teeter 1978). Though this life expectancy is short, the wear is predictable and control of the slurry feed rate is excellent. For these reasons, the company has decided to continue operations with the same type of pump. At most other sites there have been no major problems in feed pump operation.

PIPING, VALVES, AND SPRAY NOZZLES

This discussion emphasizes design details related to the piping, valves, and spray nozzles in limestone scrubbing applications. Since many of the design considerations of these components are integrally related to their materials of construction, this discussion focuses on materials and summarizes some of the information presented later in Section 3 and in Appendix G. This information provides guidance for the project manager in reviewing the proposals of scrubber system suppliers.

Piping, valves, and nozzles have been subject to moderate material failures and mechanical problems, mostly attributable to the mode of use of the component, difficulty of repair, unavailability of materials, and lack of standby units or bypass capability (Rosenberg et al. 1980). Although the "best" designs for this application are not yet identified, much has been learned by review of the experience records of full-scale operation and correlation of reported failures with the service environment.

Piping

The choice for slurry piping has been predominantly rubber-lined carbon steel pipe. Alloyed metal pipe is expensive and often not as resistant as the carbon steel pipe to erosion and corrosion. Reinforced fiberglass pipe has been discarded in some plants, although it is being used in lieu of rubber-lined carbon steel at other plants that have been dissatisfied with metals or elastomers (O'Keefe 1980).

Piping constructed of various materials has proved moderately successful in limestone scrubbing systems, and with proper maintenance has given fairly reliable service at a reasonable investment cost.

Typical piping applications in a limestone scrubbing system are listed below:

Type	Service
Carbon steel	○ Steam and condensate return ○ Instrument air ○ Industrial/service water ○ Noncorrosive slurry ○ Vacuum piping ○ Miscellaneous water service ○ Lube oil piping

(continued)

Type	Service
Rubber-lined carbon steel	° Severely abrasive slurry with acid corrosives
Reinforced plastic, typified by fiberglass reinforced polyester (FRP)	° Nonabrasive or mildly abrasive slurry. Severely abrasive or corrosive conditions in straight piping runs
316L or 317L stainless steel	° Mildly abrasive slurry with acid corrosives

Rubber-lined carbon steel piping is used frequently in the scrubber area and the limestone slurry preparation area. It is used because of its superior resistance to abrasion by solid particles carried in the slurries and because it is resistant to corrosion by acid liquids and by solutions of certain inorganic salts. Rubber-lined steel pipe is not easily damaged, but care must be taken not to heat the rubber above its maximum operating temperature.

FRP pipe is used in similar service because it is resistant to chemical attack and is moderately resistant to erosion by solid particles. This piping also can be damaged by overheating. It is used at most locations as reclaimed water piping either for pond return or thickener overflow service. These streams are essentially solids-free. Some problems with FRP pipe have been reported: it has been difficult to heat trace (although this problem has recently been corrected), it is subject to rupture, and it will leak if threaded fittings are used. Flanged or shop-fabricated joints are preferable to field-cemented joints because many pipefitters are not skilled in making FRP joints.

Type 316L stainless steel piping along with FRP is primarily chosen for spray headers located inside scrubbers. Stainless steel 317L is gaining favor because it provides greater corrosion resistance as a result of its additional molybdenum content.

Some important design considerations for slurry piping are increasing the sidewall thickness at points of high erosive wear and the use of long radii for pipe turns. The use of too many reducers and sharp bends in piping lines are design errors that have accelerated wear and failure at these points. The abrasive slurry can shear off the rubber lining in chunks that can be carried to the spray nozzles and spray headers (Laseke 1979). The bottoms of the pipe sections tend to erode first, but rotation extends pipe life fourfold or more. Therefore the joint and pipe support should permit easy turning. Provisions for automatic clear water flushing and drainage are absolutely necessary for good operability. All slurry lines must be sized to maintain a minimum velocity (7-11 fps) adequate to prevent solids from settling in the lines during minimum flow and abrasion during maximum flow.

Valves

Frequent problems with isolation and control valves have been caused by mechanical or material failures. Therefore, most designers agree that the number of valves should be kept to a minimum. Rubber-lined valves are the most common, although many stainless steel valves are used and trials of polyethylene have been promising. Designers have successfully modified conventional valves used in mining and mineral processing for the erosive/corrosive service of limestone slurry systems, but their size, weight, and price are substantially greater than those of conventional valves.

Deep erosion can occur where a valve causes an abrupt change in flow direction. Areas that can fill with slurry will block the opening and closing of valves. The scouring effects on seats and disks make valves leak quickly. These characteristics have ruled out the use of several conventional valve types in this application. A variety of valve types and materials of construction are in use, however, and the chief factors that dictate valve selection are corrosion and erosion resistance, minimum resistance to flow, and positive shutoff in long-term use. Hand-operated valves are of the overhead chain wheel type or of the gear assistance type. Following are typical hand-operated valves used in a limestone scrubbing system:

Eccentric plug valves
Lined butterfly valves
Flex check valves
Wafer knife-gate valves } clear liquor only
Wafer gate valves

Air-operated or electric-operated control valves are very important for control of the various liquor streams in accordance with the process control scheme. They are also used for isolation of equipment such as pumps. Typical control valves used in a limestone scrubbing system are pinch valves and plug valves.

Globe valves are used to control the service water, mist eliminator wash, and miscellaneous water flows. A few globe-type valves with favorably formed body passages simulating plug valves have seen service in slurry control.

Butterfly valves are used as on-off control valves for slurry service lines and in recirculation slurry pump discharge piping. They range in size from 2 to 20 inches and can handle flow rates up to 10,000 gpm. When used as isolation valves in abrasive slurries, they are typically rubber-lined carbon steel. In the throttled position, butterfly valves can damage downstream piping because their discharge flow is asymmetric.

Knife-gate valves are used for shutoff and also for modulating service for clear liquor like streams. The knife-gate valve uses a shearing action in which its thin disk knifes through any deposit on the valve seats. The seat rings can be elastomeric, but the disk is usually stainless steel.

Plug valves are used for isolation service in water and/or slurry lines. Eccentric plug valves, in which the plug wipes past the seat and is clear of flow in the open position, have given excellent results in limestone scrubbing slurry service.

Quarter-turn valves, in which the stem rotates without the sliding that permits slurry into the stem packing, are preferred for slurry service.

Spray Nozzles

Spray nozzles provide the liquid distribution pattern that is essential to scrubber operation, especially in a spray tower scrubber. A sufficient number of spray nozzles and source spray banks and headers must be used to cover the cross section of the scrubber. Spray nozzles should be selected to provide droplets of about 2500 microns (mass median diameter), which results in minimum droplet entrainment while providing enough surface area for SO_2 absorption (Saleem 1980).

The nozzle must be nonclogging and abrasion-resistant. Hollow-cone and full cone nozzles cast of silicon carbide and/or ceramic meet these requirements. Liquid is atomized by centrifugal action induced by tangential entry or swirl vanes. The nozzle chamber has few internals which can become clogged. They can be made in various sizes to suit liquid flow requirement at a normal operating pressure of 10 psig. Full cone nozzles are normally used in tray or packed scrubbers.

Wear, plugging, and installation problems with spray nozzles have been reported. Wear occurred with plastic nozzles made of Noryl resin (Laseke 1979), but replacement nozzles of ceramic proved successful. Nozzles made of extremely hard alumina and silicon carbide have also been used successfully. Stainless steel appears to be preferred for wash spray nozzles used in mist eliminators (Rosenberg et al. 1980).

The number of sprays and spray banks installed in spray towers can vary depending on the required SO_2 removal efficiency. One to six banks are used. When each spray bank is fed by a separate pump, a great degree of flexibility can be built into the scrubber by allowing the use of individual pumps depending on need.

Early experience has led to a great improvement in the spray nozzles, spray banks, and headers now being offered. These components now have a relatively low incidence of problems and they are amenable to rapid repair or replacement.

DUCTS, EXPANSION JOINTS, AND DAMPERS

The ductwork, expansion joints, and dampers of the limestone scrubbing system, along with the fans, scrubber, mist eliminator, reheaters (if installed) and stack, constitute the flue gas handling portion of the system. A ductwork assembly consists of ducting, dampers, stiffeners, turning vanes, flanges, transition pieces, access doors, test ports, gasketing, expansion joints, and duct support bases. The assembly is field-erected as part of a structural steel installation contract or as part of the overall limestone scrubber supply contract. Designs for all components are well standardized. Design details for these various components are briefly discussed in the following subsection. Again, as in the foregoing discussion, this discussion focuses on materials of construction as a critical design factor. Because of the volume of steel required and the alloys used for dampers, the cost of ductwork is a major part of the total FGD system cost, sometimes representing as much as 10 to 15 percent.

Ductwork

Ductwork in an FGD system is usually made of carbon steel plates 3/16 or 1/4 inch thick, welded in a rectangular cross section. The ductwork is supported by angle frames that are stiffened at uniform intervals. The following factors should be considered in designing limestone scrubber ductwork:

- Pressure and temperature
- Velocity
- Flow distribution
- Variations in operating conditions
- Materials of construction
- Materials thicknesses
- Pressure drop (ΔP)

The ductwork must be designed to withstand the pressures and temperatures that occur during normal operation and also those that occur during emergency conditions such as an air heater outage.

The maximum permissible gas velocity through the ductwork should be designed so that ductwork furnished with the scrubbing system is compatible with that provided under other powerplant contracts. Typically the maximum gas velocity should be 3300 to 3600 ft/min.

Ductwork associated with the scrubbing system is subject to a variety of conditions, depending on location within the system. The following list identifies the basic variants:

- Inlet ductwork

(continued)

- Bypass ductwork (all or part of the flue gas)
- Outlet ductwork (with reheat and without bypass)
- Outlet ductwork (with reheat and with bypass for startup)
- Outlet ductwork (without reheat and without bypass)
- Outlet ductwork (without reheat and with bypass for startup)

Outlet ductwork has been a major problem, particularly in units with duct sections that handle both hot and wet gas. Materials of the outlet duct range from unlined carbon steel to Hastelloy C-276. Environmental differences related to the location and use of reheaters and the type of coal burned are factors that affect the use of different materials.

The primary material of construction for ducts is carbon steel plate with various linings selected for their cost-effectiveness in resisting the effects of corrosive or abrasive gases. The gas environments can be categorized as relatively mild, moderate, and severe. Typical lining materials that can adequately serve each environment are as follows:

Mild: Wet corrosive gases with or without liquids at temperatures up to 180°F with no abrasion.

Glass-flake filled plastic coating is satisfactory. The polyester resin can successfully withstand the chemical and immersion environment. The glass flakes provide mechanical strength and resistance to permeation by creating a labyrinthine path for liquid. This lining type requires sand blasting, priming, and two hand-troweled coatings. The total thickness is 60 to 80 mils.

Moderate: Wet corrosive gases and impingement by abrasive slurry liquid or sprays at temperatures up to 160°F.

An aggregate-filled, fiberglass-reinforced plastic resin system is satisfactory. It requires sand blasting, priming, a base coat overlaid with fiberglass woven cloth, and finally a top coat filled with a sand-like aggregate to resist abrasion. The total thickness is approximately 1/3 in. The fiberglass cloth gives good control of the troweled thickness, minimizes the thermal expansion coefficient, resists mechanical drainage, and serves as a backup abrasion-resistant barrier. The resin used is the same as that in the glass-flake-filled plastic resin systems.

Severe: Alternate exposure to hot dry gas (at temperatures up to 700°F) and to saturated gas or liquid.

A heavy castable lining, such as an aluminous cement with silicate binders, is applied by guniting to a thickness of 1-1/4 in. This lining is resistant to the chemical, thermal, and immersion environment.

In addition to carbon steel with the various linings, ducts have been made from high-grade alloys and from carbon steel base plate cladded with the alloys to serve as a lining. Acid-proof brick has been used in ducts located in severely abrasive environments. Thickness is 2-1/2 in. or more.

The inlet ductwork from the precipitator to the scrubber is unlined. The scrubber outlet ductwork is lined for severe service because it is exposed to both hot dry flue gas and saturated gas. Table 3-14 lists ductwork applications in a typical limestone system containing four scrubber modules, with suggested materials of construction.

The choice of linings or the use of acid-proof brick or high-nickel alloys should be determined so as to insure the reliability of various duct service applications. The project manager should seek out the experience record of the scrubbing system supplier and weigh this record against cost-effective economics of alternative choices.

TABLE 3-14. TYPICAL DUCTWORK APPLICATIONS AND MATERIALS OF CONSTRUCTION

Application	Materials
Fan inlet duct: conveys flue gas from discharge duct of boiler ID fan to FGD booster fan; can contain turning vanes	1/4- to 3/8-in.-thick carbon steel plate; externally stiffened, unlined, insulated
Fan discharge duct: conveys flue gas from booster fan outlet damper to scrubber inlet manifold	1/4-in.-thick carbon steel plate; externally stiffened, unlined, insulated
Scrubber inlet manifold: receives flue gas from booster fan outlet duct for distribution to scrubbers	1/4-in.-thick carbon steel plate; externally stiffened, unlined, insulated
Scrubber inlet duct: conveys flue gas from manifold to scrubber; usually provided with manholes	1/4-in.-thick carbon steel plate; externally stiffened, unlined, insulated
Bypass duct: conveys flue gas from inlet manifold through bypass damper	1/4-in.-thick carbon steel plate; externally stiffened, unlined, insulated
Reheat duct: conveys air from discharge of reheat fan to stack inlet duct through gas reheater	1/4-in.-thick carbon steel plate; externally stiffened, unlined, insulated
Stack inlet duct: conveys gas from scrubber outlet manifold to stack; also accepts hot air from reheater duct and provides chamber for mixing with wet flue gas; severe service	1/4-in.-thick carbon steel plate lined with 1-1/4 in. aluminous cement; covered with 2-in. mineral fiber blanket insulation

Expansion Joints

Two basic types of expansion joints are available: metallic and elastomeric. The following factors should be considered in designating the type and construction of scrubbing system expansion joints:

- ° Movements to be accommodated
- ° Temperature
- ° Pressure

(continued)

- Location and attendant variations in operating conditions
- Exposure to particulates

The layout of the ductwork and the location of the expansion joint will determine the magnitude and type of movements a joint must accommodate. Metallic joints may be used for axial and/or angular movements. Elastomeric joints may be used for all types of movements.

For determining the maximum expansion and contraction movements, the maximum and minimum ambient temperatures expected and the normal sustained operating temperature should be considered, as well as the maximum excursion temperature and its duration. Maximum positive pressure and negative pressure (vacuum) should be a part of the design basis. Elastomeric joints should be designed for the normal operating temperature; their service life is shortened by exposure to high temperature during emergency conditions. Suppliers can provide guidance in determining life expectancy factors for various conditions.

Elastomeric expansion joints should be provided with internal baffle plates to shield the joint fabric from impingement by particulates. Metallic expansion joints do not require baffles and should be left exposed to facilitate cleaning and inspection.

Metal expansion joints have been satisfactory in dry service, but even stainless steels can corrode if exposed to concentrated acid and/or high-chloride condensate. Operators at several installations have replaced metal expansion joints with elastomer joints because the metal has corroded.

In a scrubbing system, expansion joints are generally U-shaped and constructed of an elastomer with fabric (fiberglass or asbestos) reinforcement. The greatest problem with joints downstream from the scrubber has been with the metals used for attachment purposes, rather than with the joint fabric. Selection of the expansion joint depends on the service temperature. Suppliers suggest the following guidelines:

- At 250°F or below, fabric-reinforced neoprene or Viton
- At 300°F or below, fabric-reinforced chlorobutyl rubber or Viton
- At 400°F, layered asbestos

A typical limestone scrubber could be designed with the following materials of construction for expansion joints in various locations:

- Asbestos and Viton in all ductwork leading to the scrubbers and bypass duct
- Asbestos and chlorobutyl rubber in ductwork between the scrubber and the stack
- Asbestos and neoprene in the reheat fan outlet ductwork in cold air service

(continued)

°   Elastomer/fiberglass or Teflon in reheat ductwork in hot air service

Dampers

Dampers are used for flow control and isolation of the flue gas stream. They are used in three locations: the inlet duct to the scrubber, the outlet duct from the scrubber, and the bypass duct. (In installations that have no bypass, all of the flue gas goes through the scrubbers.) Table 3-15 lists the basic damper types installed in scrubbing systems. Designs of isolation dampers, which prevent the flow of the gas into the scrubbing system, include the slide gate (guillotine), the single-blade butterfly, the multiblade parallel (louver), and the two-stage louver with a pressurized seal air system to maintain positive pressure between stages. The seal air system provides superior sealing over other damper designs, and minimizes the potential for gas leakage during periods of scrubber maintenance.

Control dampers are used to balance the flow between the scrubbers and to regulate the amount of flue gas through the bypass duct in systems with bypass reheat. Control dampers are of the multiblade opposed (louver) type.

Proper materials of construction and adequate mechanical design are necessary to ensure reliable operation of dampers. In addition to materials, the factors to be considered in selecting dampers are the damper location and service function, and the pressure and temperature of the gas flow. Care must be taken to minimize areas of fly ash deposition. Mechanical problems with dampers caused by deposition of solids from the flue gas have outweighed any materials problems.

TABLE 3-15. BASIC DAMPER TYPES

Generic	Specific	Common designs
Louver	Opposed blade multilouver	Single louver Double louver Double louver/seal air compartment
	Parallel blade multilouver	Single louver Double louver Double louver/seal air compartment
Guillotine	Top-entry guillotine Top-entry guillotine/seal air Bottom-entry guillotine Bottom-entry guillotine/seal air	
Blanking plate	Isolation	Steel plate
Butterfly	Pneumatic operator	

The damper frames are usually channel-type, of either rolled or formed plate. The material and weight of the frames should be determined on the basis of stress resulting from seismic loading, total size and weight of the damper, and operating conditions. The blade deflection should be less than 1/360 of the blade span. A detailed stress analysis should be performed on the blade for a specific design. Each damper requires an activator that should be mounted out of the gas stream.

The simplest damper is the simple isolation blanking plate. It is advisable to be sure that blanking plates are available and that provisions are made in the ductwork design for their insertion. Blanking plates are essential when ducts must be isolated for protection of the maintenance crew. Should it become necessary for persons to enter any section of the ductwork or scrubber, the blanking plate will ensure isolation. When used in conjunction with positive ventilation air purge, the blanking plates do protect the safety of personnel.

The following paragraphs describe typical damper types and service applications in a limestone scrubber system.

Precipitator outlet dampers perform dry flue gas service. The usual design is a top-entry, carbon steel, double guillotine damper with a seal air blower to maintain zero flue gas leakage across the damper. Included with the damper is the blower motor, motor operator, torque limiter, flexible 317L stainless steel seals, and limit switches. This assembly serves as an isolation damper.

The bypass damper can be located in a bypass duct between the scrubber inlet and outlet manifolds. If the bypass duct is typically round and $\leq$ 10 feet in diameter, a butterfly-type damper can be used. Each butterfly is moved by a pneumatic operator. This is a modulating damper that controls flue gas flow.

The reheat damper is located in the duct branches that come from each scrubber outlet and from the stack inlet duct. It is situated between the reheat fan and the stack inlet. This is a modulating damper for control of flue gas flow to the reheater. All materials of construction in the gas stream can be 317L stainless steel except seals, which need to be Inconel 625. Again this service dictates a butterfly damper in a round duct. A pneumatic operator that strokes the damper in 5 seconds should be provided.

The scrubber inlet damper serves as an isolation damper. It is a double-bladed guillotine type, equipped with a seal air blower to pressurize the sealing space and thus ensure against leakage past the damper. The dual blades can be carbon or stainless steel. The damper seals can be Type 316L or 317L stainless steel, Hastelloy G, or Inconel 625. This damper should be capable of being opened or closed in approximately 60 seconds by use of a dual chain and sprocket drive.

The scrubber outlet damper may be a combination isolation and modulating control damper. This unit can be a combination guillotine and multilouver damper assembly. The guillotine is served by a seal air blower. All materials of construction in the gas stream are stainless steel (316L or 317L) to provide protection against corrosion by wet flue gas. Where corrosion has been a problem, high-grade alloys such as 904L, Inconel 625, Hastelloy G, and JESSOP JS 700 have been used to replace components. Inconel 625 is preferred for the seals. An operator for the dual chain and sprocket drive should close the guillotine damper in 60 seconds; a separate operator should close the multilouver damper in 30 seconds.

The expected electrical loads needed to energize the various damper motors and seal air blowers are as follows (500 MW, four scrubber system basis):

Scrubber inlet damper motor to drive chain and sprocket	10 hp
Scrubber inlet damper blower	20 hp
Scrubber outlet damper for the guillotine	10 hp
Scrubber outlet blower	20 hp
Scrubber outlet louver damper	1 hp

The smaller butterfly dampers are moved by pneumatic operators.

All dampers should be inspected initially for proper installation, e.g., for flow direction and free movement. During operation of the scrubbing system, they should be inspected for cleanliness, working freedom of positioners, and positive closure.

On startup and in commissioning of gas flow, the scrubber inlet isolation dampers must be opened and the scrubber outlet louver damper must be adjusted to balance the gas flow equally across as many scrubber modules as are in service. The bypass and reheat dampers must be balanced to give the desired exit gas outlet temperature. The amount of hot air supplied for reheat is controlled by modulating the reheater inlet butterfly damper.

On shutdown the scrubber inlet and outlet guillotine isolation dampers must be activated to shut off individual scrubbers and the damper seal air blowers must be operating.

The 60-second interval for activation of the isolation dampers will meet emergency shutoff requirements.

TANKS

This subsection discusses the major tanks in the liquid circuit of the limestone scrubbing system. The tanks are used for storage, for mixture and reaction, and for collection and recirculation of slurry, makeup water, wash water, and other fluids.

Scrubber Effluent Hold Tank (EHT)

The EHT is also known as the reaction, recycle, or recirculation tank. The EHT must provide sufficient retention time to relieve supersaturation of the liquor and thereby minimize scaling. The EHT may be a separate tank or it may be the bottom part of the scrubber vessel that serves as a reservoir.

The critical design factor is to size the EHT so that it provides enough holdup time to ensure optimum limestone utilization and precipitation of gypsum. Failure to provide sufficient holdup time increases the supersaturation and can cause scaling. The exact holdup time required is a function of the degree of supersaturation that is allowed to take place during SO_2 absorption and also of the tank design (Saleem 1980).

If the tank is an integral part of the scrubber, it must be designed for the same pressure. Typically 10 minutes is adequate holdup time to ensure that crystallization occurs in the open EHT vessel rather than in the piping and spray headers. Eight-minute retention time has been proven as adequate for slurry holdup at the Shawnee test facility. Additionally, the Shawnee work has shown that plug flow tanks are more efficient as reaction tanks than single back mixed tanks (Borgwardt 1975).

In addition to holdup time, control of the solids content of the recirculation slurry is essential to maintain the amount of seed crystals needed for scale control. Seed crystals provide a large surface area on which the dissolved gypsum deposits preferentially in the hold tank rather than on the walls or internals of the scrubbing vessel. For a fly-ash-free limestone scrubber, normal operation with 8 to 15 weight percent solids in the EHT is typical. Where the scrubber is part of a system that also removes fly ash, a normal concentration of solids in the hold tank would be 15 percent.

Specifications for construction of the EHT should be in accordance with applicable portions of recognized tank codes such as API 650.

Some hold tanks include strainers to prevent large chunks and solid particles from the slurry circuit from plugging the spray nozzles. For example, strainers were installed after startup in the Sherburne County FGD systems. These strainers consist of large, semicircular perforated plates fitted around the suction side of the slurry recirculation pumps that send the slurry to the spray nozzles. Each strainer is equipped with an oscillating and retractable wash lance for periodic backwashing. Laseke (1979) gives a detailed account of this in-tank strainer design.

Effluent hold tank designs also have included baffles to aid in mixing of the slurry by the agitators, which are both center- and side-wall mounted. Intricate baffling configurations are sometimes used to provide compartmentalized zones and separate mixing areas.

The scrubber supplier should provide design information and background data to assure the project manager that the EHT is adequate and correctly sized.

Limestone Slurry Feed Tank

The limestone slurry feed tank is designed to provide surge capacity for the limestone slurry before it is fed from the storage tank to the scrubber. A typical design would provide about 2 hours of retention capacity. The slurry feed is 40 to 60 percent solids, containing finely ground particles of limestone. The current trend is toward finer grinding, with specifications ranging from 60 to 90 percent through 325 mesh. Since agitation equipment is costly and consumes power continuously, some designs equip a portion of the storage with agitation only by compressed air, to be used only when needed.

As with the EHT, the limestone feed tank should be constructed in accordance with applicable portions of recognized tank codes.

Thickener Overflow Tank

The thickener overflow tank acts as a surge tank and stores the clear supernatant liquid from the thickener to be pumped back to the EHT.

The pH of the thickener overflow should be between 6 and 8. Because pH excursions may occur, however, specification of proper materials of construction is important. Unless there are upsets in thickener operation, serious erosion should not occur.

The thickener overflow tank often provides surge capacity to achieve water balance in the system. Therefore, the tank should be sized to allow for system swings. Adequate NPSH for the overflow pump must also be considered.

Mix Tank (Optional)

The mix tank is sometimes used to mix a fixation agent with the spent slurry (sludge). Thickener underflow is fed to the tank, the fixation agent is added, and the mixed product flows or is pumped to a sludge pond. Erosion is the major consideration. High-torque agitators are needed to achieve proper mixing of the slurry. The rapid movement of the slurry and fixation agents against the tank walls heightens abrasive action. So as to minimize maintenance, the tank walls and bottom should be rubber-lined.

The tank must be sized to allow proper mixing. If slurry is pumped from the tank, the NPSH requirements of the pump must be considered.

Mist Eliminator Wash Tank

In systems using mist eliminator wash tanks, simple design and construction materials have been adequate. The tank should be sized to provide 3- to 6-minute surge capacity. Nominal sizes are in the 2000 to 6000 gallon range.

AGITATORS

Agitators are used to provide slurry mixing in the EHT and the slurry feed tank so as to keep the solids suspended. Additionally, all other tanks and/or sumps in the system that serve as vessels for slurry should be equipped with agitators to keep solids in suspension.

The agitators are standard models supplied by commercial manufacturers. Some are top-entry agitators center-mounted in the tanks; some EHT's are equipped with agitators mounted on the side wall. Commercial manufacturers have worked closely with the scrubbing system suppliers to perform model studies on mixing characteristics, especially for the EHT; they have provided design details on baffling and the use of both top and side-wall entry agitators.

A limestone scrubbing system in a typical 500-MW case could use as many as 20 agitators ranging in horsepower demand from 3 to 250 Hp.

Relatively few problems with materials or mechanical reliability have been reported. Additionally, agitators can be repaired or replaced rapidly in limestone slurry service.

MATERIALS OF CONSTRUCTION

Some of the foregoing discussions have touched upon materials of construction that are integral to the design of specific components. This discussion summarizes the principal overall design considerations in selection of materials for a limestone FGD system. Additional information and design guidance are given in Appendix G.

Basic Classification of Materials

A systematic grouping of the various construction materials used in limestone FGD systems is shown in Table 3-16. The materials are grouped into two basic types: (1) base metals and (2) plastics, ceramics, and protective linings. Within each group the materials are further categorized according to generic, specific, and common or trade name classifications. The generic classification of base metals includes carbon steels, stainless steels, and high-grade alloys; the plastics, ceramics, and protective linings are grouped generically as organic and inorganic materials. The specific classifications designate distinct types of materials within each generic grouping. The common or trade name classifications designate special or proprietary materials within each specific grouping.

It should be emphasized that Table 3-16 lists only those materials that have been used or tested in limestone FGD systems. For example, the base metals grouping lists only the wrought alloys because they are used to the virtual exclusion of cast alloys. Also, the martensitic stainless steels

TABLE 3-16. BASIC TYPES OF CONSTRUCTION MATERIALS

Generic	Specific	Trade or Common Name
Base Metals*		
Carbon steel	AISI 1110	
	High strength low alloy (HSLA)	Cor-Ten
Stainless steel	Ferritic	Type 430
		E-Brite 26-1
	Austenitic	Type 304
		Type 304L
		Type 316
		Type 316L
		Type 317
		Type 317L
High alloy	Iron base/nickel-chromium-copper-molybdenum**	Carpenter 20
		Uddeholm 904L
		Ferralium
	Iron base/nickel-chromium-molybdenum**	Haynes 20
		Jessop JS 700
		Allegheny AL-6X
		Nitronic 50
	Nickel base/chromium-iron-copper-molybdenum***	Incoloy 825
		Hastelloy G
	Nickel base/chromium-iron-molybdenum***	Hastelloy C
		Hastelloy C-276
		Hastelloy C-4
		Inconel 625
Plastics, ceramics, and protective linings†		
Organic	Natural rubber	Black natural rubber
	Synthetic rubber	Neoprene
		Chlorobutyl
	Polyester	
	Mica flake-filled polyester	
	Alumina flake-filled polyester	Heil and Ceilcote linings
	Glass flake-filled polyester	
	Fiber-reinforced polyester	
	Mat-reinforced polyester	
	Vinyl ester	Plasite
	Inert flake-filled vinyl ester	
	Glass flake-filled vinyl ester	
	Epoxy	Carboline
	Glass flake-filled epoxy	
	Mat-reinforced epoxy	
	Fluoroelastomer	Colbrand CXL-2000
Inorganic	Acid-resistant brick and mortar	Prefired brick
	Abrasive-resistant brick and mortar	Aluminum oxide
		Silicon carbide
	Hydraulically bonded cement	Calcium aluminate cement
	Chemically bonded cement	Gelled silicates

*Wrought alloys only.
**Arranged according to increasing iron content.
***Arranged according to increasing nickel content.
†The various manufacturer grades available for certain trade or common name materials are not listed because of their manufacturer specificity and numerous types.

are excluded because they do not see widespread use in limestone FGD systems.

Performance Characteristics

Base Metals. To date, four major test programs have been performed for evaluation of base metals with respect to performance and corrosion-resistance in wet scrubbing systems. These programs were sponsored by International Nickel Company (Inco), the High Technology Materials Division of Cabot corporation, Combustion Engineering, and the EPA/TVA, whose tests were done at the Shawnee facility. Test results are summarized in Appendix G.

Evaluation of results of these test programs and of operating experience in the industry has led to the widespread use of Type 316L stainless steel. Under certain stringent operating conditions, in which this alloy may undergo localized attack, nickel-based alloys with higher molybdenum and chromium contents are superior. Although more expensive initially, these high-grade alloys may be economically justified in severe scrubber environments. It should be noted that the use of high-grade alloys demands careful fabrication. The welding recommendations of the alloy producer should be followed precisely.

Organic Linings. Among the basic organic linings, the resins, polyester, bituminous, epoxy, vinyl ester, furan, and rubber linings are the most commonly used in utility FGD systems. Table 3-17 summarizes the resistance properties of these materials in various environments. Programs of systematic testing with the various organic linings have indicated that type of coating, thickness of coating, application techniques, and degree of surface preparation are important variables affecting performance. The quality of the coating application, especially rubber lining, is critical for good service and long life. The applications contractor must be selected very carefully.

Inorganic Linings. Bricks, ceramics, and concrete are the inorganic linings in common use. The bricks used most often in FGD systems are red shale, fire clay, and silicon carbide. Red shale should be used where minimum permeation of liquor through the brick is required and thermal shock is not a factor. Fire clay should be used where minimum absorption is not required and thermal shock is a factor. Silicon carbide brick is used where high resistance to abrasion is required.

Selection of ceramics is governed by their physical properties, which result in objects of relatively thick cross-section, heavy weight, and lack of resistance to impact. Since ceramic shapes are usually made by extrusion or casting, the shapes that are symmetrical (nozzles, cylindrical scrubbers, ductwork, piping) are most suitable. Where ceramics are used, it is nearly impossible to make changes, adaptations, or repairs in the field that will provide resistance equivalent to that of the as-fired material.

Unlike the ceramics, concrete need not be fired as a formed shape to achieve strength and resistance but achieves these values by the process of hydration. To meet strength requirements, concrete is usually reinforced

TABLE 3-17. ORGANIC LININGS: BASE MATERIALS AND RESISTANCE TO SOME ENVIRONMENTS

Name	Base material	Resistance to environment						
		Abrasion	Heat	Acid	Alkali	Solvent	Water	Weather
Bituminous	Coal tar	F	P	F	F	P	F	F
Chlorinated rubber	Chlorine, natural rubber	F	P	G	G	P	E	F
Natural rubber	Fluorine, ethylene	E	F	G	G	P	E	E
Fluorocarbons	Fluorine, ethylene	E	E	E	E	E	E	E
Epoxy	Epichlorohydrin, Bisphenol-A	G	G	G	E	G	G	G
Furan	Furfural alcohol	F	E	F	G	P	F	F
Coal tar epoxy	Coal tar, epoxy	F	F	G	G	P	G	G
Phenolic	Phenol, formaldehyde	F	F	G	G	P	G	F
Polyesters	Phthalic acid, Bisphenol-A	G	G	E	F	G	G	G
Polyurethanes	Compounds containing isocyanate and hydroxyl groups	E	G	G	G	G	G	E
Vinyls		F	P	E	G	P	E	E

E – Excellent – May be used under all conditions.
G – Good – May be used under all conditions.
F – Fair – May be used under certain conditions.
P – Poor – Should not be used under any conditions.

with steel; more recently, fibrous materials have been used for reinforcement. In a limestone FGD system, concrete is used in presaturators, tanks, and piping.

Basis for Selection of Materials

The following discussion briefly reviews the major factors involved in selection of construction materials for service in a limestone FGD system. The chief considerations are operating conditions, location within the FGD system, material characteristics, safety factors, and economic factors.

Operating Conditions. The physical and chemical parameters of temperature, pressure, pH, flow rates, and the presence of trace elements are the important process-related considerations.

Operating temperature governs the selection of materials because corrosion increases with increasing system temperature. Nonmetallic materials such as plastics are temperature-limited. Maintaining system temperature as close as possible to ambient is therefore always desirable.

System pressure determines the requirements for vessel materials and reinforcements.

The pH level of the recirculation slurry should be known accurately, because corrosive properties of the fluid depend primarily on its pH.

The abrasive effects of flue gas containing fly ash particles are proportional to velocity of the gas. Where high gas flow rates are to be maintained, abrasion-resistant lining is needed, especially at the inlet to the scrubber. High liquid flow rates can wear away metallic oxide coatings, exposing bare metal to the corrosive slurry. Conversely, low liquid flow rates can allow solids to settle, with consequent potential for fouling. Process design of ductwork, piping, and nozzles should therefore accommodate optimum fluid velocities.

A complete elemental analysis of the coal, limestone, and fly ash (especially where fly ash is to be collected in the FGD system) is very important with respect to corrosion because of the presence of chlorides and fluorides, which can cause pitting of metal surfaces. The source of makeup water to the FGD system should also be analyzed for corrosive agents, especially where cooling tower blowdown, well water, or brackish river water are considered as makeup water sources.

Location in the System. The operating parameters just described, especially temperature, pH, and flow rate, will vary with location in the FGD system and therefore should be identified as a function of location. Also, a range must be specified for each variable, indicating maximum, minimum, and average values. Figure 3-35 is a schematic diagram showing trouble spots common to various limestone FGD systems. Where the flue gas

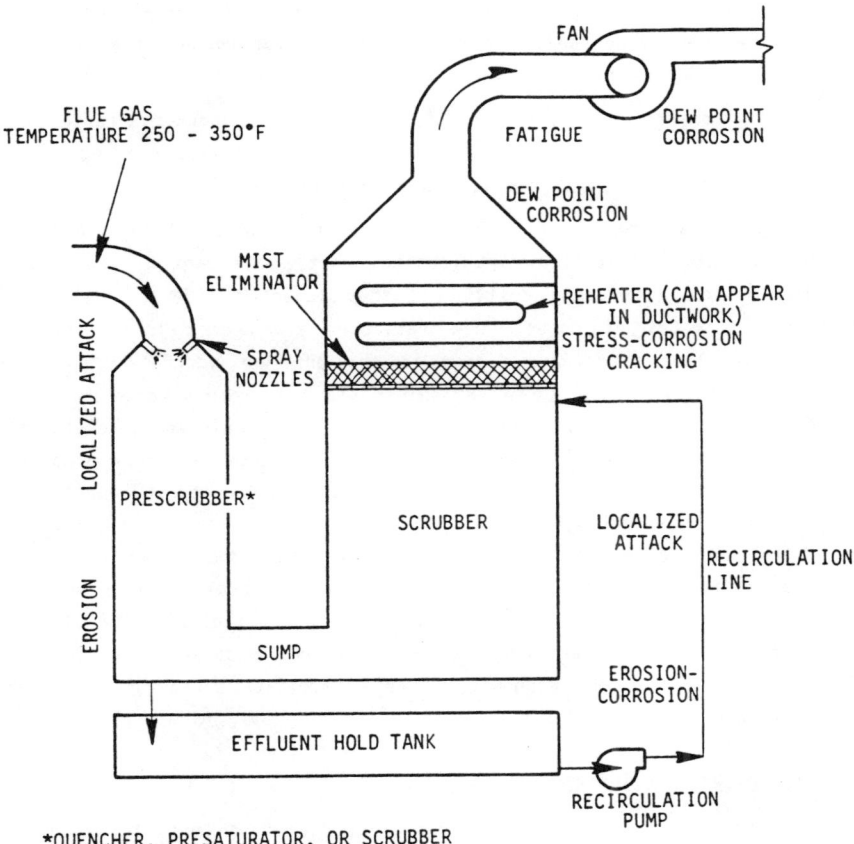

Figure 3-35. Erosion/corrosion as a function of location within a limestone FGD system.

enters and is quenched by spray nozzles, localized attack and erosion by fly ash are likely, particularly when fly ash is collected in the FGD system. Erosion-corrosion can present a problem in recirculation pumps. Corrosion of fittings and crevice corrosion may predominate inside the absorber, whereas stress-corrosion cracking occurs in reheater tubes. Corrosion by moisture takes place downstream of the mist eliminator and in stacks, and fatigue cracking has caused the failure of exhaust fans.

Material Characteristics. Temperature limitations and resistance to corrosion and erosion are the primary process design considerations. Fabricability and material strength are the primary mechanical considerations. FGD system design may involve a variety of shapes, with points of high stress. Selection of materials is especially critical when two different metals are to be welded. Quality control of materials can prevent difficulties caused by scales, cracks, scratches, or rough surfaces, all of which enhance corrosion.

Safety Factors. Fire-retardance, always an important feature, is especially so in systems using plastics and/or organic coatings. Fire retardance adds to the cost and must be expressly specified.

Economic Factors. The chief economic factors in selection of materials include the total installed cost (capital cost), the cost of replacements and maintenance (operating cost), and the projected life of the system. Accurate estimates of these three variables are needed for an economic evaluation. The recommended scheme is the "Present Worth" method, which essentially combines the three variables into one present worth value. It is difficult, however, to determine service life and maintenance costs without many years of operating experience with each material under consideration. Most of the experience accrued to date in limestone FGD applications is summarized in Appendix G.

From the spectrum of available materials, only those that satisfy the physical criteria should be considered for economic evaluation. The physical and economic factors are complementary: the physical criteria eliminate materials from the lower range of the spectrum because of unacceptable performance, whereas the economic criteria eliminate those from the upper range because of prohibitive cost.

PROCESS CONTROL AND INSTRUMENTATION

The primary functions of a limestone FGD process control system are:

- To hold SO_2 emissions within the required limits
- To achieve continuous and reliable operation
- To achieve optimum limestone utilization and energy consumption
- To ensure boiler safety

Table 3-18 lists the major variables that are measured in limestone scrubbing systems so as to fulfill these functions. The methods of controlling these measured variables are discussed later in this section.

Control of process chemistry is a major factor in successful operation of a full-scale FGD system. Following is a brief review of some essential findings that have emerged from the development of process control technology.

- Virtually all operating full-scale systems regulate reagent feed rate by controlling slurry pH. A pH electrode probe activates a signal that regulates the position of control valves to control the rate of limestone feed. This procedure, however, has proved

TABLE 3-18. MAJOR VARIABLES IN A LIMESTONE FGD CONTROL SYSTEM

Stream/equipment description	Measured variables
Flue gas at boiler outlet Flue gas at scrubber inlet Flue gas at scrubber outlet	Flow rate, temperature, pressure SO_2, O_2
Limestone feed to ball mill Fresh water to ball mill Limestone slurry to effluent hold tank Fresh slurry hold tank Makeup water	Flow rate, limestone composition Flow rate, dissolved solids Flow rate, slurry solids, solids size, particle size distribution Slurry level Flow rate and water level
Spent scrubbing slurry Fresh slurry to scrubber Bleed slurry to thickener Scrubber EHT	pH, slurry solids pH, slurry solids, flow rate Slurry solids, flow rate Slurry level
Thickener overflow Thickener underflow Sludge to disposal	Flow rate, suspended solids Flow rate, slurry solids Sludge solids, flow rate

unreliable at times and fluctuations in boiler load may cause pH instability.

º The pH probe should monitor the scrubber EHT because changes in boiler load are first noted there. To avoid cracking of the probe, a small dip tank in a screened location in the EHT is preferred.

º Sufficient operating experience has been gained to identify most of the limestone feed control problems. These problems have been resolved mostly through design modifications or new operating and maintenance procedures. For pH control, dip-type electrode probes, which are inserted into a slurry tank, are preferable to in-line probes because they are easier to clean and calibrate. In-line, flow-through probes, located in a section of piping, are generally subject to more wear and abrasion and generally require more frequent maintenance. Flow-through sensors must be removed and kept wet if the sample line is drained when the module is shut down. The dip sensor stays wet despite shutdown.

º Other limestone feed control systems have been used or are being evaluated on full-scale systems. One type involves feed control based on the inlet flue gas flow rate and SO_2 concentration, with trim provided by slurry pH. Another type involves control of limestone feed rate based on the outlet SO_2 concentration as the control variable. Only partial success has been reported for both systems because of the difficulty in obtaining accurate and consistent readings from SO_2 gas analyzers, particularly in high-sulfur coal applications.

An EPRI publication (Jones, Slack, and Campbell 1978) covers the subject of process control in great detail and is recommended as a guide in design of control systems for limestone scrubbers.

Limestone FGD System Control Loops

This discussion concerns five of the principal control loops of a typical limestone scrubber unit: limestone feed control, slurry solids control, gas flow control, bypass reheat control, and makeup water control at the thickener overflow tank.

Limestone Feed Control Loop. Limestone slurry feed rate is one of the factors that determines the extent of limestone dissolution in the EHT and thus pH of the slurry in the tank, which in turn affects the SO_2 removal. The pH level can be used in a simple feedback control loop to regulate the flow of limestone feed to the hold tank. The advantage of this system is its simplicity. The disadvantages are the nonlinearity and sluggishness of the response and other limitations of pH as the controlled variable. Proper pH control through limestone addition can prevent scaling and plugging, optimize reagent utilization and SO_2 removal, and thus improve the reliability of the scrubbing system.

All operational systems control the limestone feed rate with a feedback loop using pH as the controlled variable. Many of these systems report problems with arrangements that rely solely on pH to control limestone feed. An improved system incorporates feed-forward control of the limestone feed rate by measurement of the SO_2 concentration at the boiler outlet. Instruments measuring the SO_2 content of a dry gas have operated more reliably in the field than have SO_2 analyzers on a wet gas stream. From measurements of SO_2 concentration and of the gas flow rate, the mass flow rate of SO_2 can be computed. The limestone feed rate is then set in proportion to the quantity of SO_2 entering the scrubber.

The primary advantage of this system is that it responds immediately to changes in SO_2 concentration and gas flow rate and assists the pH feedback system in control of limestone feed. Feed forward of the inlet SO_2 concentration does operate well when properly limited (trimmed) by an indicator of slurry pH level. A schematic of this arrangement is shown in Figure 3-36.

Slurry Solids Control Loop. The solids content of recirculation slurry must be controlled to provide adequate crystallization surface area for precipitation and to minimize erosion and solids buildup. A density feedback control system for recirculation of slurry solids is shown in Figure 3-37.

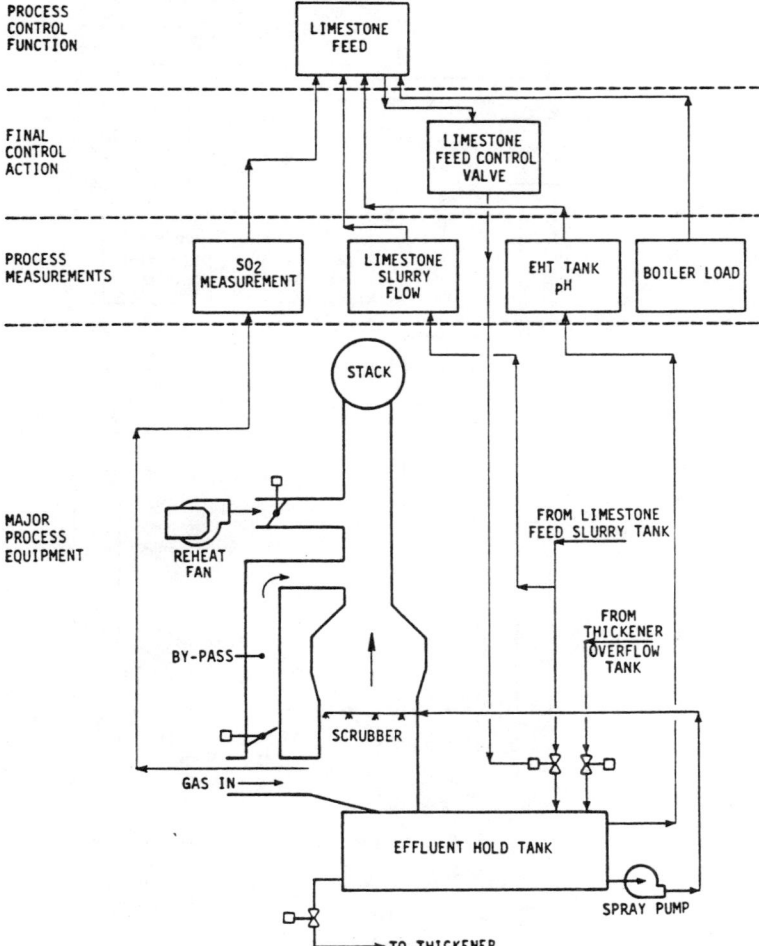

Figure 3-36. Limestone feed control loop.

In the primary loop, the density sensor activates the density controller, which changes the flow rate of the bleed stream to the thickener. In the independent secondary loop, the level sensor adds more makeup water/ thickener overflow if the level falls below a set point. The advantage of the system is its simplicity; the disadvantage is a higher rate of wear of the control valves.

Gas Flow Control. Gas flow rate to the scrubber is determined by boiler operation, and the scrubbing system must respond to boiler load changes. If multiple scrubbing modules are used, a dependable system must be provided to balance the flow rates to the parallel modules. Improperly

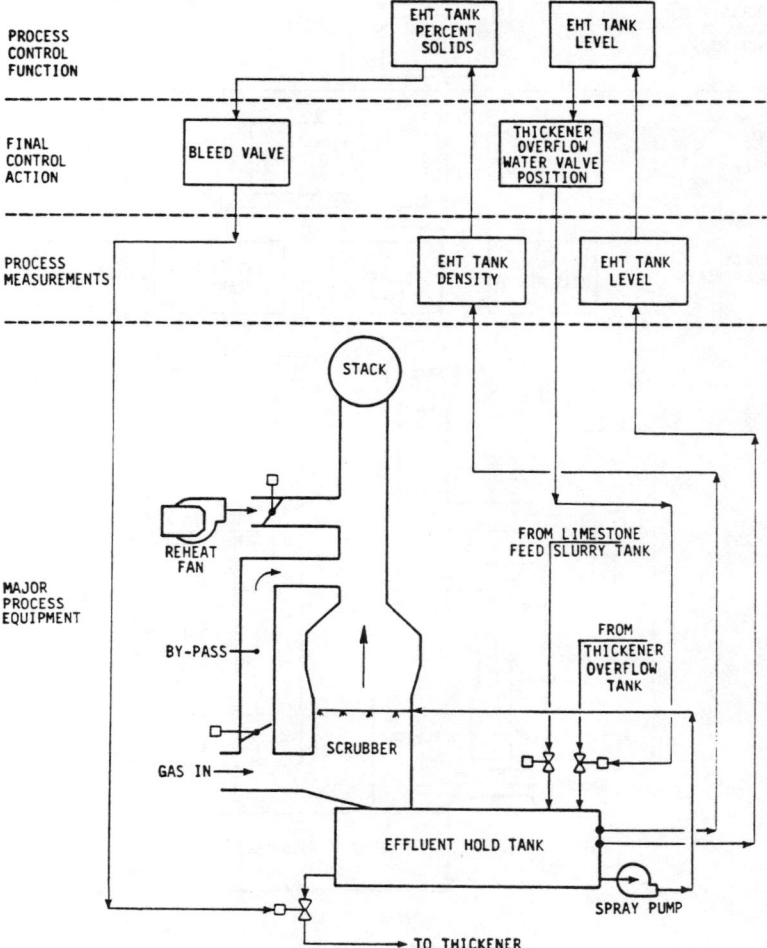

Figure 3-37. Slurry solids control loop.

balanced gas flow results in high gas velocities in one or more scrubbers, which will lead to a reduction of SO_2 removal, an increase in moisture carryover to the mist eliminator and downstream equipment, an increase in particulate loading, and possible erosion damage to downstream equipment.

The most difficult problems of gas flow control arise in protection of the boiler/scrubber system from explosion or implosion damage on tripout of the boiler or on loss of a scrubber or fan. When the boiler shuts down, there is a sudden increase or decrease in gas flow rate, depending on the safety requirements of the boiler. Although interlocks are used to achieve simultaneous shutdown of the boiler and the scrubber, a pressure or vacuum surge can develop in the boiler and ductwork if the dynamic response to this condition is unfavorable. Similarly, loss of a fan will produce a pressure surge in the opposite direction that will trip the boiler but may also create potentially damaging surges.

If the scrubber system is designed to scrub only part of the flue gas and the remainder is routed through a bypass damper, connection of the bypass damper and the fans to the boiler flame safeguard system is an acceptable solution. Then, in the event of an emergency boiler shutdown, the fan can be shut down and the damper opened. If the fan fails, the damper can be opened and an operator can then conduct an orderly shutdown of the boiler.

Makeup Water Control. Control of the overall system water balance is critical and varies significantly from site to site. The objective is to maintain a stable water inventory in the system and not to discharge any excess water. This is usually accomplished with recycled thickener overflow rather than new water. The scrubber water system should be operating in the closed-loop mode so that all the water discharged from the system is contained either in the dewatered sludge or as water vapor in the flue gas.

Figure 3-38 depicts a system that uses measurements of thickener overflow tank level to control system makeup water at the thickener overflow tank. The thickener overflow tank then provides water for makeup at the EHT to maintain proper slurry solids content and also for use as mist eliminator wash water. By eliminating the direct use of fresh makeup water for mist eliminator wash, this control approach maintains the water balance during periods of washing the mist eliminators.

Bypass/Reheat Control Loop. Figure 3-39 shows a combination bypass/reheat system in which a portion of the hot flue gas from the boiler bypasses the wet scrubber to be used for reheat and additional reheat is supplied by a supplemental indirect hot air reheat system.

Bypass reheat offers the advantages of low capital investment and economical operation, but the maximum use of bypass reheat is limited by the constraints of SO_2 emission standards. Therefore, as in this case, supplemental reheat must be added.

The SO_2 concentration and temperature of flue gas at the stack are measured to balance both the bypass damper and the reheat fan damper and control both SO_2 and temperature at the system outlet.

Instrumentation

In early limestone FGD systems, sensors caused most of the major problems with control instrumentation (Jones, Slack, and Campbell 1978). The performance of measuring devices has been improving in newer limestone FGD systems and has contributed to an increase in system availability. This

160   Handbook for FGD Scrubbing with Limestone

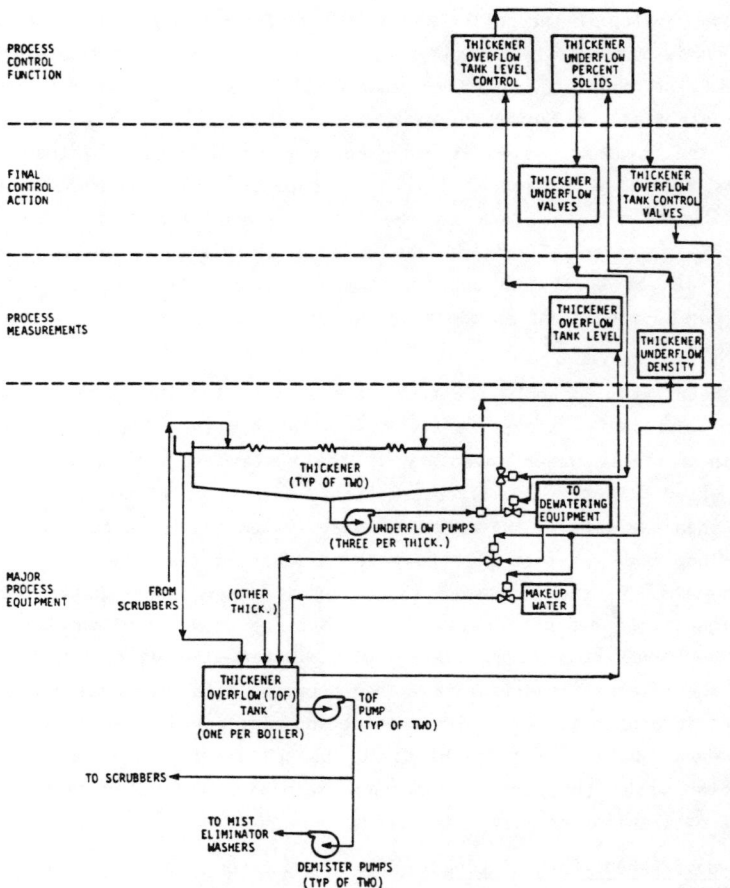

Figure 3-38. Makeup water control.

discussion presents basic information on common instrument applications in a limestone FGD system and suggests suitable hardware. The major applications include measurements of pH level, slurry solids content, liquid/slurry flow rate, liquid/slurry level, and SO_2 in the flue gas. Information is also given concerning SO_2 analyzers for continuous monitoring to meet the requirements of NSPS.

Measurement of pH Level. Recycle slurry pH is used in all limestone scrubbers to control the limestone feed rate. On-line determination of pH in the slurry of a limestone scrubber is difficult. Dependable pH readings require the use of multiple, nonlinear controllers. The following factors

The FGD System   161

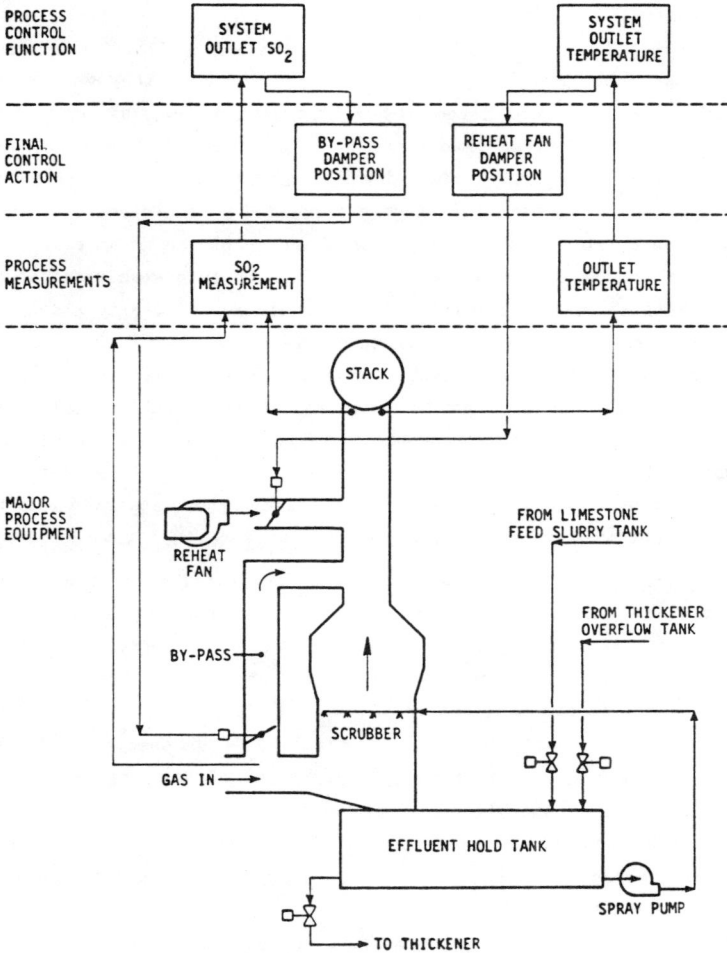

Figure 3-39. Bypass/reheat control loop.

should be considered in the design and specification of a pH electrode probe and controller:

- Probe location and maintenance
- Probe type (dip-type or flow-through)
- Nonlinearity

Electrodes are fragile devices, easily damaged by the action of an abrasive slurry. Scrubber slurry can form a deposit on an electrode, which may act as an electrical insulator and cause false readings.

Dip sensors are easier to maintain and calibrate when installed in an open overflow pot rather than directly in the EHT. Flow-through sensors are prone to higher erosion than are dip-type sensors; they must either be installed in a slipstream from the recirculation pump or supplied with slurry by a separate pump. Additionally, flow-through sensors must be kept wet when the FGD system is shut down. Pressure must be high enough to produce a flow rate within the range recommended by the electrode supplier. The piping should be as short as possible and should be drawn completely by gravity when the FGD system is shut down. Flow-through electrodes should be installed with valves to permit frequent maintenance. Dip-type sensors are more widely used in operational limestone FGD systems.

The pH electrodes are available with ultrasonic self-cleaning devices. Ung et al. (1979) report that the maintenance required for self-cleaning electrodes is significantly reduced by removing the screen surrounding the electrode or by increasing the mesh size of the screen. The self-cleaning electrodes generally require less maintenance than standard (non-self-cleaning) electrodes in continuous service of more than 2 weeks duration.

Measurement of Solids Content. A scrubber installation should include instrumentation for continuous control or recording of solids content, especially for the fresh limestone slurry, recycle slurry, and thickener underflow. Densinometers are used to monitor and control the percentage of solids in slurry streams and the EHT.

Slurry density can be measured directly with special differential pressure instruments, but a minimum liquid depth of 6 ft is needed to measure a span of 0.1 specific gravity unit. Ultrasonic devices directly measure the percentage of suspended solids. Vibrating reed instruments measure the dampening effect of the slurry on vibrations from an electrically driven coil.

Nuclear density meters, which measure the degree of absorption of gamma rays from a radioactive source, are preferred for this service. They have the minor disadvantage of producing a signal that is not linear with solids content unless the unit contains an electronic linearizer. The nuclear density meter can be precalibrated by theoretical calculations if an accurate chemical analysis of the slurry being metered is available. The main advantage is ease of application. The meter can be strapped to a pipe without insertion into the pipe line.

Measurement of Liquid/Slurry Flow. Measurement of liquid or slurry flow rates is vital to the optimization of system performance. The flow rate of fresh limestone slurry for pH control is an important application, as are the flow rates of recycle slurry, slurry bleed to the thickener, and thickener underflow.

Mechanical flowmeters are not suitable for the abrasive slurry of limestone scrubbers. The most acceptable meters for this service are electronic devices. Electronic measurement of flow rate can be accomplished with vortex-shedding instruments, ultrasonic transmission devices, Doppler-effect ultrasonic meters, and electromagnetic flowmeters.

The Doppler-effect ultrasonic meter is a fairly new development for slurry applications. The principal advantage of this device is that the electrodes are attached to the outside of the pipe through which the slurry is flowing; as is the case with the nuclear density meter, there is no penetration of the pipe.

The electromagnetic flowmeter, or "magnetic meter," is the best proven instrument available for the measurement of pressurized slurries. It consists of a stainless steel pipe section lined with an electrically insulating material. The magnetic meter does not require installation in straight piping. It has no operating parts in contact with the fluid, produces very little pressure drop, and is fairly accurate. A magnetic meter should be recalibrated periodically. The only disadvantage of the magnetic meter is the high cost.

Good quality magnetic flowmeters give trouble-free performance; in selection of a vendor, the availability of personnel to inspect and calibrate the meter at startup should be considered. Magnetic flowmeters come in sizes from 1/10 in. to 6 ft in diameter.

Less expensive devices are also made that use the electromagnetic principle. The only advantage of these instruments is their lower cost. Their disadvantage is uncertainty of the accuracy of calibration.

Measurement of Liquid/Slurry Level. Level controls in a limestone scrubber serve two functions: to feed makeup water into the system and to control the levels in the various hold tanks. The system should be supported by high- and low-level signals for notification of malfunction.

Controllers that operate with mechanical flow sensors are not suitable for limestone scrubber applications because the slurry contains chemicals and abrasive substances. Displacement controllers and flange-mounted differential controllers are best suited for scrubber applications. Capacitance level controllers, pneumatic bubble tubes, and ultrasonic devices are also effective.

Measurement of SO_2 Concentration. Jahnke and Aldina (1979) have discussed continuous monitors in detail; parts of their discussion are briefly summarized here. There are two basic types of continuous SO_2 monitors: extractive and in-situ. The extractive monitor withdraws a sample of the gas from a stack or duct and performs analysis in an analyzer usually situated in a housing near the sampling site. The in-situ monitor, as the name suggests, performs the analysis within the stack or duct.

Extractive monitors--

The total extractive system must remove a representative gas sample from the stack or duct, maintain integrity of the sample during transport to the analyzer, condition the sample so that it is compatible with the analytical method, and provide a means for reliable calibration at the sampling interface.

The most consistent problem with operation of the extractive SO_2 analyzer has been the difficulty of withdrawing a sample of gas, preconditioning it, and feeding it to the analyzer cell. Sampling systems can become plugged frequently and corroded very quickly. The function of the preconditioning system is to remove solid particulates and condense water. In practice, however, as water is collected it continues to absorb SO_2 and oxygen and can create a strong sulfuric acid solution. Solids preferentially collect on other precipitated solids and form scale. Careful design of the system is therefore required.

Design of an extractive monitoring system involves determination of the gas stream parameters, selection of the sampling site, selection of an analyzer, and design of the sampling interface. Gas temperature and velocity profiles, the particulate loading and the water vapor content of the gas, and the absolute pressure should be determined at all possible sampling sites. Selection of a sampling site should be based on accessibility and whether the samples obtained will be representative. If the site is at least 8 or more duct diameters downstream from a point of air in-leakage or mixing of different gas streams, the gases are generally mixed enough to give a representative sample. The analyzer selected must be compatible with the gas parameters, sampling site, intended housing or location, and the sampling interface, which is designed to precondition the gas sample.

A typical sampling interface includes a coarse in-stack filter, gas transport tubing, sampling pump, fine filter, and acid mist removal system. The coarse filter, made of sintered or fine mesh screen stainless steel, is located at the probe tip in the stack to prevent plugging of the probe and the sample lines. A sample line of ¼-in.-OD Teflon tubing may be used to transport the gas to the analyzer over a short run. Caution should be taken with longer runs of 50 feet or more. This and all exposed components of the interface must be heat-traced thoroughly to prevent condensation of moisture. It is advisable to place a fine filter upstream of this pump to reduce maintenance. A moisture trap to remove water droplets should be placed in front of the analyzer cell and should be kept at a controlled temperature to ensure consistent water vapor content in the sample cell. Some analytical methods require a drying (conditioning) system to reduce the sample dew point to 0°C. This can consist of refrigerated traps or permeation type dryers.

The use of absorption reference filters and the introduction of reference gas downstream from the sample lines can reduce the cost of automatic calibration valving and the time required to perform daily calibrations; however, periodic checks involving the introduction of reference gas through the entire sampling interface must be performed to ensure accuracy.

The extractive instruments in use today are based on a variety of principles of detection and analysis. They differ in sensitivity, susceptibility to specific interferences, complexity, ease of operation, and other operational factors, as well as in initial and operating costs. Some offer the flexibility of application to stack gas monitoring, analysis of process streams, and analysis of in-plant or ambient air. Selection from among these instrument systems must be based on information from vendors and users and on the intended application. Table 3-19 lists some of the manufacturers of these instruments and the principles of operation.

As a means of verifying the reliability of continuous monitors, the EPA has awarded a contract to GCA Corporation to finalize a monitoring system design for use in a 1-year demonstration program to be conducted at the

TABLE 3-19. SOME MANUFACTURERS OF EXTRACTIVE SO_2 MONITORS BASED ON VARIOUS OPERATING PRINCIPLES

Operating principle	Manufacturer
Nondispersive ultraviolet	DuPont CEA Instruments Esterline-Angus Teledyne
Ultraviolet fluorescence	Celesco Industries, Inc. Thermo Electron Corp.
Flame photometry	Tracor Melroy Laboratories Bendix Corp. Process Analyzers
Conductimetry	Calibrated Instruments, Inc.
Coulometry	Barton ITT Beckman Instruments, Inc.
Polarography	Dynasciences Beckman Instruments, Inc. Theta Sensors Teledyne
Spectrometry	Environmental Data Corp. Wilks Scientific Corp. Environmental Research and Technology Barringer Research, Ltd. Lear-Siegler, Inc.

Conesville Power Station operated by the Columbus and Southern Ohio Electric Company. This demonstration program will result in a set of guidelines for specifications, certification requirements, and recommended operation and maintenance practices for the use of extractive continuous SO_2 analyzers.

In-situ monitors--

"In-situ" monitors directly measure the pollutant concentrations in the stack or duct. These systems do not modify the flue gas composition by dilution and are designed to detect gas concentrations in the presence of moisture and particulate matter. Because particulates cause a reduction in light transmission, ceramic filters are used to exclude particles. Several types of in-situ monitors are available, all based on absorption spectroscopy.

In comparison with an extractive monitor, the in-situ monitor offers greater flexibility in site selection, reduced calibration time, and fewer components for maintenance. Also, a single in-situ monitor can measure several gases and opacity. The disadvantages of in-situ monitors include problems with optical systems and failure of complicated electronic components. An in-situ monitor can sample only one stack or duct at a time. Where a number of stacks must be monitored, the use of multiple probes leading into a single extractive system might be a cost-effective choice.

SO_2 Monitoring for NSPS Compliance. The foregoing discussion has dealt with measurement of SO_2 as a process control factor, i.e., as an indicator for use in process operation. The utility operator should also consider available instrumentation for SO_2 monitoring within the context of NSPS requirements. Under the NSPS promulgated in 1979, the owner or operator of a new electric utility steam generating unit must install, calibrate, maintain, and operate a continuous monitoring system for SO_2 concentrations and record the output of the system. The EPA has presented performance specifications and test procedures for SO_2 and NO_x continuous monitoring systems. Any of the instrument systems used for emissions measurements must meet these performance specifications at the time of installation and throughout their operation. A summary of the current specifications is presented in Table 3-20.

Diluent concentrations--

SO_2 monitor data alone do not give accurate information on emission rates. If only a volumetric measurement is desired, such as for control of limestone feed rate, the monitoring of only SO_2 concentrations may be sufficient. If emission rate information is required, however, diluent concentrations must be measured. This can be done by measuring the concentrations of oxygen (O_2) or carbon dioxide (CO_2) at the point of SO_2 analysis.

If diluent monitoring is required, the extractive sampling system

TABLE 3-20. CURRENT EPA PERFORMANCE SPECIFICATIONS FOR CONTINUOUS MONITORING SYSTEMS AND EQUIPMENT

Parameter	Specification
Accuracy[a]	$\geq$ 20 percent of the mean value of the reference method test data
Calibration[a]	$\geq$ 5 percent of each (50 percent, 90 percent) calibration gas mixture value
Zero drift (2-hour)[a]	2 percent of span
Zero drift (24-hour)[a]	2 percent of span
Calibration drift (2-hour)[a]	2 percent of span
Calibration drift (24-hour)[a]	2.5 percent of span
Response time	15 minutes maximum
Operational period	168 hours minimum

[a] Expressed as sum of absolute mean value plus 95 percent confidence interval of a series of tests.

interface used for SO_2 sampling can be used by taking a side stream of the conditioned or preconditioned sample. Just as with the SO_2 monitors, care must be exercised in evaluating the compatibility of sample conditioning with the analysis method. If in-situ SO_2 monitoring methods are being used, a separate in-situ diluent monitor will be needed.

In evaluation of the monitoring systems, it should be recognized that multiple use of the extractive monitor interface tends to increase its cost effectiveness over that of in-situ monitoring.

Recording systems--

The strip-chart recorder is used most frequently in continuous monitoring applications because of the ease of reading the recorded data and compactness of the display. The strip-chart recorder provides a continuous analog record. Low-cost digital recording devices are available for recording and processing of emission data.

Control Philosophy

In a control loop the objective is to maintain one variable at a fixed value. In an integrated control system the objective is to maintain the overall output from the plant within limits. An integrated FGD control system consists of many control loops that interact. In the design of each control subsystem, the interactions must be considered as part of an overall control philosophy.

The earliest control concept was to maintain a slurry pH value that controlled the limestone feed rate and thus the stoichiometric ratio. Control by feedback of pH value alone proved inadequate because of high limestone stoichiometric rates and the following limitations:

- The response of pH to a change in limestone feed rate is extremely nonlinear.
- There is an inherent time lag caused by obtaining equilibrium in the scrubber EHT.
- The pH value is not suitable as a variable on which to base limestone feed rate when limestone utilization is ≤ 90 percent.

Recent commercial experience by Peabody Process Systems, however, has shown that simple feedback control of pH has proven to be satisfactory (see process control writeup of Colorado Ute). The reason the Peabody process is able to overcome these pH feedback limitations is the unique use of the hydrocyclone in a single loop system. The hydrocyclone permits very high ($\geq 90\%$) limestone utilization to be achieved and this results in a typical pH set point value of 5.5 maximum. Peabody's process has measured low stoichiometric ratios (1.02) and this supports the successful performance of feedback pH in their system (Ostroff 1981). Simple feedback pH can successfully be used to control limestone utilization or limestone stoichiometric ratio and thus limestone feed rate.

The shape of the titration curve of an acid/base neutralization is nonlinear. Greater quantities of a base are needed to change the pH of a solution from 5 to 6 than to change it from 6 to 7. A standard controller, however, is a linear device. It is adjusted so that if the pH value is doubled, the rate of limestone addition is also doubled. If the acid/base neutralization is buffered, as it is in a limestone scrubber, the titration curve is even less regular. It shows "plateaus" where the mixture absorbs quantities of limestone and there is little change in pH. As the degree of buffering changes, the shape of the curve changes. Buffered solution in a scrubber, therefore, is less amenable to standard linear control than is an unbuffered solution.

Several manufacturers have recently produced nonlinear controllers specifically designed for control of pH. Within a pH band near the set point, the controller output signal is amplified only slightly. As pH change increases, greater amplification makes a greater change in limestone feed rate to drive the pH back more rapidly into the acceptable range. Controllers of this type are more suited to this application than are standard linear instruments.

Since dissolution of limestone and precipitation calcium sulfite/ sulfate are not instantaneous, an EHT generally retains the mixture of scrubber effluent slurry and fresh limestone slurry for about 10 minutes before it is recycled to the scrubbing vessels. Thus, a time lag of several minutes is inherent in a system using pH of the recirculation slurry as the measured variable.

The point of pH measurement is an important consideration (Jones et al. 1978). The primary choice for location of the probe is in the EHT itself.

As a means of improving pH feedback, a simple pH cascade control loop can be used to regulate limestone feed rate. The feed rate is controlled primarily by pH level of the spent slurry. In a secondary loop, the controller is activated on the basis of a measurement of pH of the fresh slurry. In the secondary loop an operator may manually adjust the set point according to the pH reading. The overall control is improved by reducing the time lag.

For a given size of the EHT and given L/G ratio, pH of the recycle slurry depends on the limestone slurry feed rate and the amount of SO_2 removed. Because of the complex interrelationships among the variables, the optimum pH set point for the control system is not easily identified, and an operating range of pH values is usually defined. The upper limit corresponds to the maximum limestone feed rate beyond which poor utilization of reagent and sulfite scaling may occur. The lower limit corresponds to the minimum feed rate required for the desired degree of SO_2 removal. Therefore, the optimum limestone feed rate should be determined not only by the pH, but also by total mass flow rate of SO_2 at the inlet or outlet of the scrubber.

The use of outlet SO_2 concentration as a controlled variable for adjustment of limestone feed rate is theoretically advantageous because the response of limestone feed to a measured error in SO_2 content is more nearly linear than the pH response. The time lag will be the same as in pH control because it is set by residence time of the EHT. The sensor used in direct SO_2 control is an SO_2 analyzer of the types described earlier. Care should be taken to assure the reliability of such a sensor for use in a control loop.

Application of this control method still requires pH control. As mentioned earlier, the pH level of recycled slurry must be maintained within a range that will cause neither low SO_2 removal (and sulfate scaling) nor excessive limestone addition (and sulfite scaling). Thus, recycle pH can be used to limit (or trim) the feedback loop. The limestone feed rate would be controlled by the loop controlling outlet SO_2 so long as the pH remains within limits. If pH reaches its high limit, the high-limit pH controller would prevent further addition of limestone. Similarly, the low-limit pH controller would prevent the SO_2 controller from closing the limestone feed valve too far.

The use of inlet SO_2 concentration to improve limestone feed control is also desirable and is being applied commercially. This system, as described earlier, is proving itself as the preferred method of limestone feed control. The advantage of measuring the SO_2 content of a dry gas has enhanced the reliability of this method.

Another technique, developed at the Shawnee test facility, merits attention. Both the inlet and outlet SO_2 concentrations are measured, and the difference is computed. The differential, which is a measure of the amount of SO_2 removed, is then used in a feed-forward control loop (trimmed by the recycle slurry pH), similar to the inlet SO_2 feed-forward loop discussed earlier. This control scheme is superior to the latter in that both the inlet and the outlet SO_2 concentrations are considered in the control loop. The SO_2 removal efficiency, however, is established by the initial design operating parameters and is not to be considered a control variable.

Improvements in the reliability of wet SO_2 analyzers should affect the manner in which scrubbers are controlled. With the emphasis on stricter emission limits and shorter emission averaging time, outlet SO_2 emissions must be monitored continuously. Thus, the new control system specifications should have provisions for outlet SO_2 feedback or inlet SO_2 feedforward control.

Operational Systems

The process control systems currently used at Sherburne 1 and 2 (Northern States Power), Lawrence 4 and 5 (Kansas Power and Light), and Craig 1 and 2 (Colorado Ute) are described in detail. The descriptions are based on EPA-sponsored survey reports and on information obtained from the scrubber supplier of the Craig system.

Process Control at Sherburne (Laseke 1979a). A dedicated computer monitoring and control system is used for process control at the Sherburne scrubber plant. This system represents a refinement over the original concept, which relied on only enough parameters to schedule the modules in and out of service to meet load requirements. The old system, operated from a separate control room for the scrubber system, monitored 3 analog and 26 digital inputs with no visual display. The refined system is operated from the main control room and is interfaced with the boiler controls. The computer now monitors 134 analog and 48 digital points, with display on a cathode ray tube. All routine operations are performed by the computer and logic network.

Process control is maintained by regulating limestone feed as a function of slurry pH, regulating waste discharge as a function of slurry solids content, and regulating water feed as a function of liquid levels in the tanks. The efficiency of particulate removal is maintained by controlling gas-side pressure drop across the venturi rod decks. Pressure drop of 12 in. H_2O across the rod decks is maintained by raising or lowering the lower rod decks with a manual scissor jack. Regulation of the vertical space between the rods controls the gas-side pressure drop, which in turn affects the amount of particulate removed by the scrubber.

The efficiency of SO_2 removal is maintained by controlling slurry pH level. The pH level is measured with a Universal Interloc pH sensor located in a fiberglass flow-through chamber that receives slurry from a slipstream tapped off the spray pump. Based on these measurements, the rate of limestone feed into the reaction tanks is manually controlled. The pH control range is 5.0 to 5.5. As the pH swings above or below this range, the amount of fresh limestone slurry is reduced or increased manually. Thus optimum removal efficiency is attained while avoiding the scaling or plugging that results from loss of chemical control. Addition of limestone to the slurry circuit ensures a minimum SO_2 removal efficiency of 50 percent while preventing possible corrosion and scaling caused by pH swings. The additive slurry feed to each EHT is controlled by an Invalco valve with a Norbide valve plug and seat (boron carbide ceramic) for abrasion resistance.

To control the level of slurry solids circulated through the scrubber system, an effluent pumping service removes a spent slurry bleed stream from each EHT. One effluent bleed pump is provided for each EHT. The spent slurry is discharged at a rate of 150 gpm per module and is sent to the thickener via the slurry transfer tank. This bleed stream is automatically controlled by a Texas Instruments nuclear density meter (standard clamp-on type) working in conjunction with a Leeds and Northrup controller. A meter inside the EHT of each module monitors the solids level of the circulating slurry. When the level exceeds 10 percent, a control valve is activated that allows a bleed stream to flow to the thickener until the proper level is reestablished. Two control valves are placed in series in the bleed line, the first manually set and the second controlled. The control is not critical, however, and may be made manual.

Maintenance of a 10 percent solids level in the scrubbing slurry is vital to the process chemistry of this system. The level of sulfate in the slurry is controlled by selective desupersaturation in the EHT and continual bleed from the spray pump discharge to the thickener. Desupersaturation of sulfate requires the presence of enough calcium sulfate solids in the slurry to act as seed crystals, promoting crystal growth of calcium sulfate in the process. This process is aided by the use of a forced oxidation system that sparges air into the EHT's, oxidizing all the remaining sulfite to sulfate. The sulfate then crystallizes on the seed crystals and is removed from the system for disposal in the fly-ash pond. As a consequence, a safe sulfite level is maintained in the slurry being returned to the scrubbers; critical supersaturation, which can produce uncontrolled gypsum scale formation on equipment internals, does not occur. The amount of calcium sulfate seed crystals is 2 to 3 percent of the 10 percent solids level in the slurry, which is sufficient to control sulfate saturation levels.

A water balance system controls the water returned from the holding and recycle basins to maintain the proper liquid levels in the EHT and makeup tanks. Bubble-tube level controllers are used in these tanks to control the flow of makeup water to the tanks. Flow rates in all slurry lines are measured by magnetic flowmeters (Foxboro).

Concentrations of SO_2 in the flue gas are measured by DuPont ultraviolet analyzers (Photometric 460). Each scrubbing system is equipped with four analyzers, each having a four-stream sample train. The monitors analyze the inlet and outlet values of SO_2 for each module, the overall content of each generating unit, and the total flow to the stack.

<u>Process Control at Lawrence (Laseke 1979b)</u>. The process control networks of the Lawrence limestone scrubbing systems rely on a significant amount of instrumentation to provide total automatic control of process chemistry. Included are SO_2 gas analyzers (DuPont Photometric 460) for all gas inlet and outlet streams, magnetic flow meters (Foxboro) for all liquid slurry streams (recirculation, bleed, and feed lines), pH meters (Uniloc) for all EHT's, and nuclear density meters for all the collection tanks and EHT's. This instrumentation is the basis of a control network that maintains particulate and SO_2 removal efficiencies at desired levels while preventing the loss of chemical control. The effect of the Lawrence control network on performance of these major functions has been very successful.

Particulate removal is maintained by controlling gas-side pressure drop across the rod-decks situated in the throat area of each module through regulation of the vertical spacing between the two rows of rods in response to gas flow. This arrangement maintains a set gas-side pressure drop (9.0 in. H_2O) across the rods and ensures particulate removal efficiency at 0.1 lb/10^6 Btu of heat input to the boiler.

Sulfur dioxide removal is maintained by regulating the flow of limestone to the scrubbing systems as a function of coal sulfur content. A signal based on coal feed rate is used to indicate the inlet SO_2 content of the flue gas, and this signal regulates the limestone feed rate. The coal flow signal is related to inlet sulfur conditions only when the sulfur content of the coal is constant; an operator-selected stoichiometry bias allows correction of the limestone demand signal to account for change in the coal sulfur content. The sulfur in the coal is usually fairly constant; therefore, the coal flow signal provides an accurate indication of the inlet sulfur for all boiler loads. This allows accurate variation of the limestone feed rate for the correct stoichiometry rate throughout the load range. The system can operate at design removal efficiencies without loss of chemical control, which can lead to scale formation or corrosion.

In the Lawrence 4 scrubbing system, discharge of spent slurry occurs in the liquid staging system, where the slurry from the EHT's (one per module) is bled to the collection tanks (one per module). Spent slurry that accumulates in the slurry circuit of the rod-deck scrubber is discharged from the collection tanks by variable-drive, effluent bleed pumps. The solids content in the EHT's is controlled at the 5 percent level by a constant gravity-flow bleed stream, which discharges to the collection tanks; solids in the collection tanks are controlled at the 8 to 10 percent level by varying the flow of the effluent bleed pump. The effluent bleed stream is transferred to the thickener, where the slurry is concentrated to 30 to 35 percent solids before it is discharged to the sludge ponds. Solids content in the collection tanks and thickener is monitored by nuclear density meters in the spray lines.

Spent slurry is discharged from the Lawrence 5 scrubbing system in a manner similar to that described for Lawrence 4. Notable differences are (1) the lack of selective liquid staging and thickening in Lawrence 5, which is equipped with only one EHT for all the scrubbing modules, and (2) the direct transfer of the effluent bleed stream to the sludge ponds without a preceding thickening step. Solids content of the EHT is controlled at the 10 percent level by cycling the effluent bleed pump on and off.

In the water balance system at Lawrence, fresh water, thickener overflow water, and pond return water are used to compensate for water loss in the process. Procedures for maintaining water balance in the two systems differ because of the presence of additional liquid-staging and thickening equipment in Lawrence 4. For Lawrence 4, fresh water is used to slurry the limestone prepared in the ball mill. Dilution water, which is added to the slurry to dilute the solids content of the mill effluent, originates from the recirculation tank, which receives pond return water and thickener overflow. This water is used to maintain liquid levels in the hold tanks and also for mist eliminator wash and tank strainer wash.

The water balance network is essentially the same as for Lawrence 4, except that because Lawrence 5 contains no thickener, only pond return water is used as the dilution water. As in the Lawrence 4 system, dilution water is used to maintain liquid level in the EHT and also to wash the mist eliminator and tank strainers.

Formation of surface scale is minimized in the EHT's of the Lawrence scrubbing systems by controlled desupersaturation, which is effected by providing calcium sulfate seed crystals for crystal growth sites, providing and maintaining adequate solids levels in the slurry circuits, and providing adequate mixing and retention time in the hold tanks.

Precipitation of calcium sulfate is maintained by providing enough seed crystals for crystal growth sites in the slurry circuit and by controlling saturation below the critical supersaturation level. Sufficient seed crystals are maintained by controlling the solids content of the slurry circuits (5 percent solids in the Lawrence 4 EHT's; 10 percent solids in the Lawrence 4 collection tanks and the Lawrence 5 EHT).

Process Control at Craig (Ostroff 1981). The system at Craig differs from the Sherburne and Lawrence installations in that it is a pure SO_2 removal system; particulate matter is removed by electrostatic precipitators upstream of the FGD equipment. There are two generating units, each with a dedicated FGD system. Each FGD system consists of four scrubbing modules complete with auxiliaries, one thickener, and one thickener overflow tank. The two FGD systems share a common limestone preparation area and a common final dewatering (centrifuge) installation. Each scrubbing module contains a liquid hydrocyclone system to remove unreacted calcium carbonate from the process slurry. The calcium carbonate is returned to the EHT for reuse; the decarbonated slurry is used for scale control in the scrubber and for waste slurry from the process. This system provides sufficient alkali for SO_2 removal at a high utilization rate.

The process is controlled with both analog and digital equipment. Digital control is accomplished with a programmable system controller (manufactured by Modicon), which controls the start/stop sequence of every major subsystem within the FGD system. Analog control is accomplished with a Bailey 820 Analog Control System, which adjusts the controllable parameters for optional system operation.

Sulfur dioxide removal is controlled primarily via the pH of the recirculation slurry. The analog controller was calibrated during unit startup to provide the required SO_2 removal. The system can handle minor perturbations in entering SO_2 concentration. More significant deviations may call for a change in the pH set point or the starting of an additional recycle pump. Changing the pH set point and starting (or stopping) a recycle pump are initiated by the operator; after that, the sequence of events involved in the pump startup (or shutdown) is controlled by the digital system. The coal used at this installation has been fairly consistent with respect to sulfur content, and the pH control system has handled adequately the small changes that have occurred.

The number of modules required is determined by boiler load. Signals from the boiler control panel are converted to an equivalent gas flow and to the number of required modules. The result of this "electronic" calculation is displayed on the FGD control panel. The starting (or stopping) of a module and the choice of module are operator-initiated activities, but once initiated, the required sequence of events is controlled by the digital system.

The waste product is removed by a simple overflow arrangement. Liquid is added to the system as limestone slurry in response to the pH signal; liquid is added as clear water for density control, for mist eliminator wash, for pump seals, and for closed-loop water balance. The system overflow is the decarbonated product from the liquid hydrocyclones. It flows by gravity from the EHT's and is collected in a common sump, then pumped to the thickeners. This arrangement precludes the use of complex, interconnected control loops for slurry density control and waste product purge. The amount of waste slurry produced varies with boiler load and coal composition, and the thickener feed rate is constant (at the pump capacity). Since these two flow rates are not equal, thickener overflow water is added to the waste slurry on the basis of sump level to provide a thickener feed of constant flow but variable density. This eliminates the necessity for frequent pump starts and protects the pump against running dry.

The thickener underflow pumps are operated continuously to prevent plugging of the transfer line. Removal of thickener underflow is controlled by slurry density. When the density is low, the system is operated in a "recycle mode," in which thickener underflow is returned to the thickener feed well, where the sludge blanket builds up to the desired density and thickness. When this occurs, a signal alerts the FGD operator that the thickener underflow may be operated in a "production mode," in which the slurry is transported to the centrifuges for final dewatering and disposal. The change from the recycle to the production mode is operator-initiated; after that, the digital system controls starting of the centrifuges and diversion of the flow. When the sludge blanket in the thickener has been depleted, the thickener is returned to the recycle mode and the centrifuge system is sequentially shut down by the digital controller.

Three centrifuges feed a common removal point. Trucks are used to remove the sludge to the disposal site. A control system at the loading site synchronizes the operation of the centrifuges with the availability of a truck. When a truck is in position to be loaded, the centrifuge accepts thickener underflow and discharges sludge to a conveyor system and to the truck. When a truck is not in position to be loaded, the thickener underflow is briefly returned to the recycle mode but the centrifuge system is not stopped.

The process makeup water is cooling tower blowdown, introduced in response to a system-water inventory signal. Liquid is stored in the alkali storage tanks, in the EHT's, in the thickener, and in the thickener overflow tanks. Since the EHT's and the thickeners are overflow devices and are always full, the water inventory is determined by measuring the contents (level) of the alkali storage tanks and the thickener overflow tanks. The cooling tower blowdown water is blended with decarbonated slurry and introduced into the modules. This arrangement allows the makeup water to equilibrate chemically with the FGD system and to isolate the reactive, alkali-containing, recycle slurry from the mist eliminating equipment.

Scaling is controlled by (1) providing sufficient residence time and agitation in the EHT to allow desupersaturation to occur; (2) providing a high enough concentration of seed crystals for nucleation sites; (3) providing enough liquid flow to minimize the SO_2 make per pass; (4) controlling the system pH; and (5) minimizing internals within the scrubber modules.

REFERENCES FOR SECTION 3

SCRUBBERS

Calvert, S. 1977. How to Choose a Particulate Scrubber. Chemical Engineering, 48(18):54-68, August 28, 1977.

Industrial Gas Cleaning Institute. 1976. Basic Types of Wet Scrubbers. Publication No. WS-3, June 1976.

PEDCo Environmental, Inc. 1981. Flue Gas Desulfurization Information System. Maintained for the U.S. Environmental Protection Agency under Contract No. 68-01-6310, Task Order No. 6. February 1981.

Saleem, A. 1980. Spray Tower: The Workhorse of Flue Gas Desulfurization. Power, 124(10):73-77.

MIST ELIMINATORS

Burbank, D. A., and S. C. Wang. 1980. EPA Alkali Scrubbing Test Facility: Advanced Program - Final Report (October 1974 to June 1978). EPA-600/7-80-115. NTIS No. PB80-204241.

Conkle, H. N., H. S. Rosenberg, and S. T. DiNovo. 1976. Guidelines for the Design of Mist Eliminators For Lime/Limestone Scrubbing Systems. EPRI FP-327. December 1976.

Industrial Gas Cleaning Institute. 1975. Scrubber System Major Auxiliaries. Publication No. WS-4. November 1975.

Ostroff, N., and T. D. Rahmlow. 1976. Demister Studies in a TCA Scrubber. MASS-APCA. Drexel University. April 23, 1976.

PEDCo Environmental, Inc. 1981. Flue Gas Desulfurization Information System. Maintained to the U.S. Environmental Protection Agency under Contract No. 68-01-6310, Task Order No. 6.

REHEATERS

Choi, P. S. K., S. A. Bloom, H. S. Rosenberg, and S. T. DiNovo. 1977.

Stack Gas Reheat for Wet Flue Gas Desulfurization Systems. EPRI report No. FP-361. Prepared for Battelle Columbus Laboratories for EPRI Research. February 1977.

Muela, et al. 1979. Stack Gas Reheat - Energy and Environmental Aspects. In: Proceedings: Symposium on Flue Gas Desulfurization, Las Vegas, Nevada, March 1979. Volume II. EPA-600/7-79-167b. NTIS No. PB80-133176.

Perry, R. H., et al. Chemical Engineers Handbook. Fifth edition. McGraw Hill, New York 1973. p. 12-2.

FANS

Green Fuel Economizer Co. 1977. "Heavy Duty Fans." Technical Bulletin No. 195. Beacon, New York.

Green, K., and J. Martin. 1978. Conversion of Lawrence No. 4 FGD System. In: Proceedings: Symposium on Flue Gas Desulfurization Held at Hollywood, Florida, November 1977. Vol. 1. EPA-600/7-78-058a. NTIS No. PB-282 090.

Laseke, B. A. 1978a. EPA Utility FGD Survey: December 1977 - January 1978. EPA-600/7-78-051a. NTIS No. PB-279 011.

Laseke, B. A. 1978b. Survey of Flue Gas Desulfurization Systems: Cholla Station, Arizona Public Service Co. EPA-600/7-78-048a. NTIS No. PB-281 104.

Laseke, B. A. 1978c. Survey of Flue Gas Desulfurization Systems: La Cygne Station, Kansas City Power and Light Co. EPA-600/7-78-048d. NTIS No. PB-281 107.

Laseke, B. A., et al. 1978. EPA Utility FGD Survey: February-March 1978. EPA-600/ 7-78-051b. NTIS No. PB-287 214.

Melia, M., et al. 1978. EPA Utility FGD Survey: June-July 1978. EPA 600/7-78-051d. NTIS No. PB-288 299.

Miller, D. M. 1976. Recent Scrubber Experience at Lawrence Energy Center, Kansas Power and Light Co. In: Proceedings: Symposium on Flue Gas Desulfurization Held at New Orleans, Louisiana, March 8-11, 1976. EPA-600/2-76-136a. NTIS No. PB-255 317.

Wells, W. L., G. T. Munson, and E. G. Marcus. 1980. Actual and Projected Materials Problems in Limestone and MgO Scrubber Processes. Paper presented at 7th Energy Technology Conference and Exposition, Washington, D.C., March 24-26, 1980.

THICKENERS AND MECHANICAL DEWATERING EQUIPMENT

Heden, S. D., and J. H. Wilhelm. 1975. Dewatering of Powerplant Waste Treatment Sludges. Paper presented at the 36th Annual Meeting of the International Water Conference, Pittsburgh, Pennsylvania.

Knight, R. G., et al. 1980. FGD Sludge Disposal Manual, Second Edition. EPRI-CS-1515.

Morasky, T. M. 1978. State-of-the-Art of Sludge Fixation. EPRI-FP67.

Morasky, T. M. 1979. Lime FGD Systems Data Book. EPRI-FP-103.

Rabb, D. T. 1978. Selected Topics from Shawnee Test Facility Operation. EPA Industry Briefing at Research Triangle Park, North Carolina.

Robbins, J. 1979. Private communication to PEDCo Environmental, Inc. from BIRD Machine Company regarding centrifuge at Mohave.

U.S. Environmental Protection Agency. 1979. Process Design Manual for Sludge Treatment and Disposal. EPA-625/1-79-011.

Wilhelm, J. H., and R. W. Kobler. 1977. Private communication from EIMCO Process Machinery, Division of Envirotech.

SLUDGE TREATMENT

Knight, R. G., et al. 1980. FGD Sludge Disposal Manual, Second Edition. EPRI-CS 1515.

PUMPS

Dalstad, J. I. 1977. Slurry Pump Selection and Application. Chemical Engineering, 84(9):101-106.

Doplin, J. H. 1977. Select Pumps to Cut Energy Cost. Chemical Engineering, 84(2):137-139.

Karassik, J., et al. 1976. Pump Handbook, McGraw-Hill Book Co. pp. 3-47 to 3-69.

Kruger, R. J. 1978a. Experience with Limestone Scrubbing: Sherburne County Plant, Northern States Power Co. In: Proceedings: Symposium on Flue Gas Desulfurization, Hollywood, Florida, November 1977. Volume I. EPA-600/7-78-058a. NTIS No. PB-282 090.

Kruger, R. J. 1978b. Personal communication to PEDCo.

Laseke, B. A., Jr. 1978a. Survey of FGD Systems: Cholla Station, Arizona Public Service Co. EPA-600/7-78-048a. NTIS No. PB-281 104.

Laseke, B. A., Jr. 1978b. Survey of FGD Systems: La Cygne Station, Kansas City Power and Light. EPA-600/7-78-048d. NTIS No. PB-281 107.

O'Keefe, W. 1980. Flue Gas Desulfurization and Coal's Upswing Direct Your Attention to Slurry Pumping. Power, 124(5):25-35.

Reynolds, J. A. 1976. Saving Energy and Costs in Pumping System. Chemical Engineering.

Teeter, R. 1978. Personal communication to PEDCo.

Wilhelm, J. H. 1977. Personal communication with Mr. Wilhelm of EIMCO Process Machinery, Division of Envirotech.

PIPING, VALVES, AND SPRAY NOZZLES

Laseke, B. A., Jr. 1979. Survey of Flue Gas Desulfurization Systems: Sherburne County Generating Plant, Northern States Power Co. EPA-600/7-70-199d. NTIS No. PB 80-126287.

O'Keefe, W. 1980. Flue Gas Desulfurization and Coal's Upswing Direct Your Attention to Slurry Pumping. Power, 124(5):25-35.

Rosenberg, H. S., et al. 1980. Operating Experience With Construction Materials for Wet Flue Gas Scrubbers. Combustion, 52(1):23-36.

Saleem, A. 1980. Spray Tower: The Workhorse of Flue Gas Desulfurization. Power, 124(10):73-77.

TANKS

Borgwardt, R. H. 1975. Increasing Limestone Utilization in FGD Scrubbers. Presented at the 68th AIChE Annual Meeting, Los Angeles, California.

Laseke, B. A., Jr. 1979. Survey of Flue Gas Desulfurization Systems: Sherburne County Generating Plant of Northern States Power Co. EPA-600/7-79-199d. NTIS No. PB 80-126287.

Saleem, A. 1980. Spray Tower: The Workhorse of Flue Gas Desulfurization. Power, 124(10):73-76.

AGITATORS

Rosenberg, H. S., et al. 1980. Operating Experience with Construction Materials for Wet Flue Gas Scrubbers. Combustion. pp. 23-36. July 1980.

PROCESS CONTROL AND INSTRUMENTATION

Gruenberg, N. R. 1979. Instrumentation and Control for Double Loop Limestone Scrubbers. Power Engineering, 83(6)72-75.

Jahnke, J. A., and G. J. Aldina. 1979. Handbook for Continuous Air Pollution Source Monitoring Systems. EPA-625/6-79-005. NTIS No. PB-300 930.

Jones, D. G., A. V. Slack, and K. S. Campbell. 1978. Lime/Limestone Scrubber Operation and Control Study. EPRI-RP 630-2.

Laseke, B. A., Jr. 1979a. Survey of Flue Gas Desulfurization Systems: Sherburne County Generating Plant, Northern States Power Co. EPA-600/7-79-199d. NTIS No. PB-80-126287.

Laseke, B. A., Jr. 1979b. Survey of Flue Gas Desulfurization Systems: Lawrence Energy Center, Kansas Power and Light Co. EPA-600/7-79-199b. NTIS No. PB 80-125628.

Ostroff, N. 1981. Private communication to PEDCo Environmental, Inc. on Peabody's simple feedback pH control system.

Ung, C., A. Acciani, and R. Maddalone, 1979. The Use of pH and Chloride Electrodes for the Automatic Control of FGD Systems. EPA-600/2-79-202. NTIS No. PB 80-138464.

Section 4

Procurement of the FGD System

This section deals primarily with procurement of the FGD system, including the basic procedures to be followed in preparing the purchase documents and evaluating the respondents proposals. Guidelines are given for management of the total project effort following award of contract so as to ensure that the contract is fulfilled, i.e., management of the activities of vendors, A/E consultant, and utility staff in the installation, startup, and testing of the system. The scope of the section thus encompasses the major project activities from the start of procurement to the point of system operation. The sequence is depicted in Figure 4-1, and the text discussion also follows that sequence.

Because each utility follows preferred or regulated procedures in administration of purchasing activities, the information presented here is intended to supplement rather than to replace those practices. Although there may be considerable variation in the format and nomenclature of documents generated by individual purchasers, the overall procedures described here are considered typical of those used in successful procurement of a limestone FGD system.

COMPETITIVE BIDDING

The competitive bidding process has become routine in business practice. So routine, in fact, that the purchasers may lose sight of the real objective, which is to obtain the best possible product from a qualified bidder at the lowest evaluated cost. Achieving this objective may be easy in theory but is sometimes difficult in practice.

The key to successful procurement by competitive bidding lies in obtaining comparable proposals. For proposals from competitive bidders to be truly comparable, each must be prepared against a baseline document, which is the purchase specification.

THE PURCHASE SPECIFICATION

As the medium through which the utility communicates with the prospec-

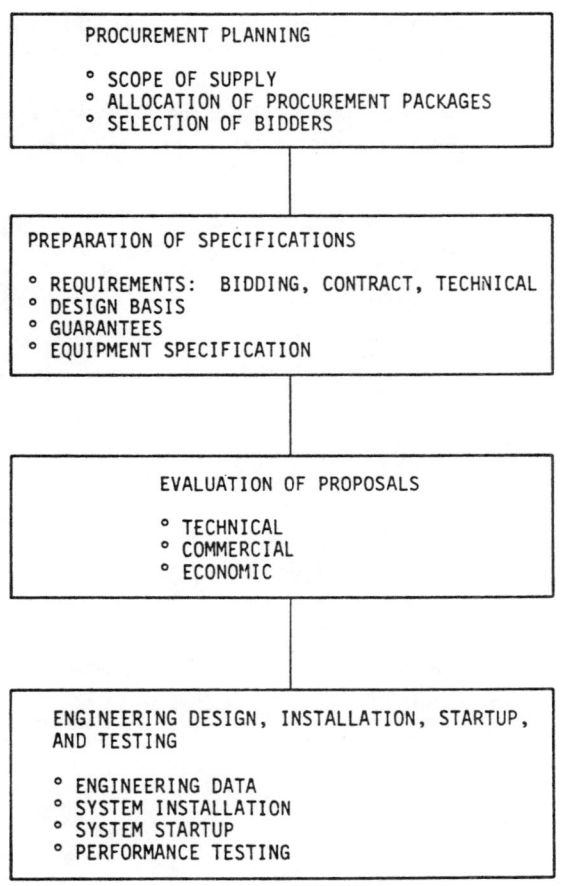

Figure 4-1. FGD system procurement sequence.

tive bidders, the purchase specification should clearly define the requirements to which the bidders will respond. The requirements are of three basic kinds:

1. Bidding requirements, which provide guidance and instructions for preparation and submission of the proposal. Typically, the bidding requirements include detailed instructions to bidders, pricing forms, and specification of data to be submitted with the proposal.

2. Contract requirements, which consist of contract forms and contract regulations. Because each utility probably follows preferred contract procedures, including the forms and regulations, the discussion in this section is limited to the significance of these documents in the purchase specification.

3. Technical requirements, which identify the design basis, the

performance requirements, and the guarantees for the FGD system, and present the detailed equipment specifications and erection requirements.

These three basic components of the purchase specification are depicted in Figure 4-2 and are discussed in detail later in this section.

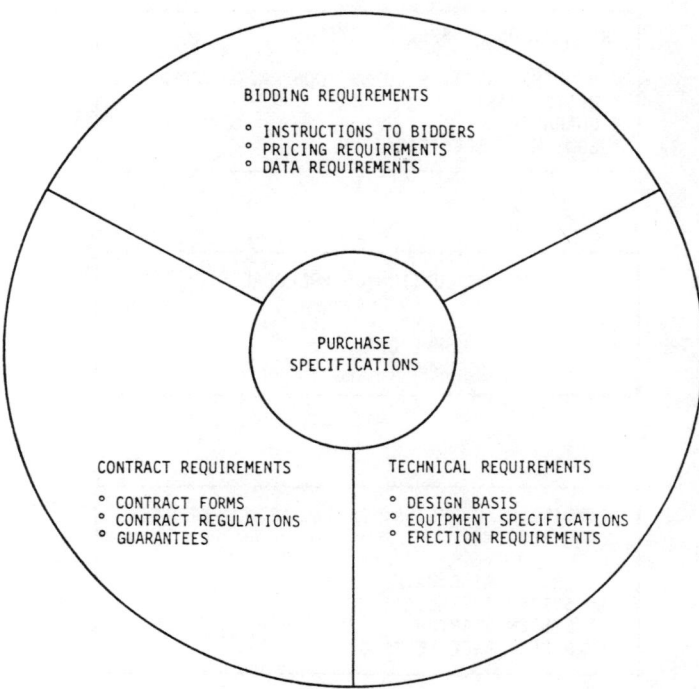

Figure 4-2. Primary components of purchase specifications.

PROCUREMENT PLANNING

Before the project team prepares specifications for purchase of a limestone FGD system, they must make certain decisions regarding the detailed system design, scope of supply, procurement package breakdown, and selection of bidders. As discussed earlier, preliminary engineering studies must be made to determine the conceptual system design, including the most advantageous scrubber type, additive type, and methods for processing, transportation, and disposal of the solid waste. These fundamental decisions must be made prior to requesting proposals.

Scope of Supply

The scope of supply for a limestone FGD system encompasses a variety of equipment. The following list of items to be supplied is based on the primary functions of specific subsystems of the FGD system:

- Scrubber modules

    Presaturators
    Scrubbers
    Scrubber purge system

- Limestone receiving, conveying, and storage subsystem

    Hoppers
    Conveyors
    Feed bins

- Limestone slurry preparation and storage subsystem

    Feeders
    Ball mills
    Storage tanks
    Mixers
    Classification system

- Liquid flow subsystem components

    Scrubber recirculation pumps
    Other pumps
    Piping and piping support
    Valves
    Tanks
    Agitators
    Liquid hydrocyclones

- Gas flow subsystem components

    Booster fans
    Ductwork and support
    Dampers - isolation and control
    Expansion joints
    Hoppers
    Mist eliminators
    Flue gas reheaters
    Soot blowers
    Test ports
    Insulation

- Sludge thickening subsystem

    Thickener
    Flocculant feed system
    Thickener overflow tank and pumps

- Sludge dewatering subsystem

    Filter or centrifuge
    Filtrate/centrate return system

- Sludge stabilization/fixation subsystem

    Lime feeding
    Flyash feeding
    Blending

Precure
Transportation

○ Instrumentation and control subsystem

In addition to the foregoing subsystem equipment items, the scope of supply includes

- ○ Foundations, structural support steel, and access provisions
- ○ Buildings to enclose equipment
- ○ Electrical distribution system
- ○ Model and model tests
- ○ System installation
- ○ System startup
- ○ Performance testing

Allocation of Procurement Packages

The project team must allocate these items of equipment and materials into logical procurement packages. The allocation depends on the utility's plant design approach and on the capability of the engineering staff in scrubber system design. The focus on capability logically carries over to allocation of the procurement packages to individual suppliers; that is, the utility project staff must become aware of the specific qualifications of each potential supplier.

Basing the allocation solely on function may not be the most economical or manageable procedure. Depending on the available resources and personnel, any of the following approaches may be used to produce a workable set of procurement packages. None of these, however, is considered optimum for all situations.

One approach is to purchase the entire FGD system from a single supplier, a "turnkey" or "design-construct" firm that would be responsible for all equipment, materials, and work required to produce a complete operational system. Essentially the opposite approach is to perform extensive detailed engineering design, purchase the components, and construct the system, maintaining primary control over the entire effort.

A third approach, and probably the most common, is to group entire subsystems or major elements of the subsystems into several major procurement packages based on the technical expertise required. This procedure places responsibility for the design of crucial components with suppliers who are qualified by experience and technical competence. Under this plan the utility can purchase and install the less critical components in a routine manner.

A typical allocation of procurement packages under the third approach would consist of four packages, as follows:

- Limestone receiving and conveying equipment. These would be purchased from a conveying equipment manufacturer.

- Scrubber modules, mist eliminators, flue gas reheaters and soot blowers (if required), booster fans, limestone slurry preparation and storage subsystem, sludge thickening subsystem and portion of liquid flow subsystem. These would be purchased from a scrubber system supplier (a possible variation would be separate procurement of limestone slurry preparation and/or sludge thickening subsystems).

- Sludge dewatering subsystem, sludge stabilization and/or fixation subsystem, and portion of liquid flow subsystems. These would be purchased from a supplier with established technical expertise in this specialty.

- Ductwork, dampers, and expansion joints. These would be purchased from a scrubber system supplier or acquired under a separate contract for ductwork or structural steel.

Obviously, other breakdowns of the systems for procurement are possible.

The procurement package allocation described above does not encompass all items needed to complete the FGD system. Additional services and work to be procured include the following:

- Site preparation and underground utilities
- Electrical construction
- Painting (other than special coatings and linings)
- Testing
- Underground piping between subsystems (if not provided by the system suppliers)

This work is usually performed under construction contracts.

Selection of Bidders

Selection of qualified bidders is vital to the competitive bidding process. Selection of qualified firms to receive a request for proposal is typically accomplished as follows:

1. Prepare a list of potential bidders based on FGD surveys (Smith et al. 1980b).

2. Request that each bidder submit a pre-bid qualifications statement, which should include related experience, a customers list, organization and manpower capabilities, and financial status.

3. Verify the experience and reputation of each bidder by inspecting the bidder's fabrication facilities and by talking with utility personnel involved with operating FGD systems.

4. Evaluate the qualifications statements and establish the list of qualified bidders.

The qualified bidders will receive the purchase specifications and will be asked to submit proposals for evaluation.

The following are limestone FGD system suppliers whose units are currently in service in domestic utility plants (Smith et al. 1980b).

Babcock & Wilcox Co.

Chemico Air Pollution Control

Combustion Engineering, Inc.

Peabody Process Systems

Pullman Kellogg

Research-Cottrell

Riley Stoker/Environeering

UOP, Air Correction Division

PREPARATION OF SPECIFICATIONS

This subsection describes a typical purchase specification document. The guidelines given here are intended to supplement a utility's normal procedures for preparing specifications.

Prospective suppliers of the FGD system must be given specific information so that they can submit proposals that are cost-effective and responsive to a utility's needs. The bid requests must be specific and detailed, so that proposals received from the various vendors are of similar content and scope and thus are readily comparable. The specifications must provide all pertinent available information concerning design of the limestone FGD system. They should also allow for enough flexibility that each bidder can apply his own technology; this could provide both technical and economic benefits for the overall project.

As a minimum, the following items must be transmitted to prospective system suppliers to provide guidance for proposal preparation.

- Scope of supply

- System equipment redundancy and interfaces with other systems

- Site constraints

- Requirements for performance guarantees that ensure compliance with applicable regulations at all operating conditions

- Requirements that establish quality baselines for equipment and materials to reflect the utility's operating and maintenance philosophy

- Description of construction site and restrictions on field activities

Procurement of the FGD System   187

- Schedule and time constraints
- Equipment erection requirements that establish quality standards for the field construction work
- Economic evaluation factors (cost of electricity, water, steam, etc.)
- Fuel, makeup water, and limestone composition

All of the above information must be presented to prospective suppliers in a logical, clear, and concise manner. As an aid in formulation of the total purchase package, the following discussion deals in order with the three major elements described earlier--the bidding, contract, and technical requirements.

Bidding Requirements

As shown in Figure 4-3, general instructions to bidders should state

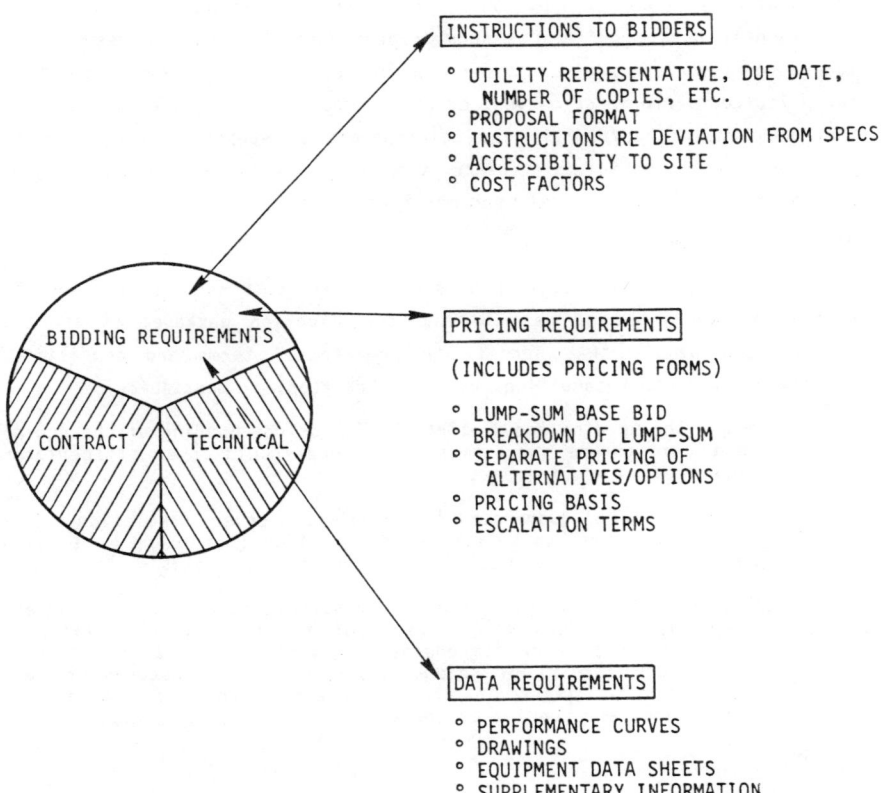

Figure 4-3. Bidding requirements for proposal preparation.

the requirements for proposal preparation, including the following types of instruction: to whom, where, when, and in how many copies to submit the proposal; style and format to be used; the information required; methods of indicating any intended deviations from the specifications; the utility's provisions for site accessibility; procedures for seeking clarification of the specification and requirements; and the technical and cost factors that will be used in proposal evaluation.

Pricing requirements should be set forth clearly, with pricing forms on which the bidder will list cost information. Proposal pricing forms should provide for a breakdown of the lump sum price into material and erection prices. This will permit the application of separate cash flow schedules and/or price adjustment indices during the utility's economic evaluation of proposals. The forms should define the basis of pricing and should require an indication as to whether the prices are firm or are subject to escalation. The forms should also request an exact procedure for calculating the effects of escalation (if applicable) on the contract amount.

Proposal data requirements should specify the information needed from bidders for technical and economic evaluation of proposals. Requested items should include performance curves, drawings, equipment data sheets, and such other supplemental information as descriptions of support and maintenance operations, anticipated maintenance schedule, startup and shutdown procedures, mass balance diagrams, and materials lists.

Contract Requirements

Copies of legal contracts to be signed by the successful bidder and the utility are submitted as part of the specification package, as are the contract regulations that specify the commercial terms and conditions, liabilities, and other conditions by which the supplier must abide.

- Bond--The bonding requirement should include bonding of the equipment performance guarantees. The bond should cover all performance testing requirements.

- Guarantee--The guarantee period should cover completion of a performance test to be conducted some time (for example, 1 year) after initial operation and testing of the equipment.

- Payments--Payment provisions may be structured to provide progressive payments at crucial stages of the project. For example, payments can be made for engineering work, at start of construction, upon receipt of shipments, at various milestones in construction, at start of system operation, and upon passage of performance and compliance tests.

Technical Requirements

A full statement of the utility's technical requirements is the key to eliciting substantive information from bidders in their proposals. As shown in Figure 4-4, the technical requirements may be grouped into three major

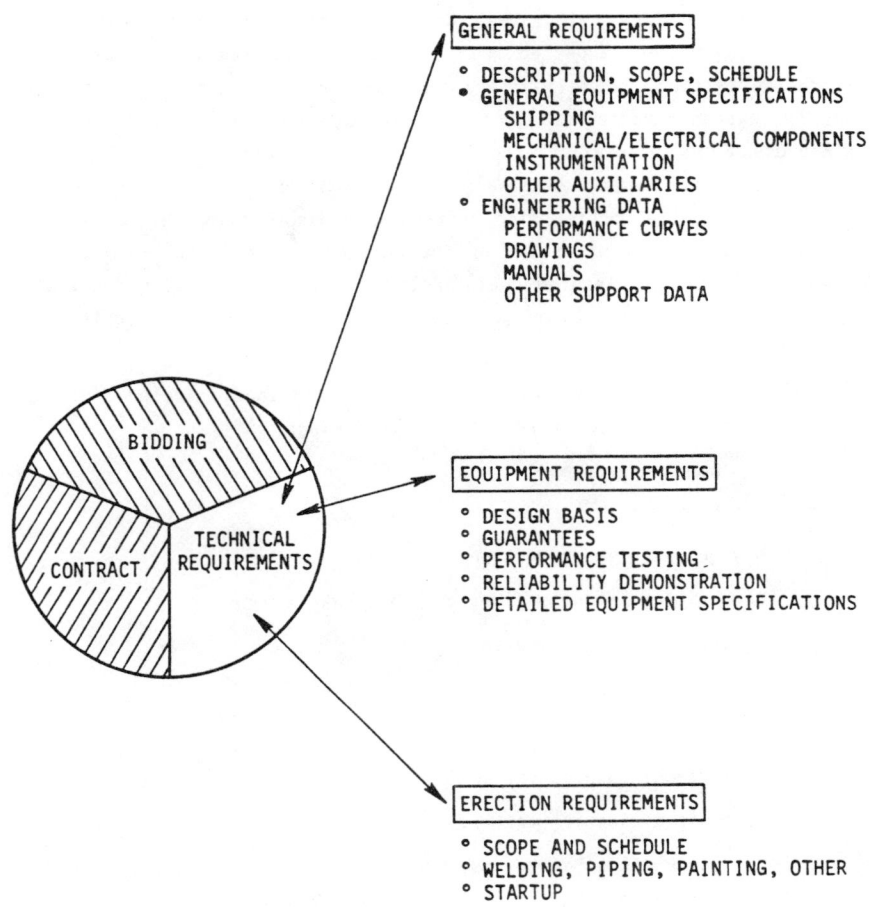

Figure 4-4. Technical specifications consisting of general, equipment, and erection requirements.

categories: general, equipment, and erection requirements.

General requirements pertain to the overall limestone FGD system rather than to specific components. These should begin with a general description of the project and its scope, including the project schedule. Specifications are given regarding shipping procedures and details of the mechanical and electrical components, including instrumentation and other auxiliary items. Instructions should be given also on procedures for submission of detailed engineering data after award of contract; e.g., such items as performance curves, drawings, and manuals that are developed during the construction/installation phase.

Equipment requirements are the heart of the purchase document. It is here that the utility delineates the design basis of the limestone FGD system, the guarantees expected from suppliers, requirements for guaranteed performance testing and reliability demonstration, and detailed specifications for system components. These critical portions of the purchase package are discussed in detail in the following subsections.

Erection requirements outline the scope and schedule of field erection work, describing the construction facilities and specifying the services to be provided by the supplier and by the utility. Specifications are given for welding, piping, cleaning, painting, and other installation functions. Erection activities will culminate in system startup, for which requirements should be stated.

Design Basis. As a minimum, the Design Basis should include information and specifications pertaining to the following:

- Equipment arrangement
- Scrubber type
- Design operating conditions
- Composition of fuel, limestone, and makeup water
- Equipment sizing criteria
- Required redundancy
- Construction criteria
- Performance data curves
- Economic evaluation factors

Although the Project Manager may not be directly involved with development of the detailed information included in the Design Basis portion of the specifications, the Manager is ultimately responsible for results and therefore should review the design information provided for the bidders. As an aid in ensuring that all necessary design basis information is transmitted to the prospective suppliers, a checklist of important items is presented in Figure 4-5.

Guarantees. Vague process guarantees for FGD system equipment should be avoided. Such guarantees sometimes prove to be nonbinding because they are not specific in covering the possible range of operating conditions.

Specific guarantees provide two advantages for the utility. They allow an in-depth comparison of the strength and scope of guarantees presented in the various proposals and also allow predictions of the operating costs to be based on guaranteed performance parameters.

Following is a suggested list of significant guarantees to be obtained for a limestone FGD system:

Design Basis Checklist

Equipment arrangement
    State whether the unit is new or retrofit. If retrofit, specify the following:
        Space limitations (arrangement drawings, elevations, plans, and ductwork schematics).
        Locations and sizes of existing fans, ducts, and stack.
        Materials of construction of existing equipment, if any.
    Instruct bidder to minimize the amount of FGD system ductwork.
Design conditions and expected ranges
    Coal and ash properties
        Proximate and ultimate analyses
        Fly ash alkalinity analysis
    SO_2 inlet loading and emission limitations
        SO_2 content of total flue gas stream
        Minimum design SO_2 removal efficiency
        Guaranteed maximum allowable SO_2 content
          of the effluent gas stream
    Boiler characteristics
        Boiler type
        Coal burn rate
        Heat input rate
        Excess air in flue gas
    Flue gas conditions
        Unit generating capacity
        Flue gas flow rate (lb), temperature, and pressure
        Emergency operating conditions
        Variability in gas flow and temperature
        Bypass limitations
        Reheat requirement and mode
    Particulate control
        Method of particulate control
        Inlet and outlet particulate loading
        Particulate emission limitation and opacity limitation
        Expected maximum particulate content in the inlet flue gas stream
        Guaranteed maximum allowable particulate content of effluent gas stream
        Possibility of operating at reduced particulate control efficiency
    Limestone supply
        Limestone analysis
        Quantity required
        Limestone grind
    Makeup water supply
        Source
        Water analysis
        Allowable water discharge under local regulations
    Water disposal
        Proposed disposal site
        Desired quality of final product (solids content,
          pH, leaching characteristics, impact strength)
Equipment sizing
    Minimum number of modules acceptable
    Spare module requirements
Equipment construction criteria
    Scrubber design pressure
        Maximum positive
        Maximum vacuum
    Maximum flue gas temperature at scrubber inlet   (continued)

Figure 4-5. Checklist for preparation of system design basis to bidders.

Figure 4-5 (continued)

```
        Makeup water pressure
        Seismic criteria
        Location (indoor and outdoor)
        Wind and snow loads (if applicable)
        Grade elevation
        Barometric pressure
        Ambient temperature
            Minimum
            Maximum
        Indoor temperature
            Minimum
            Maximum
    Performance data and curves requested
        Mass balance diagrams (showing flow rates, pressures and temperatures of
            flue gas, makeup water, additive, slurry, sludge, chemicals, etc.)
            At maximum continuous capacity of steam operator
            At expected average operating capacity
            At reduced conditions
        Scrubber performance curves
            Pressure loss through scrubber versus inlet gas flow
            Water droplet content in flue gas leaving mist eliminator versus load
            $SO_2$ removal efficiency versus $SO_2$ concentration and gas flow
        Pump characteristic curves
            Head
            Power requirements
            Efficiency
            Net positive suction head required versus capacity
        Fan performance curves
            Static pressure
            Power requirements
            Fan efficiency
            System resistance versus capacity
```

o SO_2 removal

o Particulate outlet loading

o Mist in the outlet gas stream

o Power consumption

o Reheat energy consumption

o Limestone consumption

o Water consumption

o Waste streams

o Turndown ratio and rate of unit load change

o System availability*

---

*Availability is defined as the number of hours the FGD system is available (whether operated or not) divided by the number of hours in the period, expressed as a percentage. Operation of the FGD system is often largely outside of the system supplier's control. For this reason, flat guarantees of availability are rare; an availability guarantee usually contains limitations on operating conditions and other factors.

A checklist for specifications requesting these guarantees is provided in Figure 4-6.

Figure 4-6. Checklist for specification of guarantees required from prospective bidders.

Guarantee Checklist

General
    Specify EPA requirements for test port locations.
    Specify sampling procedure.
    Specify analysis procedure.
    Specify data reporting procedure.
    Specify assignment of financial responsibility for sampling and analysis.
SO_2 removal
    Request a guarantee for SO_2 removal over entire range of scrubber design operating conditions.
    Specify EPA test procedures.
    State all conditions that require compliance testing.
Particulate outlet loading
    Specify that quantity of particulate leaving scrubber system is not to exceed the quantity entering sscrubber system.
Mist in the outlet gas stream
    Request guarantee for maximum quantity of entrained water droplets leaving the mist eliminators. Since this is a difficult guarantee to verify, adequate design data should be requested for use in technical evaluation of the design.
Power consumption
    Request guarantee for FGD system power consumption on a 24-hour time-averaged basis.
Reheat energy consumption
    Request guarantee for maximum energy consumption.
    Specify location and methods of energy and gas flow measurement; specify test interval.
Limestone consumption
    Request guarantee for lb limestone consumed per lb SO_2 removed.
    Specify test method for limestone feed rate.
Water consumption
    Request guarantee for closed-loop operation: net amount of water consumed.
Waste streams
    Specify acceptable waste streams (whether plant will use landfill, settling pond, or other).
    Request guarantee for limits of wastewater quality.
    Specify the sampling and analytical methods and level of boiler operation at which waste streams will be measured.
Turndown ratio
    Request guarantee for ratio of maximum flue gas flow rate to minimum flue gas flow rate.
    The ratio should be consistent with the lowest expected steam generator load at which power can be produced efficiently.
    Identified by system as well as individual scrubber module.
    Request that bidders state FGD system lag time as boiler load increases and whether operating conditions increase in stepwise or continuous manner.
Availability guarantee
    Request demonstration of operation for a specified time period (e.g., 60 days).
    Request demonstration of operation at rated capacity and minimum load capability of FGD system.

Detailed Equipment Specifications

This subsection identifies some of the significant factors to be considered when specifying the components of the FGD system. The equipment specifications should be organized so as to list together the requirements for similar items insofar as is practicable. For example, one section should cover all pump requirements, another section all piping requirements, and so on for tanks, valves, agitators, ductwork, and similar classes of equipment.

The objective of the specifications is to control the quality of equipment and services without limiting the supplier's freedom to apply his expertise. Specifications for systems or subsystems must be performance-oriented. Where a component must interface with another system or subsystem, detailed specifications are needed to ensure that the component does not prevent the interfacing items from achieving the guaranteed performance.

Detailed information is given in Section 3 regarding the functions of FGD system equipment and the preferred materials of construction. That information should be considered as the technical basis for formulation of the equipment specifications. The discussion that follows considers the major equipment items and focuses on the means of transmitting the required technical information to the prospective suppliers. Guideline sheets are given for several of the major components, indicating what the utility should specify and what information the bidder is asked to provide in the proposal.

Scrubber. The chief factors to be considered in specifying requirements for the scrubber include the following:

- SO_2 removal
- Acceptable scrubber types
- Pressure drop
- Scaling and plugging
- Corrosion/erosion
- Maintenance

Guidelines for specifying the scrubber requirements are given in Table 4-1.

Provisions that simplify maintenance must be specified because they are generally extra cost items that may not be included in competitive bidding. Repair of coatings or linings, for example, may be a difficult and time-consuming process. Thus the use of highly alloyed stainless steels, though initially more expensive, may be justified in some applications because it may reduce maintenance costs and the number of unplanned shut-

TABLE 4-1. SPECIFICATION GUIDELINES: SCRUBBERS

Utility specifies the following:

    SO_2 removal requirement
    Acceptable scrubber types
    Acceptable materials of construction
    Acceptable coatings or linings
    Acceptable types of packing materials for mobile beds
    Nozzle materials
    Maintenance provisions

Bidder provides this information:

    Configuration and overall dimensions
    Inlet and outlet dimensions and other internal dimensions as required to define cross-sectional areas of various stages
    Materials of construction and thicknesses
    Lining or coating types and manufacturers
    Packing type, size, and material
    Nozzle locations, size, type, and material
    L/G ratio
    Gas pressure drop across scrubber
    Estimated weights empty and for normal operating and flooded conditions
    Detailed maintenance provisions

downs. It is essential that the supplier provide for rapid and easy access to the scrubber for cleanout and repairs. There should also be a drain system that allows complete drainage of the scrubber for inspection and maintenance. Provisions must be made for removal of deposits from the scrubber interior, including ready access as well as proper location of cleanout doors.

    Mist Eliminator. The following factors must be considered in specifying FGD system mist eliminators:

- Droplet and particulate collection efficiencies
- Configuration, bulk separation, stages, spacing, and geometry of separation devices
- Corrosion/erosion
- Pressure drop
- Structural integrity
- Wash system

Spray pressure
Top and bottom wash flow rates
Ratio of fresh to return water
Number of sprays per unit area
Intermittent and sequential washing frequencies
Piping layout

- Maintenance features

Associated factors to consider are the scrubber system design and operating conditions, system construction, scrubbing medium, solids content of the slurry, sulfur content of the coal, and chloride content of the water and coal.

Collection efficiency is implicitly specified when allowable particulate emissions are stipulated as part of the design basis requirements. The scrubbers must not increase the particulate load. Specifying a percentage elimination efficiency probably is impractical because compliance cannot be readily determined. Attempts to design to a specified efficiency may compromise reliability by a reduction in washability of the mist eliminator, which could lead to plugging and higher pressure drops. The supplier should provide enough information on the proposed design to permit comparison with units now in operation.

Where a utility wishes to install vertical mist eliminators (horizontal flue gas flow), the specifications must clearly so stipulate. A vertical installation probably will increase the initial capital expenditure and may require more space for ductwork and the scrubber module.

<u>Flue Gas Reheaters</u>. The following factors must be considered in specifying FGD system reheating equipment:

- Design temperature increment and allowable range
- Type of reheat strategy
- Energy source or heating medium
- Materials of construction (for in-line reheat)
- Soot blowing requirements (for in-line reheat)

<u>Soot Blowers</u>. Soot blowers are required for removing deposits from ductwork downstream of the scrubber and from the reheater. The following factors must be considered in specifying FGD system soot blowing equipment.

- Choice of cleaning medium (compressed air or steam)
- Compressor type and receiver capacity, if applicable

Choice of the cleaning medium is determined by economic considerations because compressed air and steam offer comparable cleaning service.

<u>Limestone Receiving and Conveying Equipment</u>. The following factors should be considered in specifying limestone receiving and conveying equipment:

○   Primary and backup methods of transporting limestone to the plantsite

○   Relative locations of receiving station, storage piles, and slurry preparation area feedbins

Mechanical conveying systems are used to move limestone. Specification guidelines are given in Table 4-2.

TABLE 4-2. SPECIFICATION GUIDELINES: LIMESTONE RECEIVING AND CONVEYING EQUIPMENT

Utility specifies the following:

    Limestone type, grindability, size range, and analysis

    Conveying rate required

    Conveying distance, elevation change, relative locations

    Type of conveying system

    Emission limits for fugitive emissions from the conveying system; type of dust collector

Bidder provides this information:

    Configuration and overall dimensions of major system components

    Manufacturer and model numbers of system components

    Construction details for major mechanical equipment items such as conveyors, feeders, and dust collectors

    Power requirements

Feedbins. The following factors should be considered in specifying limestone feed bins.

○   Space available for storage

○   Configuration of the bins

○   Feedbin capacity considered necessary to permit routine daily operating procedures

Specification guidelines for feed bins are given in Table 4-3.

Limestone Feeder. The following factors should be considered in specifying limestone feeding equipment:

○   Type of feed measurement best suited for the system installation: volumetric or gravimetric

○   Required accuracy of measurement of limestone usage

○   Manual or semiautomatic control

○   Type of conveying mechanism (belt conveyors, screw conveyors, oscillating hoppers, or vibrating hoppers)

○   Provisions for shutdown and necessity for cleaning and maintenance

TABLE 4-3. SPECIFICATION GUIDELINES: LIMESTONE FEEDBINS

Utility specifies the following:

    Required capacity of feedbins in terms of time period of operation or other criteria

    Materials of construction and minimum thicknesses

    Configuration criteria such as maximum diameter or maximum height, rectangular cross-section, offset hopper construction

Bidder provides this information:

    Configuration and overall dimensions of feedbins

    Materials and thicknesses

Specification guidelines are given in Table 4-4.

TABLE 4-4. SPECIFICATION GUIDELINES: LIMESTONE FEEDER

Utility specifies the following:

    Limestone type and size range

    Acceptable type(s) of feeder

    Accuracy requirements for measurement of limestone usage

Bidder provides this information:

    Configuration and overall dimensions of equipment

    Manufacturer, model, and construction features of equipment

    Power requirements

    Control systems

    Maintenance provisions

    Feeder capacity in terms of excess capacity over the anticipated limestone demand rate or other criteria

---

    <u>Limestone Ball Mill</u>. Since the properties of limestone slurry affect the removal efficiency and economics of the FGD system, specification of the ball mills is important to the operation of a scrubbing system. The following factors should be considered in specifying limestone ball mills:

- Limestone type
- Ball mill type
- Grinding media
- Mean particle size of product
- System reliability

Guidelines for specification of ball mills are given in Table 4-5.

TABLE 4-5. SPECIFICATION GUIDELINES: LIMESTONE BALL MILL

Utility specifies the following:

    Limestone type and size range
    Acceptable types of ball mills
    Hours per day of operation

Bidder provides this information:

    Configuration and overall dimensions of equipment
    Manufacturer, model, and construction features of equipment
    Product particle size distribution
    Power requirements
    Reliability of system
    Ball mill capacity in terms of excess capacity over the anticipated limestone slurry demand rate

---

Thickener and Flocculant Feed. The thickener plays an important role in water balance, sludge characteristics, and sometimes in chemical reactions. The following factors must be considered:

- Corrosion
- Solids concentration
- Clarification
- Recirculation
- Solids removal

Flocculant is usually added to the thickener by a reciprocating pump, which can be of a piston or diaphragm type and is invariably associated with a check valve in the discharge piping.

Vacuum Filter. The following factors should be considered in specifying a vacuum filter (Knight et al. 1980):

- Materials of construction
- Drying time
- Barometric legs
- Filter medium
- Final solids content

A 10-second drying time should be specified to maximize the cake solids content of the filter cake without causing sludge cracking. To keep filtrate from going through the vacuum pump, a barometric leg should be installed to protect the pump by trapping liquid before the suction of the vacuum pump. The filter medium should be cheap, durable, and noncorrosive, and it should allow easy cake discharge.

Centrifuges. The following factors should be considered in specifying a centrifuge for limestone slurry applications.

- Materials of construction
- Rotational speed
- Conveyor
- Final solids content

All materials that contact liquid in the centrifuge should be made of corrosion-resistant materials. The tips of the conveyor should be made of tungsten carbide to reduce abrasive wear.

Rotational speed should be midrange, 3000 rpm or less, to gain some of the benefits of high-speed clarification while preventing excessive abrasion and difficult solids discharge. If centrifuge speeds are too high, the conveyor and the bowl will lock. The screw conveyor within the bowl should turn at the minimum speed required to remove solids without causing excessive turbulence. Since the scrubber caking rates may vary, a variable-speed conveyor should be specified.

Mix Tank or Pug Mill. The following factors should be considered in specifying a mix tank or a pug mill (Knight et al. 1980):

- Corrosion/erosion
- Tank size
- Quantity of material
- Degree of blending
- Additive

In the event of low pH swings, the equipment could be stopped to prevent corrosion. Erosion is the major specification consideration. To achieve proper mixing of the slurry, high torque agitators are needed. The rapid movement of the slurry and fixation agents against the tank walls heightens abrasive action. For reduced maintenance, the tank walls and bottom should be rubber-lined.

Fixated Sludge Conveying System. The chief factors to be considered in specifying sludge conveying equipment are the type of system (belt or screw conveyor) and the materials of construction.

For belt conveying, the weight per cubic foot of material to be handled should be determined accurately and specified in an "as-handled" condition, rather than taken from published information. Belts can be quickly damaged by high temperatures; a high-priced belt may prove most economical in the long run. The elastomers available for belt construction include Neoprene, Teflon or Teflon-coated rubber, Buna-N rubber, and vinyl rubbers.

Capacities of screw conveyors are generally restricted to about 10,000 ft^3/h. Serviceable materials range from cast iron to stainless steel. Additional information is given by Knight et al. (1980).

Additional Equipment Items. Many of the other equipment items to be specified for the limestone FGD system are common to most major engineering installations--pumps, piping, ductwork, and the like. In specification of these items, it is most important to define the conditions of service so that the potential supplier has a firm grasp of the service limits to which the equipment will be exposed. For a limestone FGD system, a chief objective is to provide designs and materials that will withstand the corrosive/erosive action of gases or slurries. The supplier must be informed of the range of pressures and temperature encountered in normal operation and also those that could occur during emergency conditions; the probable duration of excursion conditions is important also. These conditions must be clearly specified in the design basis presentation. Some considerations for specification of these additional equipment items are discussed briefly here; detailed information on design and materials is given in Section 3.

Specification of pumps and piping depends largely upon the type of service. Equipment that handles slurry transfer and makeup must withstand erosive conditions but need not be acid-resistant. Slurry recirculation and discharge service is the most severe; service involving reclaimed water and fresh water makeup generally is not severe. All pumps, however, should have capability for remote start/stop and manual local start/stop.

Valves are used for isolation and control functions within the system, but because of potential plugging and mechanical problems their number should be limited. Rubber-lined valves are the most common, although many stainless steel valves are used; functional problems are seldom related to materials.

As with other equipment, specifications for tanks are determined by the service function. The EHT must allow adequate retention time (typically 8 to 10 minutes) for completion of the slurry reaction with the SO_2. Specifications for the thickener overflow tank should allow for some system swings in the demand for recycle water.

A principal factor in specification of ductwork is compatability among all parts of the system, regardless of supplier. The maximum permissible gas velocity through the ductwork should be specified; typically this is 3300 to 3600 ft/min. The minimum permissible thickness of ductwork materials in a given application must be consistent with that of the remainder of the unit. Typical ductwork thickness ranges from 3/16 to 1/4 inch. Acceptable materials for ductwork in a limestone FGD system are discussed in Section 3 (Materials of Construction).

Specifications for dampers should emphasize the importance of mechanical design provisions that will minimize fly ash deposition. Mechanical problems with dampers caused by deposition of solids have outweighed any problems attributable to materials.

In specification of expansion joints the utility should provide information with which the supplier can determine the magnitude and types of movements that the joints must accommodate. Maximum and minimum ambient temperatures are important, along with operating and excursion temperatures.

The chief factors to be considered in specifying system instrumentation are the overall control philosphy, the parameters to be measured or used for control functions, the acceptable types of instrumentation for each service, and the degree of redundancy required for reliable operation.

EVALUATION OF PROPOSALS

The proposal evaluation process is considered in terms of three categories: technical, commercial, and economic evaluation. A detailed example evaluation of a proposal for an FGD system on a hypothetical 500-MW powerplant is given by Smith et al. (1980a).

Technical Evaluation

The technical portion of each supplier's proposal should be reviewed thoroughly, mainly to determine compliance with the specification. When noncompliance is indicated, the supplier must be contacted for clarification.

The important technical information supplied by each manufacturer on the equipment data sheets should be summarized in tabular form for ease of comparison. Table 4-6 is an example of a technical data summary sheet.

Most of the data furnished are used for determining compliance with the specification. Each proposal is compared with the specifications, and price additions or credits are applied where the systems are deficient or overdesigned. Examples of deviation from specifications would be changes in materials of construction or failure to supply required dampers. Where the proposal is otherwise adequate, the reason for such deviations is sometimes ascertained by contact with the supplier.

The technical expertise and construction experience of the prospective suppliers should be evaluated and compared. Obvious indexes of the credibility of potential system suppliers are the numbers of limestone FGD systems each has placed in operation, those under construction, and those for which contracts have been awarded.

TABLE 4-6. TECHNICAL DATA SUMMARY SHEET

	Bidders:			
	A	B	C	D
Scrubber ΔP, in. H_2O				
Through scrubbing stage	___	___	___	___
Through total all scrubber elements	___	___	___	___
Superficial velocity, ft/s				
Through scrubber	___	___	___	___
Through mist eliminator	___	___	___	___
Water droplet carryover past mist eliminator, lb/h	___	___	___	___
Overall system SO_2 removal, percent	___	___	___	___
Slurry recycle system				
Liquid to gas ratio-scrubber,[a] gal/min per 1000 acfm	___	___	___	___
Slurry recycle-scrubber, gal/min	___	___	___	___
Solids in recycle slurry, percent	___	___	___	___
Scrubber tank retention time, min	___	___	___	___
Additive system				
Limestone additive, lb/h	___	___	___	___
Limestone grind, maximum particle size	___	___	___	___
Limestone purity, percent	___	___	___	___
Stoichiometric feed rate (based on SO_2 removal), ratio	___	___	___	___
Solids in additive slurry, percent	___	___	___	___
Additive slurry flow rate, gal/min	___	___	___	___
Freshwater requirements, gal/min				
Additive system freshwater	___	___	___	___
Scrubber makeup freshwater	___	___	___	___
Vacuum filter freshwater	___	___	___	___
Mist eliminator wash freshwater	___	___	___	___
Pump seal freshwater	___	___	___	___
Total freshwater	___	___	___	___
Sludge disposal system				
Waste slurry from scrubbers, gal/min	___	___	___	___
Solids in waste slurry, wt. percent	___	___	___	___
Underflow from thickener, gal/min	___	___	___	___
Solids in thickener underflow, wt. percent	___	___	___	___
Filter cake production, dry tons/h	___	___	___	___
Solids in filter cake, wt. percent	___	___	___	___
Structural factors				
Overall dimensions of each module, ft	___	___	___	___
Shell material	___	___	___	___
Mist eliminator material	___	___	___	___
Temperature increase after reheat, °F (ΔT)	___	___	___	___
Heating tube material	___	___	___	___
Distance between uppermost spray bank and bottom of mist eliminator, ft	___	___	___	___
Distance between top of final stage of mist eliminators and the bottom of the reheater, ft	___	___	___	___

[a] Based on saturated flue gas conditions at scrubber outlet.

Commercial Evaluation

Each bidder may take numerous exceptions to the commercial terms of the specifications, such as terms of payment and escalation factors. Any exceptions should be clearly stated in terms compatible with those specified. For comparison purposes, a summary of significant commercial terms specified and those offered by each bidder should be tabulated. A commercial data summary sheet is shown in Table 4-7.

TABLE 4-7. COMMERCIAL DATA SUMMARY SHEET

	Bidders:			
	A	B	C	D
Payment Terms -- Material and Shop Labor				
Percentage paid each month for work performed	___	___	___	___
Percentage paid after completion of the initial performance guarantee test	___	___	___	___
Percentage paid upon satisfactory completion of the 1-year performance guarantee test	___	___	___	___
Payment Terms - Erection				
Percentage paid upon completion of work	___	___	___	___
Percentage paid upon completion of performance guarantee tests	___	___	___	___
Escalation				
Portion of material, erection, and labor subject to escalation	___	___	___	___
Time period of escalation	___	___	___	___
Value of each index or the date of the base value of each index	___	___	___	___

In addition to the key items shown in the summary sheet, the commercial evaluation should consider in a qualitative way the extent to which each bidder defines responsibilities for various aspects of the contracted work. Performance guarantees, for example, should not only stipulate the guarantee period and completion dates for performance tests, but also should cover contingencies, e.g., procedures to be followed if the initial performance tests are unsatisfactory, and at whose expense additional tests would be run. Details of shipping procedures should be clear-cut: who pays transportation charges, what is the F.O.B. site, and so on. Project staffing proposed by the supplier should be identified by function if not by name: technical service rep, erection supervisor, project engineers.

Delays in the project schedule, whether caused by the utility, the supplier, or other factors, can lead to substantial expense; likewise, as the project develops, certain work not specified in the contract may become desirable or necessary. Although both purchaser and supplier may be reluctant to make firm commitments regarding such unknown quantities, each supplier should address such eventualities as a basis for commercial evaluation of his proposal.

Economic Evaluation

Economics is a key element in the evaluation of bids. Various capital and operating costs must be assessed. One proposal may indicate the lowest operating costs and another, the lowest capital costs. A systematic approach for evaluating the overall economic impact of each proposal is a necessity. The FGD suppliers should be provided with a list of economic evaluation factors.

Capital Investment. So that the capital investment required for each FGD system proposal can be compared on an equal basis, the as-received proposal prices for equipment, materials, and erection are adjusted by adding the following:

- Technical cost adjustments
- Balance of plant costs
- Commercial cost adjustments

Technical cost adjustments are price additions or credits applied to a proposal when the systems are technically deficient or overdesigned relative to the intent of the specifications.

Balance of plant costs are equipment and material costs that are outside of the manufacturer's scope of supply but that would be incurred in completing the proposed system; therefore, these costs must be included in a comparison of prices. The following are several examples of balance of plant costs:

- Cost of material and labor for concrete foundations and piling
- Erected cost of structural steel for scrubber module and building support, platforms and stairs, and wall panel and roofing material
- Cost additions or deductions to provide ductwork to selected reference points on the inlet and outlet of the system
- Cost additions for installed electric wiring, conduit, starters, cable trays, circuit breakers, and other miscellaneous electric equipment
- Cost additions for a comparatively larger induced-draft fan as a result of greater system static pressure

The commercial costs calculated for each proposal are based on the supplier's terms of payment, the price adjustment policy, and the economic criteria given in the specifications. Commercial costs include the costs of escalation and interest during construction. The escalation should be based on each bidder's delivery and erection schedule, if one is given, or on an estimated schedule. The escalation factor should incorporate the value of each index or the date of the base value of each index that the bidder proposes to use. The escalation is calculated from the base proposal price plus all differential technical adjustment and balance of plant costs. Interest during construction should be based on the escalated bid price. The interest is calculated from the date of payment to the date of commercial operation.

Annual Operating Costs. Operating costs are levelized over the expected plant life and then capitalized for use in the evaluation. The operating costs include the costs of additive, air, water, electrical demand

and energy (recirculation pumps, ball mills, and miscellaneous pumps), and sludge disposal. The economic criteria used for this evaluation should be clearly stated in the bid specification so that the bidders are fully aware of the economic penalties associated with the operation of their systems. This will enable the bidders to optimize their proposed systems and make key decisions regarding the tradeoffs associated with capital versus operating costs.

Comparisons of operating costs are normally based on the information provided by the bidders in their proposals. Suppliers vary, however, in the degree of conservatism applied in estimating operating cost parameters. Therefore, it is prudent for the utility Project Manager or consulting engineer to carefully review, and if necessary adjust, the indicated operating costs to ensure that they are reasonable.

Total Evaluated Costs. The total evaluated costs of the FGD system should include the total evaluated bid price (technical and commercial cost adjustments), the balance of plant costs, and the differential operating costs.

Selection of Successful Bidder

Selection of the FGD system supplier is based on more than economics. It should be recognized that many factors that contribute to a successful FGD installation can be assigned no specific economic (dollar) value. These are such items as effluent hold tank retention time, carryover of water droplets through the mist eliminator, and overall system layout. Although such factors are difficult to evaluate economically, they should be considered carefully after the economic advantages and disadvantages of each offering are assessed. Selection of the "best" of the proposed FGD installations is also based on system operation and maintenance, on the stated basis of guarantees, and on schedule considerations. Because of these and other factors significant to success of the overall project, the economics are not necessarily controlling but must be factored into a broader judgemental process.

Before a contract is awarded, all of the technical and commercial exceptions indicated by the successful bidder should be clarified and resolved.

INSTALLATION, STARTUP, AND TESTING

Following the award of contracts, the utility project team will work closely with suppliers on detailed engineering design, fabrication, installation, startup, and performance testing--all of these activities being prerequisite to operation of the FGD system. Good lines of communication must be established with the system supplier and the A/E consultant.

The supplier provides engineering information for the design and purchase of interfacing auxiliary equipment. Fabrication and erection must be closely monitored to ensure that the installed equipment conforms with specifications. Performance testing is conducted to demonstrate compliance with applicable flue gas emission regulations and to demonstrate that the equipment meets the performance guarantees. During startup and initial operation of the system, procedures are established for operator training, recordkeeping, and maintenance. Particularly during early operation, system reliability must be monitored to confirm satisfaction of contractual performance requirements. Again, it is emphasized that communication among all parties concerned is essential from preliminary design through operation of the system.

Engineering Data

The supplier should be required to provide as a minimum the following engineering data:

- Complete, detailed mass balance diagrams for maximum capacity and average operating capacity, as well as specified intermediate and lower load conditions. These conditions should include the minimum expected percentage of steam generator capacity. These data would support detailed engineering analysis of such items as minimum acceptable flow velocities in slurry pipelines.

- Scrubber performance curves. These curves should include superficial gas velocity through the scrubber and mist eliminators versus load. They should indicate the recommended points of changeover to increase or decrease the number of modules in operation, L/G versus SO_2 removal efficiency, expected SO_2 removal efficiency versus coal sulfur content, and SO_2 removal efficiency versus boiler load. The supplier should provide a chart relating modules and pumps in service to boiler load and sulfur content.

- Pump characteristic curves for each pump furnished under the specifications. The curves should indicate head, power requirements, efficiency, and net positive suction head required versus capacity.

- Fan performance curves for all fans. The curves should indicate static pressure, power requirements, fan efficiency, and system resistance to gas flow versus capacity.

System Installation

Installation of the FGD system, one of the major power plant systems to be constructed, should be managed as part of the overall power plant construction activities. Written procedures for system installation should be prepared in advance of actual installation and should be used to control the project field organization, project documentation, and lines of communication with the utility and FGD system construction contractor. The procedures should include standard forms for use in control of activities. The following are some of the major requirements for effective management of FGD

system installation at a large utility power plant:
- Resident FGD system management and field engineering personnel
- Interface with other system construction contractors
- Management of startup
- Contract administration of the construction contract and material supply contracts
- Management of inventory control, storage, and maintenance of material and equipment
- Document control to support field management operations
- A quality assurance program to ensure effectiveness and compliance with design specification
- Progress and status reports to utility management

System Startup

The FGD system supplier must be informed of the overall power generation project schedule and of all other interfaces between his work and the work of other contractors on site. A detailed schedule should be developed for FGD system erection and subsystem shakedown. This schedule should be followed from the start of construction activities.

Weekly and monthly progress reports should include information regarding the labor force, weather, conference memorandums, effects of change orders, and other significant information related to progress of the construction. Also essential are inspection of FGD equipment and materials received at the project site, and followup activities such as handling of loss claims and correction of manufacturing errors.

Well-planned startup and shakedown of the major FGD subsystems and components can minimize any adverse impacts of construction errors or design deficiencies. Representatives of the major subsystem equipment suppliers should be present during initial startup of their equipment.

Performance Testing

Sampling and analysis of the flue gas stream are performed to demonstrate compliance with emission standards and with performance guarantees. Emission compliance testing consists of measurements for concentrations of SO_2, particulate, and nitrogen oxides. Testing of the FGD system performance will depend upon the guarantees established in the specifications and in contract negotiations with the equipment vendor. This testing commonly consists of measurements of SO_2 removal efficiency, particulate emission rate and/or removal efficiency, and scrubber additive usage; it may also include such items as flue gas stream pressure loss, water usage, and sludge generation rate.

The purpose of the tests determines to some degree the selection of sampling locations. The EPA provides guidelines for location of gas stream measurements. These should be considered prior to construction of the power generation unit and the FGD system.

A written test protocol should describe clearly the responsibilities of all participants, the operational requirements for the power generation unit and FGD system, and methods to be used in testing and analysis. An unbiased statistical method should be used to reject poor data. The protocol should identify the goals of the test program so that the end uses of the test data are well established. This is especially important in a combination of tests to determine system performance and to demonstrate regulatory compliance.

One person should coordinate all field testing activities. This person can thus coordinate operation of the boiler and FGD system with the testing schedule.

All tests should be performed at steady-state boiler load conditions. Load should be stabilized for at least an hour before testing begins. The FGD system operations should be maintained at constant conditions during the tests, and for gas stream sampling the FGD system operation also should be stabilized for at least 1 hour before the test. For sampling of the liquid/slurry streams, a much longer stabilization period may be required to reach steady-state conditions. Testing should be interrupted during any major upsets of the boiler or FGD system. Testing need not be interrupted because of minor deviations from specified conditions because most of the gas stream measurements are time-averaged.

Sulfur dioxide removal efficiency is determined by simultaneously measuring SO_2 concentrations at the inlet and outlet of the scrubber system. Similarly, particulate removal efficiency is determined by particulate measurements at the inlet and outlet of the particulate removal system, which may also be the SO_2 scrubber. Where a separate particulate removal system is located upstream of the scrubber, particulate measurements at the outlet of the scrubber may be required to ensure that the scrubber is not generating particulate.

Limestone utilization or stoichiometric ratio may be guaranteed at designated emission performance levels, and confirming performance tests are required. Limestone stoichiometry is defined and described fully in Appendix A. The limestone stoichiometric ratio may be calculated from measurements of the limestone feed rate and the SO_2 removal rate. Determination of limestone stoichiometries by performance testing has been found to be only ±10 to 20 percent accurate because of inherent discrepancies in flow measurement of liquid/slurry and gaseous streams and because of the resulting inaccuracy of sampling techniques.

The limestone stoichiometric ratio can also be measured directly by chemical analyses of the molar ratio of total sulfur to calcium in the dry scrubber sludge and by analyses of the ratio of carbonate to calcium in the sludge. Because composition of the solids is time-averaged over long periods and can be determined accurately, chemical analysis of the sludge solids provides the best means of establishing limestone utilization. These chemical analyses also provide a valuable check on the measured SO_2 removal efficiency when used in conjunction with data on limestone feed and SO_2 concentration of the incoming flue gas.

Confirming performance tests of makeup water usage may also be needed when usage is guaranteed at designated emission performance levels. Water usage by the scrubber is determined by measuring selected process stream flow rates and densities and performing a water balance determination. A good water mass balance must satisfy the requirement that the amount of water in the incoming flue gas stream, the limestone slurry feed, and the makeup water streams, together with any accumulation in the scrubber system, equals the amount of water in the output streams, consisting of moisture in the flue gas outlet and in the scrubber sludge.

REFERENCES FOR SECTION 4

Knight, R. G., E. H. Rothfuss, K. D. Yard, and D. M. Golden. 1980. FGD Sludge Disposal Manual, 2d ed. EPRI-CS 1515.

Smith, E. O., W. E. Morgan, J. W. Noland, R. T. Quinlan, J. E. Stresewski, D. O. Swenson, and C. E. Dene. 1980a. Lime FGD System and Sludge Disposal Case Study. EPRI CS-1631.

Smith, M., M. Melia, N. Gregory, and M. Groeber. 1980b. EPA Utility FGD Survey: July-September 1980. NTIS No. PB 81-142655. EPA-600/7-80-029d.

Section 5

Operation and Maintenance

The mark of a successful scrubber installation is reliable operation. Achieving reliability in the chemical processes involved in a limestone scrubber may present new and unfamiliar operating and maintenance concepts for a utility staff. Even though redundancy has been designed into the system, poor operational practices and inadequate maintenance could lead to reduced unit output or could curtail power production entirely. Fortunately, together with the improvements incorporated into scrubber design in recent years, the experience gained by utility personnel in operation and maintenance of limestone FGD units can further enhance system performance and availability.

The roles of operating and maintenance personnel must be clearly delineated, but with enough flexibility to optimize overall system performance. Operators should be able to perform the more basic maintenance so that downtime is minimized. The maintenance staff must be equipped to respond rapidly to component failure. Just as the design and procurement phases of the project focus on system operability and maintainability, so the goal of the operation and maintenance functions is to achieve a high level of scrubber availability for the life of the unit.

This section addresses the operational and maintenance requirements associated with a limestone FGD system, including standard operating practices, routine startup and shutdown, and operating modes for system upset conditions. The discussion deals with the major components of the gaseous and liquid flow paths. The size, duties, and training needs of an operating crew are reviewed.

Maintenance practices are reviewed in detail. Even with a well-executed preventive maintenance program, unscheduled maintenance will be needed. Familiarity of the staff with typical malfunctions, proven troubleshooting techniques, and spare parts requirements will reduce maintenance time and will improve the reliability and availability of the system. The requirements for maintenance personnel, in terms of numbers, duties, experience level, and training are also discussed.

STANDARD OPERATIONS

Planning of system operations begins when the system is in the conceptual design stage and continues through procurement and construction. Since the goal is to achieve a high degree of reliability and availability, the equipment and the controls must be arranged in a manner easily understood by the operating staff.

With stringent requirements on plant emissions, the utility must make a strong commitment to scrubber operations, including adequate staffing. Operators should be assigned specifically and solely to the scrubber system during each shift. Considerations of scrubber operation must be incorporated into the unit's power generation schedules and even into the purchasing of coal.

Many of the first-generation FGD installations were required to control SO_2 from widely varying unit loads (cycling and peak), with different coals (low-sulfur western, high-sulfur eastern, and blends). Often, too, the control systems in the early installations demanded a response beyond the capability of the FGD system. Resulting variations in the reagent feed rate, loss of chemical control, and many chemical and mechanical problems caused numerous forced outages and low reliabilities. Building on experience gained in operation of those first-generation systems, suppliers and designers are now providing better design configurations and construction materials.

Among several general tendencies in recent FGD system designs are an increased degree of flexibility and reliability. Specifically, design trends that increase availability are toward the use of spare modules and spare ancillary components, and away from the use of interdependent systems (i.e., systems in which major unit operations are affected by difficulties in upstream components).

Some of the current difficulties with limestone FGD systems relate to poor operating practices, unnecessarily complex operating procedures, or both. In some cases, although the equipment has been correctly installed, it rapidly deteriorates and breaks down. The user blames the supplier for selling inferior and poorly designed equipment; the supplier blames the user for improper operating practices; and both may well be right. Both can benefit from the operating experience with similar units. The operating characteristics of a limestone FGD scrubber can be established during the initial startup period, which is also a time for finalizing operating procedures and staff training.

Operation at steady-state conditions is the goal of every scrubber designer and operator. The system must be monitored and controlled to ensure proper performance, even at steady state. During periods of changing load or variation of any system parameter, additional monitoring is required. The roles of variables and of components in system operation are discussed in the following sections.

Varying Inlet SO_2 and Boiler Load

The number of scrubber modules in service is governed by the flue gas flow rate. As boiler load is increased, additional modules are placed in service and, conversely, modules are removed from service when boiler load is reduced. With each change in load, the operator must check the scrubber to verify that all in-service modules are operating in a balanced condition.

Although the changes are not as rapid as boiler load changes, system operating parameters may vary with time. For example, the SO_2 concentration in the inlet flue gas may change because of variations in the coal. The scrubber system should be able to accommodate and compensate for such changes. Operator surveillance of system performance is needed, however, to verify proper system response. In suitably designed systems, pumps can be added and removed from service as the SO_2 concentration increases or decreases.

Effective control of the FGD system requires communication between FGD operators and boiler operators. It is important that the FGD system operator be aware of impending load changes. The limestone feed rate should be reset according to these changes, and the operator should not rely solely on the pH controller, which responds slowly.

Verification of Flow Rates

The operating staff should routinely monitor and record readings from all instruments used to measure flow of the different process streams. Deviations from anticipated values can indicate potential problems, either in the scrubbing system or with specific instruments. As the only means of obtaining performance data, the routine monitoring of instrumentation is a principal task of the operator. On the basis of previous experience, the operator may interpret the data or may recognize the need for unscheduled maintenance. The operator should keep in mind that steady-state conditions may well fluctuate with time in a manner that does not affect the overall performance of the system.

The easiest method of verifying liquid flow rates or evaluating pump/nozzle erosion is for an operator to determine the discharge pressure in the scrubber slurry recirculation header with a hand-held pressure gauge. (Permanently mounted pressure gauges frequently plug in slurry service.) The discharge pressure of the recirculation pumps should be determined manually on a periodic basis. An increase in discharge pressure usually indicates plugging of some spray nozzles. A decrease in discharge pressure indicates wear of either the spray nozzle orifices or the pump impellers, or both, in which case maintenance work should be scheduled.

Flow in slurry piping can be checked by touching the pipe. If the piping is cold to the touch at the normal operating temperature of 125° to 130°F, then the line is plugged. It is difficult to verify flow in individual slurry pumps discharging to a common manifold. Where slurry pumps are piped in parallel, an operating pump can be used to backflush a plugged pump. With the plugged pump shut down, and the pump discharge valve open, the pump suction valve should be slowly opened. Pressure from the common discharge manifold then forces slurry backwards through the plugged pump until it spins free in the backward direction. The shaft torque will be the same as in the driving mode so that threaded impellers should not unscrew if this procedure is followed correctly.

Unplugging long runs of slurry lines is difficult. Overnight pressurization of lines from one end is sometimes effective when lines are plugged with slurry but not with scale. This method is especially useful when someone is available throughout each shift to rap on the pipe or otherwise provide the vibration needed to help reslurry thixotropic material. Water from a high-pressure fire protection system can also be used to unplug slurry lines. Dismantling or replacing the piping is usually a last resort.

Surveillance of Scrubber Operations

Visual inspection of the scrubber section and hold tanks can identify scaling, corrosion, or erosion before they impact the generation output of the unit. Visual observation can identify leaks, accumulation of liquid or scale around process piping, or discoloration on the ductwork surface resulting from inadequate or deteriorated lining material. When these checks are coupled with timely maintenance, many costly repairs may be avoided. Thus, routine surveillance of the scrubber system is needed to spot potential problems in time to implement corrective action.

The scrubber module must be designed to withstand the demands imposed on it, chiefly high temperatures and erosive/corrosive environments. The site-specific combination of the projected demands governs the selection of materials of construction. Carbon steel, alloys, and all types of coatings or liners are subject to deterioration or scale formation. System operations must be conducted in accordance with the design. For example, to equalize the effects of erosion, all modules should be subjected to an equal number of operating hours. Combining such operating practices with timely maintenance will maximize the service life of the overall limestone scrubber system. Station outages should be used as an opportunity for maintenance personnel to enter the scrubber modules for a visual inspection.

Mist Eliminators

In limestone FGD scrubbers, mist eliminators have been subject to buildup of slurry solids and chemical scale in the narrow passages with a resulting restriction of gas flow. When scaling occurs in an FGD system, it is usually noticed first in the mist eliminator as an increase in pressure drop. Continued uncontrolled growth could lead to shutdown of the scrubber module. Scaling may be tolerated for as long as 6 months at one installation, whereas at another it may cause problems after 1 week.

Many techniques have been employed to improve mist collection and minimize operational problems. The mist eliminator can be washed with a spray of process makeup water or a mixture of makeup water and thickener overflow water. Most limestone FGD systems are designed with a two-stage mist eliminator, which allows more washing on the first stage. Washing of both sides is recommended; washing may be continuous or intermittent, or both. Intermittent washing with a high liquid flow rate may remove hard scale; continuous washing is necessary to limit scale buildup. An intermittent wash sequence may be based on a number of factors:

1. Time interval and unit load maintained since the last wash.
2. Pressure drop across the mist eliminators.
3. System liquid inventory level (too much washing can cause water balance problems).
4. Experience with a specific mist eliminator design.

In addition to the mist eliminator design features discussed in Section 3, several specific operating procedures affect the performance of mist eliminators. The range of flue gas flow rates over which a module operates is one example. If a scrubber is operated below, or even above, the design L/G range, problems are likely. For these reasons, operations must be conducted within the design ranges.

Successful, long-term operation without mist eliminator plugging generally requires continuous operator surveillance, both to check the differential pressure across the mist eliminator section and to visually inspect the appearance of blade surfaces.

Flue Gas Reheat

In-line reheaters are frequently used for flue gas reheat; these units can be subject to corrosion by chlorides and sulfates. Plugging and deposition can also occur, but are more rare. Usually, proper use of soot blowers prevents these problems. Soot blowing may not remove hard scale, but if it is done frequently enough the blowing may prevent deposition of hard scale. Generally, soot blowing once every 4 hours with air or steam is adequate.

Fans, Ductwork, and Chimney

Fly ash erosion and deposition on the fan blades are the major operational problems with dry and wet fans, respectively. These conditions can cause vibration and high torque, leading to excessive noise, bearing failure, and rotor cracking.

Extensive use has been made of ductwork and chimneys lined with refractory that is coated with various materials or fabricated from alloys. Essentially all materials have been subject to acid attack when the scrubber was operated outside of the design range.

Severe problems have occurred in operation of both louver and guillotine dampers. Typical problems are corrosion, erosion, fly ash buildup that prevents opening and closing, and mechanical failure. The most serious deficiency is that all dampers leak, and zero leakage can be achieved only by using double dampers with a barrier of higher-pressure air between them. Even with this design, however, dampers may still leak in a dirty environment. When leakage is severe, it may be necessary to shut down the boiler to allow entry into the FGD system for maintenance.

Limestone Receiving, Storage, and Slurry Preparation

Operational procedures associated with handling and storage of limestone are similar to those of coal handling. At some plants, the capacity for fugitive dust collection in conveyor enclosures has been undersized, so that escaping dust presents a potential safety hazard for operating personnel. The ball mill used for limestone grinding and slurry preparation must be vented. Operation of pumps, valves, and piping in the slurry preparation equipment is similar to that in other slurry service.

Limestone Slurry Feed Control

Slurry feed requirements are generally determined from the pH level of the EHT, slurry recycle line, or scrubber blowdown stream. Frequent backflushing of sensor lines where flow-through pH elements are used and frequent calibration with buffer solutions are routine requirements for reliable operation. One method for ascertaining instrument operability is to cross-check measured values with readings on redundant instruments or with values determined from analysis of grab samples. These readings must be made immediately, with a temperature-compensating pH meter. Delay would allow the system pH to rise because of the presence of unreacted limestone. Another method is to correlate slurry pH, reagent feed rate, and outlet SO_2 concentration during the initial system tests. Then, in routine operations, the three variables can be compared to detect a malfunction in any instrument.

A recent study at the Shawnee test facility was undertaken to develop a reliable field laboratory method for determining the pH of slurry liquor. The method was to be applied in a series of short-term tests in which conditions were changed every 6 to 8 hours. In these tests the pH was to be controlled to within ± 0.2 pH unit by laboratory measurement. The study report (Bechtel 1976) recommends the following:

1. Use of easily cleanable electrodes with glass-to-glass seals.
2. Use of commercially available certified buffers for standardizing the pH meters. Buffer pH should be within 0.5 unit of system pH.
3. Storage of the glass electrodes in hot (50°C) buffer (pH 4) saturated with KCl when not in use.
4. Air conditioning of the field laboratory to safeguard the pH meters.
5. Changing of electrodes at the beginning of each shift.
6. Frequent monitoring of the agreement between values obtained with the laboratory and the in-line pH meters, with action to be initiated to correct any disagreement greater than 0.2 pH unit.

Additional methods of improving the reliability of pH sensors are indicated in Table 5-1.

TABLE 5-1. METHODS OF IMPROVING pH SENSOR RELIABILITY[a]

Dip-type sensor	Flow-through sensor
Provide enough agitation in the tank to prevent accumulation of solids on the electrode	Provide extremely short sample lines (1 to 2 ft) of at least 1 in. diameter in slipstream of the recycle line
Locate probe away from quiescent zones but provide mechanical support	Avoid installing sample taps at the bottom of horizontal slurry lines
Provide an external tank for easy access	Provide backflushing capability (also can be used for calibrating)
Feed tank by a slipstream from the recycle line	Install upstream deflector bar to prevent erosion of the pH cell

Both types:

Provide redundant sensors (100 percent redundancy)

Conduct frequent calibration (once every shift)

Ensure proper mechanical and electrical hookups during installation

[a] Jones et al. 1978.

Pumps, Pipes, and Valves

Operating experience has shown that pumps, pipes, and valves can be significant sources of trouble in the abrasive and corrosive environments of a limestone FGD system. The flow streams of greatest concern are the limestone reagent feed slurry, the EHT/spray slurry recirculation loop, and the scrubber bleed stream.

When equipment is temporarily removed from slurry service, it must be thoroughly flushed. Failure to flush will result in plugging by suspended solids. In colder climates, protection against freezing may be required.

Because pumps, pipes, and valves are vulnerable to a comparatively high failure rate when used in either abrasive or corrosive process streams, redundancy is needed for successful operation. With redundant equipment, the staff can continue operation as repairs are made. Where capital investment is of concern, an alternative to two full-capacity components is three 50 percent capacity components. Although such considerations are mostly applicable to the design and procurement phase of a project, the operating staff must be fully familiar with the design intent so as to maintain high availability for the life of the plant.

Thickener

The solids content of the thickener underflow may vary between 20 and 40 percent, depending upon thickener design and the composition of solids in the slurry. Considerable operator surveillance can be required to minimize the suspended solids in the thickener overflow so that this liquid can be recycled to the scrubber system as supplementary pump seal water, mist eliminator wash water, or limestone slurry preparation water.

If the thickener feed rate and slurry properties are constant, then proper control of flocculant dosage rates (if any) and thickener bed density can generally be achieved. Because of load-following system inputs, however, these parameters are rarely constant. Therefore, for optimum results the operator must maintain surveillance of such parameters as underflow slurry density, flocculant feed rate, feed slurry characteristics, and turbidity of the overflow.

Thickener overflow is returned to the scrubber, either by mixing with the fresh water used to wash the mist eliminator or to the hold tank for level control. A thickener upset condition that causes excessive amounts of suspended solids to be carried into the overflow can seriously upset the makeup water system.

Sludge Disposal

As it leaves the thickener, the scrubber sludge may either be discharged to a pond or subjected to a second stage of dewatering in preparation for landfill disposal. Each of these options entails a different set of operational considerations. For slurry disposal, enough lines must be installed to accommodate the anticipated range of boiler loads. When any line is shut down, it must be flushed and drained so that the slurry solids do not settle and plug the pipe. In severe climates, protection against freezing is needed. A second pipeline network is required to return the supernatant pond water to the unit for reuse. Operation of both the discharge to the pond and the return water equipment requires attention of the operating staff. In addition to normal operations, the pond site must be monitored periodically for proper water level, embankment damage, and security for protection of the public. Even after the unit has ceased operations, some type of continued care is usually mandated for pond disposal sites.

Landfill disposal involves the operation of secondary dewatering equipment such as vacuum filters, centrifuges, or settling/evaporation ponds. Operators are required to adjust the process equipment to maintain optimum conditions over a broad spectrum of boiler loads. Again, when any of the process equipment is temporarily removed from service, it must be flushed and cleaned to prevent deposition of sludge solids.

Depending upon the composition of the solids, the dewatered slurry will contain 50 to 80 percent solids. For stabilization, fly ash may be mixed with the scrubber sludge in pug mills or muller type mixers. Regardless of whether the sludge disposal system incorporates dewatering only, stabilization with fly ash, fixation with lime, or one of the proprietary processes, personnel are required to operate the equipment and to maintain the proper process chemistry.

Transport of the wastes between process units and from those units to the landfill disposal site necessitates additional operating staff. Proper placement of the wastes at the landfill, pile compaction, and shaping require both personnel and special equipment.

Process Instrumentation and Controls

Operation of a scrubber system requires more of the operating staff than surveillance of automated control loops and attention to indicator readouts on a control panel. Manual control and operator response to manual data indication are often more reliable than automatic control systems and are often needed to prevent failure of the scrubber control system.

Typical problems include mist eliminator plugging, severe scale formation caused by pH sensor malfunctions, solids contamination in thickener overflow recycle water, and pump damage caused by false level or flowmeter indications. Many of these problems can be prevented when a scrubber operator can integrate manual control techniques effectively.

For units of more than 500 MW capacity, maintenance of pH sensors can be a full-time task for one or more instrument technicians. Frequent acid washing, cleaning, and calibration with buffer solutions have been required to ensure reliable operation. In addition, frequent backflushing of all pH sensor lines has been needed at some installations having flow-through pH sensor elements.

Malfunction of pH sensors, typically resulting in a constant output signal despite changing level of slurry pH, can usually be discovered by cross-checking procedures. A good cross-checking technique is to monitor several redundant pH sensors or to determine the slurry pH periodically with grab samples and portable pH meters. The sample pH should be checked immediately.

In many scrubbing systems, especially in high-sulfur coal applications, the continuous gas analyzers that determine outlet SO_2 concentration have malfunctioned. The malfunctions have been caused by plugging of sensor taps and lines and by corrosion attack inside the analyzer. The outlet SO_2 level can be estimated on the basis of a sulfur balance around the scrubbing system. The sulfur balance can also be checked by reference to a calcium balance of reagent feed rate, fly ash collected, and sludge produced. Such methods are not recommended as primary control techniques, but are useful as cross-checks of various instruments. Usually, the outlet SO_2 concentration determined by grab samples with wet chemical analysis can be correlated with scrubber slurry pH when the coal sulfur content is known and when the unit load and slurry recirculation rate are both fixed. Thus, pH of the scrubber slurry can be used as an indication of the outlet SO_2 concentration for known operating conditions.

INITIAL OPERATIONS

Very seldom does a system perform properly when it is first placed in service. Even though stringent quality control may be exercized during the construction phase, it is usually necessary to optimize the control functions and to correct minor problems. Although the individual components may be completely checked out during construction tests, the integrated system performance can be evaluated only when the system is placed in operation.

Initial Operational Tests

Any problems in the design, performance, or operation of a limestone FGD system should be identified through complete and comprehensive testing. Such testing is normally accomplished immediately after initial startup. Where possible, single tests may be conducted for multiple purposes. For example, tests conducted to ascertain system operability can also demonstrate compliance with emission limits. Tests conducted under the proposed normal operating procedures can verify the procedures and familiarize the station staff with the installed system. The station maintenance staff should make any required repairs, execute field modifications, calibrate instruments, and optimize the control system. A log of all activities, including details of aborted or failed tests, will provide clues to specific solutions for various problems and will serve as the basis for any field changes.

Extensive documentation of startup tests may seem more bothersome than useful during initial operations. These data, however, constitute the base against which future performance trends are measured. Test data may also be useful in routine operational decisions (e.g., the maximum rate of load changes may be influenced by response of the scrubber to load changes). Correlations between measured system parameters at various stages in the process may be useful in developing techniques for equipment operation. Together with the operating log, the test data complete the detailed history of the system.

After the initial startup tests have established a norm for system operation, additional testing is conducted for two purposes: to verify performance guarantees and to demonstrate compliance with regulations. Testing for fulfillment of supplier guarantees is discussed in Section 4.

At each installation, the applicable standards for emissions of SO_2, NO_x, and particulate will be established by the required permits. At new coal-fired generating stations the pollutant emissions must be monitored continuously.

Continuous source monitors were not originally intended to demonstrate compliance with emission standards. Several states, however, are developing implementation plans that utilize continuous monitoring data. The EPA has issued the following reference methods* for sampling and analyzing regulated emissions: selection of sampling location, Method 1; SO_2 emissions, Method 6; particulate emissions, Method 5; and NO_x emissions, Method 7.

STARTUP, SHUTDOWN, STANDBY, AND OUTAGE

Startup and shutdown are two nonsteady-state scrubber operating modes that occur frequently. Furthermore, two nonoperating conditions that necessitate action by the operating staff are scrubber system standby and extended station outage. Each of these situations is of special interest to the limestone FGD system operating staff.

Scrubber Startup

Before flue gas is introduced into a scrubber module, proper preparations must be made. Generally, limestone slurry is added to the system as a "lean" stream (low slurry solids content), and reaction product solids are permitted to build up to a specified control level. Specifically, the

---

* Described in Appendix A of Part 60, Title 40, Code of Federal Regulations.

solids content and the pH of the limestone reagent in the scrubber must be brought to predetermined levels. The slurry recycle lines and sprays also must be placed in operation. All of these actions must be accomplished with enough lead time to allow the parameters to come into the design range before the scrubbing operation begins.

A prerequisite to the startup activities is initiation of the limestone grinding process to ensure the availability of limestone slurry feed. Since the duties of the operating staff are greatest during transient scrubber conditions, such as startup and shutdown, the inventory of slurry feed should be increased to maximum levels in preparation for startup.

After scrubber operation begins, the operating staff must be ready to process the scrubber bleed stream. Operation of the thickener must be monitored. If sludge disposal is to a pond and if the pipelines have been drained, they must be filled before pumping can begin. If disposal is to a landfill, the secondary dewatering equipment must be ready for operation.

When the scrubber is placed in service, the operating staff must be available to monitor system response as boiler load is increased. The equipment lineup may need to be altered accordingly. As the flue gas flow rate through the scrubber modules approaches the maximum design value, additional modules are prepared and brought into service.

Scrubber Shutdown

As boiler load is reduced in preparation for unit shutdown, the startup sequence is executed in reverse. Even with automated controls, the operating staff must closely monitor the scrubber response to the changing conditions. When the boiler load is reduced and the unit is not shut down, the scrubber operators must ensure that the system reaches equilibrium at the new conditions. Associated activities such as limestone slurry preparation and sludge processing must be adjusted to accommodate such changes in scrubber status.

System Standby

A scrubber module that is ready to process flue gas is said to be on standby. The module may have been removed from service because of a reduction in station load and now prepared for service because of boiler startup or an anticipated increase in boiler load. When a module is removed from service because of a power reduction, the scrubber bleed stream must be terminated and the bleed line flushed. Failure to stop the scrubber bleed flow would waste the limestone additive. Failure to flush the line would lead to plugging as the solids settle. When a module is brought into service, the operator must prepare the blowdown line to accept flow by checking valve position and backfilling the line if necessary.

Extended Outage

Additional operations are necessary when a scrubber module is removed from service for an extended period. The limestone slurry recycle pumps and the recycle lines should be drained and flushed. If the hold tank agitator is not left in service, the hold tank must also be drained and flushed of solids.

During an extended outage, the operating staff should conduct inspections of equipment that is normally inaccessible. By entering the shutdown scrubber module, an operator can check the conditions of the structural materials and linings for evidence of corrosion. He can also inspect tower internals, spray nozzles, and the mist eliminator for scale deposits, abrasive wear, or evidence of other developing problems. All auxiliary equipment should also be inspected. With proper operation and servicing, the limestone FGD system will provide high reliability and availability when returned to service.

SYSTEM UPSETS

The scrubber is the last stage in a series of systems in which any malfunction can impact scrubber operations. Upsets are usually associated with the boiler, the scrubber system, or the sludge disposal system.

A boiler trip will terminate the flow of flue gas through the scrubber. Except for the possible discharge of unreacted limestone slurry to the waste processing equipment, there should be no adverse impact on the scrubber. Transient conditions causing an increase in flue gas flow may produce scaling of the mist eliminator or excess liquid carryover. The effects of greater-than-design inflow of particulate to the scrubber would depend upon particulate composition. Although there would usually be no adverse effects, the solids content of the scrubber bleed stream would increase. If the fly ash is alkaline, SO_2 removal may be higher than anticipated.

Inability of the scrubber to process flue gas could lead to a boiler upset and removal of the unit from service. Failure of a single scrubber module could lead to a reduction of station output. A reduction in SO_2 removal efficiency can be caused by an inoperable slurry recycle pump, plugging of spray nozzles, or an insufficient supply of limestone slurry feed. Scaling of the mist eliminator could cause an increase in flue gas velocity and a reduction in mist eliminator performance. The high pressure differential resulting from mist eliminator scaling could cause an unnecessary increase in fan operating costs, even with no imminent threat to the station output.

The inability to process scrubber bleed for sludge disposal could impair SO_2 removal efficiency and station output. Since waste processing systems usually incorporate some spare capacity, station output should only be reduced, at worst. The solids processing system should incorporate some type of surge capacity. Operating flexibility provided by redundancy or installed surge capacity will allow the staff to work around a malfunction of sludge processing equipment without a reduction in station output.

OPERATING STAFF AND TRAINING

The size, experience level, responsibilities, and training of the operating staff are significant factors in FGD system performance. A conservative approach is to assign highly qualified personnel to the operating staff until the system response to the range of operating conditions is understood and reactions to changes in performance become routine. The number of personnel assigned to each operating shift will vary with the type of equipment and with the normal operating mode of the unit, i.e., base-load versus load-following.

In staffing, the scrubber operations and waste disposal operations must be considered separately. The permanent assignment of key personnel to specific work areas will allow them to become completely familiar with the process equipment and its chemistry. As the operating personnel gain understanding of the system, they will be able to anticipate problems before unit output becomes impaired.

In addition to the normal complement of equipment operators and supervisory personnel on the operating crew of each shift, certain specialists should be available to assist them. For example, a chemical engineer would be a valuable resource during atypical operating conditions. A chemical laboratory technician should also be available to analyze the process chemistry in the event of suspected trouble. During normal operations, this technician can monitor routine system performance through sampling and laboratory analyses. Except for the period following initial operation and testing of a newly installed system, the chemical engineer and the laboratory technician need not be dedicated full time to scrubber operations.

The time period following initial startup and operation of the scrubber system presents an excellent opportunity for training of the operating staff. When a scrubber system is first placed in operation, vendor personnel are usually available on site to ensure that the equipment is operating properly. During this period all equipment should be operated by utility staff personnel, under the guidance of the vendor representatives. Whenever possible, written procedures should be followed so that any error can be identified and corrected.

The following estimate of the operating staff personnel requirements is based on experience at full-scale, limestone FGD systems at coal-fired power plants:

Operating responsibility	Number of personnel per shift per unit
FGD scrubber	2 to 4
Limestone receiving and storage	1/3 to 1/2
Slurry preparation	1/3 to 1/2
Waste processing and disposal	2 to 10

Although some reduction in crew size may be possible where multiple units are operated at the same site, the reduction will depend on what equipment is common to all units. Requirements for waste processing and disposal personnel vary with operating schedules and method of final disposal. It should be kept in mind that the limestone FGD scrubber is only one system in series with the rest of the operating units in a powerplant and that the size and experience level of the operating staff must be correlated with the overall utility requirements.

PREVENTIVE MAINTENANCE PROGRAMS

Preventive maintenance is the practice of maintaining system components in such a way as to prevent malfunctions during periods of operation and to extend the life of the equipment. The goal of preventive maintenance is to increase availability of the FGD system by eliminating the need for emergency repair.

The term preventive maintenance is almost synonymous with periodic maintenance. The manufacturer specifies how often each component should be serviced, usually depending upon the type of duty it sees. The manufacturer's operating instructions for each component should specify both the schedule and procedures to be used in preventive maintenance activities. Such procedures may be as simple as lubrication of a pump or as complex as complete disassembly for inspection and overhaul. The upkeep of pumps, for example, normally involves lubrication and inspection of bearing wear and liner conditions. Replacement of valve seats and packing may be necessary after a specified number of hours of operation. Routine inspection of linings in chimneys, ducts, fans, and the scrubber permits early detection of damage. When corrective actions are taken before the damage becomes severe, repairs are usually less complex, less costly, and less time consuming.

The operating log also can be useful in establishing appropriate preventive maintenance schedules. The operating log and the equipment maintenance records together should indicate trends on which to base a program of systematic upkeep.

Scrubber Modules

The scrubber module is typically a passive component with no moving parts except for agitators in the EHT. Of primary concern in the scrubber module is the integrity of the structural materials, including the lining, and any scale buildup that may be occurring. Maintenance personnel should enter and inspect the scrubber module at least annually. Where there is a recurring problem with either scaling or corrosion, more frequent inspections are in order.

The EHT must also be checked for buildup of settled sludge. The structural integrity of pump suction strainers must be checked. Agitators should be inspected for not only corrosion and erosion but also for bearing wear and seal deterioration at the tank wall.

Mist Eliminators

Scale deposits typically are the chief maintenance factor with mist eliminators. The mist eliminator may be subject to nonuniform flow or a faulty wash system. Plugged or worn spray nozzles, solids deposits in the supply piping, or other malfunctions may cause a loss of wash pressure and lead to localized scaling. Wash spray pressure should be monitored. Mist eliminators should be inspected periodically and, again, if a problem is recurring, inspections should be made more often.

Reheat System

Procedures for maintaining the reheater system vary with heating method. In-line reheaters are subject to both scaling and corrosion. In addition to visual inspection, pressure testing and measurement of heat transfer efficiency are useful in quantifying the magnitude of a reheater problem.

In an indirect reheat system, the mixing chamber and the air heating equipment must be checked routinely. Scale and corrosion are the principal concerns.

Dampers, Fans, Ductwork, and Chimneys

The location of the fans and the ductwork governs the types of potential problems that could occur. Components located in the wet portion of the system, downstream of the scrubber modules, are subject to scaling and corrosion. Upstream fans and ductwork could be subject to erosion, depending on the particulate removal efficiency. All points in the system must be checked for integrity of lining materials and for damage resulting from collection of condensation products in low-lying pockets. Such pockets may occur at expansion joints or instrument taps into the duct, or at points where the sloping of straight duct runs may be inadequate. Chimneys should be inspected for effects of acid condensation or thermal cycling.

Limestone Slurry Preparation

Preparation of limestone slurry subjects the ball mill to abrasive wear. This equipment must be lubricated often during operating periods. Because the equipment sees intermittent service, it should be inspected visually each time it is placed in service. Periodic disassembly is also needed to check for excessive wear on the milling surfaces. The frequency of disassembly depends on the composition of the limestone. Conveyor and feeders should be inspected and maintained at the same time as the grinder or mill.

Limestone Slurry Feed

In addition to the pumps, piping, and valves, the slurry storage tank and associated equipment must be maintained. Even though the prepared slurry is agitated, some settling can occur. If the settling of solids becomes excessive, the transfer pump suctions and drain lines could become clogged. Tank agitators must be inspected for impeller abrasion, corrosion, and bearing wear. Maintenance of the slurry feed system is critcal because failure of this equipment strongly impacts the FGD system operation.

Pumps, Pipes, and Valves

Slurry pumps are normally disassembled at least annually. The purpose of the inspection is to verify lining integrity and to detect wear and corrosion or other signs of potential failure. Bearings and seals are checked but not necessarily replaced. The frequency of these inspections depends on the slurry solids content and also on the size and shape of the limestone particles.

Pipelines must be periodically disassembled or tested in other ways both for solids deposition and for wear. Measurements of flow capacity and pressure drop, and possibly ultrasonic inspection, could be substituted for disassembly. Slurry lines are subject to solids deposition and to abrasive or corrosive wear. Deposition is more prevalent in straight runs of piping where flow patterns are undisturbed. Wear may predominate at points of flow perturbation, such as elbows, restricting orifices, instrument taps, or other sidestream connections. Appropriate component design, e.g., flanged pipe connections, can simplify maintenance procedures.

Valves must be serviced routinely, especially control valves. Valves that are operated rarely must be exercised to ensure that they are functional. Valve operators, both motor and pneumatic, must also be functionally tested. Seats must be checked to verify leakage rates. Valve stem packing must be checked to ensure that leak-tightness does not prevent proper valve operation.

Lined valves, such as those used in slurry service, require more frequent maintenance than unlined valves. They should be disassembled at least yearly and their liners inspected. Valves with replaceable seats are easier to maintain and therefore are preferable.

Thickeners

Thickeners are usually constructed of a concrete base slab with coated carbon steel walls. The coating should be inspected periodically to prevent massive corrosion, which could cause delamination of the coating or damage to the underlying materials. Drag rakes, torque arms, and support cables must also be inspected for wear. The drag rake should be lifted when the system is shut down. Drive motors must be lubricated frequently.

Sludge Disposal Equipment

Secondary dewatering devices, mixing components, and transport equipment also must have periodic maintenance to check for abrasive wear and solids deposition. Motors and device couplings require periodic checking of lubricant level and lubrication as needed.

Vacuum filters, both drum and belt type, require periodic replacement of the filter media. Frequency is determined by the media materials, filter cake removal method, hours of operation, and abrasiveness of the sludge. In a centrifuge, both the scroll coating and the bowl surfaces are subject to wear. Materials of construction usually preclude corrosion problems, but this must be verified during maintenance. Similar considerations apply to mixing and handling equipment.

Process Instruments and Controls

All electronic equipment must be calibrated periodically. The frequency is governed by the type of service and the accuracy to which a variable must be controlled.

Numerous installation and maintenance techniques have proved beneficial in ensuring the reliability of pH sensors (Table 5-1). A consistent difficulty with pH measurement is that electrodes and amplifiers are often poorly installed or badly located. Experience has shown that where maintenance is inconvenient, the operators often neglect it to the point that pH measurement becomes unreliable. Thus, ease of access to the electrodes is very important. The electrodes should be cleaned and standardized at least once every shift. The best practice is to install dual pH metering systems so that calibration can be cross-checked continuously.

Wiring between the electrodes and the preamplifier should be as short as possible; some vendors mount the preamplifier in the electrode housing to prevent short-circuiting. This arrangement, however, has the disadvantage of placing the electronic component in a wet atmosphere, which could lead to failure of the preamplifier if the housing fails. All vendors offer either voltage or current output signals, most of which are field-adjustable for both range and span.

The electrode station should contain a workbench and a cabinet to hold spare parts, small tools, and standardizing solutions. At least two identical electrode assemblies are desirable, with valves and switches arranged for simple crossover to a set of standby electrodes when a set is checked or serviced. All amplifiers and calibration controls should be installed at the electrode station so that one person can perform the necessary adjustments. This arrangement eliminates the need for communication between control room and maintenance personnel during calibration.

Flow devices, pressure sensors, temperature monitors, and level instruments are susceptible to similar problems. A sound preventive maintenance program will minimize adverse conditions resulting from an inadequate design that cannot be changed by modifications.

Experience with process instrumentation and controls in limestone FGD system applications has shown that a good preventive maintenance program begins with daily operating procedures. Proper use of instruments will include daily flushing of most instrument lines in slurry service just before monitoring of process variables. Routine comparison of the instruments in a process stream with similar instruments in parallel streams can point out incipient failures. Operating data, especially from the startup test program, can also indicate potential problem areas.

Housekeeping

Because of the nature of the materials handled and the process equipment used in a scrubber and its support system, the work area can become dirty. Leaks and even routine use of equipment will lead to accumulation of sludge on the floors. In the limestone receiving and sludge processing areas, dust can become a significant problem. The resulting dirty environment not only affects the operating and maintenance staff adversely, but also reduces overall system reliability by accelerating the rate of wear and other malfunctions. In such an environment, both operating and maintenance personnel must follow a continuous and stringent housekeeping program.

UNSCHEDULED MAINTENANCE

Even the most rigorous preventive maintenance program will not prevent random failures, to which the maintenance staff must respond. Redundancy achieved through excess capacity and the use of multiple fractional-capacity process streams will enable the station to continue power production while repairs are effected.

Most malfunctions are correctable by unscheduled maintenance. In some situations, usually during initial system startup, design modifications may be required to bring the system into compliance with operating standards. As system operation continues, means of improving system performance may be identified. Improvements are normally incorporated during scheduled outages.

Each component of a limestone FGD system is subject to malfunctions from a variety of causes. Most such malfunctions are typical of those that occur in other systems, regardless of the application; examples are pump motor failures and malfunctions of bearings and valve operators. Some malfunctions, however, are unique to the components of a scrubber system. The discussion that follows deals with these problems and the probable responses.

Scrubber Modules

Degradation of the coating or lining of the scrubber vessel and subsequent damage to the base material are normally identified and corrected during preventive maintenance. Structural failure of spray tower trays and recycle pump suction screens have occurred as a result of excessive vibration, uncorrected corrosion damage, or high pressure differentials. These malfunctions must be repaired immediately before operations are resumed. Repairs may include patching or repairing the lining in the vicinity of the failure.

Mist Eliminator

Failure of the mist eliminator in an operating system is typically due to scaling and plugging, as indicated by excessive pressure differential. The scale may be removed either by thorough washing or by mechanical methods, in which maintenance personnel enter the scrubber and manually chip away the scale deposits. The cause of the scaling should be determined (e.g., a clogged spray nozzle, insufficient wash water flow or pressure, or improper process chemistry). If the cause is not identified, scaling or plugging is likely to recur.

Reheat System

Reheater malfunctions could be caused by tube failures in in-line reheaters, by damper problems in bypass reheat, or by nonuniform flows in indirect reheaters. The lack of sufficient reheat capability could cause condensation and corrosion in the stack.

Fans

Fans can develop vibrations resulting from deposition of scale in wet service or from erosion of blades in dry service. Before operation can resume, the cause of the vibration must be eliminated and the fan repaired and rebalanced.

Ductwork

Most problems associated with ducts develop over a long period. Deterioration of the lining and damage to the base material are usually detected and corrected during preventive maintenance. Correction may include modifications to the system design. Sudden or gross failures, such as a major leak, call for immediate repair. Temporary repair or patching may suffice until the next scheduled outage. Again, the source of the malfunction should be identified; for example, if the problem is traceable to improper process chemistry, system operation should be adjusted accordingly.

Acid condensation in a chimney can cause lining deterioration and subsequent damage to the base metal. These problems are usually identified during preventive maintenance inspections and probably require long-term solutions. Temporary repairs could become necessary, in which case the chimney should be routinely monitored to check the extent of any additional damage.

Scale can also accumulate on dampers. Unscheduled maintenance is limited to removing the scale and investigating the cause. Whenever possible, the cause is eliminated; e.g., by adjusting the process chemistry. Where the scale accumulation is attributed to a design deficiency, equipment modifications may be necessary. Flue gas leakage is one such deficiency that usually requires equipment modifications.

Limestone Slurry Preparation

Malfunctioning components such as ball mills must be repaired in accordance with the manufacturer's instructions. Some facilities have experienced trouble with plugging of the limestone feeder due to intrusion of moisture. Excessive dust has also been a problem at some facilities. Correction of these problems will probably necessitate changes in equipment design.

Pumps, Piping, Valves

Excessive wear of the impeller or separation of the lining from the pump casing is a common problem. Pieces of the lining can plug the pump discharge or flow downstream, impeding the operation of other components, such as valves and instrument sensing lines. These malfunctions are corrected by repair of the pumps.

Operation of slurry pipeline with insufficient flow velocity can cause clogging. High flow velocity or extended service can cause erosion. If pipes cannot be unclogged, they must be replaced.

Malfunction and binding of a valve operator are typically caused by wear-induced misalignment. It may be possible to continue operations with manual actuation of the valve until repairs can be made. During steady-state conditions, it may be possible to replace a valve operator without interrupting process operations. Lined valves are subject to the same types of failures as lined pumps. Use of valves constructed with replaceable seats and/or other internal parts will facilitate repairs. Leakage of the valve stem packing can often be corrected by tightening enough to terminate the leak but still prevent binding. With some types of stem leakage the packing must be replaced.

Thickeners and Sludge Disposal Equipment

The thickener underflow can become plugged because of excessive solids in the slurry. A plugged underflow or rapidly settling sludge solids will produce a blanket in the bottom of the thickener. The rake must then be raised so that the torque remains within acceptable limits. If the torque cannot be kept within limits, operation of the thickener must be terminated. The thickener must be drained and the sludge blanket removed manually.

Where secondary dewatering equipment is used in preparation for landfilling, water removal may be inadequate. An excessive concentration of sulfate solids in the sludge can lead to cracking of the filter cake on the vacuum filter. Cracking of the filter cake is not usually serious, however.

Process Instrumentation and Controls

Slurry service is probably the most severe application for instrumentation. Most malfunctions result from plugged sensing lines caused by low velocities or flow restrictions. Fouling of probes also has been a problem. Although temporary measures such as routine flushing of the sensing lines or frequent cleaning of the probes will allow operation to continue, the system should be modified to eliminate the cause of plugging/fouling. The best approach is to install instrumentation that minimizes the amount of required maintenance. Table 5-2 lists types of instruments preferred for various applications in a limestone FGD system.

TABLE 5-2. PREFERRED FGD SYSTEM INSTRUMENTATION

Process variables	Type of instruments
Slurry level	Ultrasonic sensors, admittance probe sensors, and flexible diaphragms with purge flushing
Slurry density	Nuclear density meter
Slurry flow	Magnetic flow meters; sonic (Doppler) detectors
pH level	Dip-type pH sensor located in auxiliary measuring vessel that can be isolated; equipped with ultrasonic cleaning device
Flue gas pressure	Capacitance cell electronic differential pressure transmitters with water purging
SO_2 concentration	Extractive or "in-situ"
Slurry pressure	Capacitance cell electronic pressure transmitters with diaphragm seals and water purging
Flue gas temperature	Thermocouples with stainless steel protector tubes

Troubleshooting Techniques

Troubleshooting of an FGD scrubber and ancillary equipment calls for a multiphase program along the following lines.

Phase 1: Problem Identification. This phase begins with a detailed inspection of the system. All observations (positive and negative) are listed, interpretations are developed (why things were the way they were), and finally, methods and items that will improve performance are recommended. Recommendations may call for design modifications, replacement of components or accessories, or fabrication of new equipment.

Phase 2: Implementation. After thorough analysis, the Phase 1 recommendations should be implemented by repair and by replacement with procured and fabricated components. The system is then started up and debugged.

Phase 3: Testing and Sampling. A performance test must be conducted to evaluate the effects of the work on system operation. Testing may be by stack sampling and/or measurements with the system in continuous operation.

Phase 4: Operational Troubleshooting. Certain symptoms are attributable to more than one cause. Table 5-3 lists typical symptoms, probable

TABLE 5-3. OPERATIONAL TROUBLESHOOTING CHECKLIST

Symptom	Potential cause	Recommended action
Low pressure drop (scrubber section)	Low flue gas flow rate Low liquid flow rate Eroded or dislocated scrubber internals Meters plugged	Check fan Check pump/nozzles Inspect Clean lines
High pressure drop (scrubber section)	High flue gas flow rate Plugging in ducts or scrubber	Check fan Inspect
Low pressure drop (mist eliminator)	Low flue gas flow rate Low liquid flow rate Media dislocated	Check fan Check pump/nozzles Inspect
High pressure drop (mist eliminator)	High flue gas flow rate High liquid flow rate Clogging Flooding	Check fan Check pump/nozzles Inspect/clean Inspect/drain
High temperature in stack	Too much reheat Liquid temperature too high	Check flue gas temperatures upstream and downstream of reheater Check sump temperature
Pump leaks Increase in pump pressures Reduction of pump	Packing or seals Nozzle plugging Valves closed Impeller wear	Replace Replace Open valves Replace

(continued)

TABLE 5-3. (continued)

Symptom	Potential cause	Recommended action
flow rate/pressure	Nozzle abraded Speed too low Defective packing Obstruction in piping	Replace Check motor Replace Check pipes, strainer, and impeller
Pump noise/heat	Misalignment Bearing damage Cavitation	Check Replace Check
Corrosion	Inadequate neutralization High Cl^- concentration	Check pH control Check Cl^- content of recirculation slurry
Erosion	Incompatible materials High recycled solids content	Replace Wastewater system
Scaling	Improper chemistry	Change chemistry variables
Pipe plugging	High solids content Abrupt expansion/contraction/bends	Cleaning Change pipe fittings

causes, and suggested remedies. This list should not be regarded as exhaustive of all possibilities; no checklist, maintenance protocol, or operator instruction manual can take the place of a well-trained maintenance staff familiar with the equipment and its operating history.

Spare Parts

The spare parts inventory maintained on site must be able to support the required maintenance activities. Initial inventories should be based on recommendations of the equipment vendor. Lead times required for delivery of parts from the supplier's warehouses must also be considered. Initial spare parts inventories may have to be adjusted depending upon system performance and the degree of redundancy in the installed system.

The quantity of spare parts can be reduced by standardizing sizes and types of equipment. Standardized parts also reduce the time needed for maintenance activities.

MAINTENANCE STAFF REQUIREMENTS

The maintenance staff for a limestone FGD system must include personnel from a number of disciplines. Mechanics are needed for component repairs. Electricians are also needed, as well as instrument technicians familiar with the system. These specialists are assisted by laborers and by the operating staff. Because different policies govern the tasks performed by each type of craftsman, an optimum maintenance staffing scheme cannot be outlined firmly.

Assignment of maintenance personnel to shift coverage also varies with individual facilities. Where maintenance on the back shift is performed by "on-call" personnel, the standard day-shift maintenance crew will be large. Where operating personnel can perform basic maintenance, such as instrument flushing, requirements for the maintenance staff can be reduced. The potential number of unscheduled maintenance activities (i.e., because of equipment malfunctions) must also be considered in sizing the maintenance staff.

REFERENCES FOR SECTION 5

Bechtel Corporation. 1976. pH Study at the Shawnee Test Facility. Phase II. EPA Contract 68-02-1814.

Jones, D. G., A. V. Slack, and K. S. Campbell. 1978. Lime/Limestone Scrubber Operation and Control Study. EPRI-RP 630-2.

Appendix A

Chemistry of Limestone Scrubbing

The primary objective of any FGD system is to enable the powerplant to operate in compliance with the SO_2 emission regulations. In order to achieve this objective in a cost-effective and reliable manner, a limestone FGD system must maximize both the SO_2 removal and limestone utilization and must precipitate the calcium sulfite/sulfate solids in a controlled scale-free manner. Numerous interrelated chemical variables and mechanical factors must be controlled. In an effort to simplify this subject matter, the process chemistry is considered separately from the key operational factors that affect the FGD system performance; the latter are discussed in Appendix B. The separation cannot be maintained absolutely because the chemical reactions are an integral part of process operation. Therefore, wherever it is possible the significant interrelationships are identified and related material in other parts of the manual is cited.

The intent here is to present a brief, theoretical description of the basic process chemistry involved in wet limestone scrubbing. Although the subject has been treated in the FGD literature by various authors, the treatments are often very rigorous, or are empirical and site-specific. This appendix is intended as an introduction to scrubber process chemistry; more detailed treatments of these topics are given in the references cited.

CHEMISTRY DESIGN OBJECTIVES

The chemistry of a simple limestone slurry scrubber should be designed to maximize SO_2 removal and to avoid scaling of calcium sulfite ($CaSO_3$) and calcium sulfate ($CaSO_4$) in the scrubber. This is achieved by separating the functions of the scrubber and the effluent hold tank (EHT) so that SO_2 removal takes place in the scrubber while solids precipitation occurs primarily in the hold tank.

In the scrubber, SO_2 is removed from the flue gas and, ideally, about half of the limestone is dissolved. Usually the dissolved alkalinity is insufficient to provide for SO_2 absorption as bisulfite (HSO_3^-); therefore either $CaCO_3$ or $CaSO_3$ solids must dissolve in the scrubber. Ideally, one mole of $CaCO_3$ should dissolve for two moles of SO_2 absorbed:

$$CaCO_3(s) + 2SO_2 + H_2O \rightarrow Ca^{++} + 2HSO_3^- + CO_2$$

If excessive $CaCO_3$ is present, $CaCO_3$ dissolution can lead to precipitation of $CaSO_3$ and possibly to scaling in the absorber:

$$CaCO_3(s) + SO_2 \rightarrow CaSO_3 \rightarrow\; + CO_2$$

If insufficient $CaCO_3$ is present, $CaSO_3$ will dissolve in the scrubber:

$$CaSO_3(s) + SO_2 + H_2O \rightarrow Ca^{++} + 2HSO_3^-$$

The pH levels at the scrubber inlet and outlet are indicative of the solids reactions that will occur. A high pH level (5.8 to 6.5) indicates excessive dissolution and precipitation of $CaSO_3$ in the absorber. The SO_2 removal by a given scrubber system is generally higher with excessive $CaCO_3$ at high pH and lower with little excess $CaCO_3$ at low pH. Hence there can be tradeoffs between SO_2 removal and $CaSO_3$ scaling and between SO_2 removal and limestone utilization.

Ideally, all of the crystallization of $CaSO_3$ and $CaSO_4$ and about half of the dissolution of $CaCO_3$ occur in the EHT. The stoichiometry for $CaSO_3$ crystallization is given by:

$$CaCO_3(s) + 2HSO_3^- + Ca^{++} \rightarrow 2CaSO_3(s) + CO_2 + H_2O$$

Completion of these reactions with minimum supersaturation requires adequate EHT volume and a high level of suspended solids.

Because the flue gas contains 3 to 10 percent oxygen, a substantial fraction of sulfite will oxidize to sulfate in the scrubber. Depending on SO_2 and O_2 gas concentrations, 10 to 100 percent of the SO_2 absorbed may be crystallized as calcium sulfate. With an adequate EHT design, the solution entering the scrubber will be slightly supersaturated to $CaSO_4$. In the scrubber, sulfite oxidation and calcium solids dissolution will increase the concentration of dissolved $CaSO_4$ and tend to cause $CaSO_4$ scaling. The critical concentration at which $CaSO_4$ precipitates uncontrollably is avoided by maintaining a liquid circulation rate high enough to minimize the change in concentration of $CaSO_4$ solution across the scrubber and sufficient nucleation sites for desupersaturation.

If the amount of sulfate in the waste solids is less than 15 to 20 percent of the total sulfate plus sulfite, $CaSO_4$ will crystallize in solid solution with $CaSO_3$, rather than crystallizing as gypsum ($CaSO_4 \cdot 2H_2O$), its usual form (Jones et al. 1976). As a result, the constraints on EHT design and liquid circulation rate can be relaxed without gypsum scaling.

Composition of the solution in $CaCO_3$ slurry scrubbing can be predicted approximately from solid/liquid equilibria. Most scrubber inlet solutions are nearly in equilibrium (saturated) with $CaSO_3$ and $CaSO_4$ solids. Depending on the amount of excess limestone, the solutions are at 0.1 to 50 percent saturation with respect to $CaCO_3$ with a CO_2 partial pressure of 0.1 to 1.0 atm. In the absence of soluble salt impurities the scrubber solution is primarily dissolved $CaSO_4$. Soluble impurities such as Cl^-, Mg^{++}, and Na^+ accumulate in the scrubber loop and frequently constitute the primary constituents.

Radian Corporation (Lowell et al. 1970) developed an equilibrium program for this system, which has been modified by Bechtel (Epstein 1975). Table A-1 lists the calculated concentrations of important solution species in solutions dominated by $CaCl_2$ and by $MgSO_4$. Variation of excess $CaCO_3$ affects primarily the solution pH and the concentrations of the ion pairs: $CaSO_3^\circ$, $CaSO_4^\circ$, $MgSO_3^\circ$, and $MgSO_4^\circ$.

A number of specific rate processes contribute to overall system per-

TABLE A-1. TYPICAL SOLUTION COMPOSITIONS

	$CaCl_2$	$MgSO_4$
$CaCO_3$ saturation	0.10	0.10
$CaSO_3$ saturation	1.0	1.0
$CaSO_4$ saturation	1.0	0.3
P_{CO_2}, atm	0.1	0.1
P_{SO_2}, atm	0.9×10^{-6}	0.9×10^{-6}
pH	5.64	6.1
CO_2, mmol/liter	1.8	1.8
HCO_3^-, mmol/liter	0.57	1.8
HSO_3^-, mmol/liter	2.2	6.9
$SO_3^=$, mmol/liter	0.1	1.1
$CaSO_3^\circ$, mmol/liter	1.5	1.4
$SO_4^=$, mmol/liter	6.4	20.3
$CaSO_4^\circ$, mmol/liter	6.0	1.8
Ca^{++}, mmol/liter	49.0	6.0
Cl^-, mmol/liter	100.0	100.0
Mg^{++}, mmol/liter	8.7	69.5
$MgSO_4^\circ$, mmol/liter	1.2	24.6
$MgSO_3^\circ$, mmol/liter	0.1	5.5

formance. The effects of these processes must be integrated in the scrubber and the EHT to project the relationships of SO_2 removal, limestone utilization, pH, oxidation, scaling, and other system variables. In the remainder of this appendix we describe in detail the four most important rate processes:

1. SO_2 gas/liquid mass transfer
2. Limestone dissolution
3. Oxidation
4. $CaSO_3/CaSO_4$ crystallization

Other rate processes that are not discussed specifically include CO_2 desorption and $CaSO_3$ dissolution.

SO_2 GAS/LIQUID MASS TRANSFER

The overall process of SO_2 absorption in a $CaCO_3$ slurry typically requires three rate processes in series: SO_2 diffusion through a gas film, SO_2 diffusion through a liquid film, and $CaCO_3$ dissolution. In general, any one of these processes can limit the SO_2 absorption, though if $CaCO_3$ dissolution is important, SO_2 liquid-phase diffusion is also important. The following discussion deals with the effects of gas and solution composition on SO_2 mass transfer through the gas and liquid films. This is followed by a discussion of $CaCO_3$ dissolution in the scrubber and the EHT. Appendix B describes the effects of physical parameters on gas and liquid films.

The two-film model of gas/liquid mass transfer assumes that concentrations of SO_2 are in equilibrium at the gas/liquid interface:

$$P_{SO_2 i} = H\, C_{SO_2 i} \qquad (A-1)$$

where $P_{SO_2 i}$ = partial pressure of SO_2 (atm)

$C_{SO_2 i}$ = liquid SO_2 concentration (mol/liter)

H = Henry's constant (atm/mol-liter)

The total flux of SO_2 (N, gmol/cm^2-sec) must be the same in the gas and liquid films, as given by:

$$\text{gas film:} \quad N = k_g\, (P_{SO_2} - P_{SO_2 i}) \qquad (A-2)$$

$$\text{liquid film:} \quad N = \phi k_1^\circ\, (C_{SO_2 i} - C_{SO_2}) \qquad (A-3)$$

P_{SO_2} and C_{SO_2} represent the concentrations of SO_2 in the bulk gas and liquid phases respectively. The mass transfer coefficients, k_g and k_1°, vary with agitation and species diffusivities in the respective gas and liquid phases, but are independent of compositions. The liquid-film mass transfer coefficient, $k_1^°$, must be corrected by the enhancement factor, ϕ, to account for chemical reactions that permit SO_2 to diffuse through the liquid film as bisulfite or sulfite species rather than as undissociated SO_2.

Equations A-1, A-2, and A-3 can be combined to eliminate $C_{SO_2 i}$ and $P_{SO_2 i}$ and give flux in terms of the overall gas-phase coefficient, K_g:

$$N = K_g (P_{SO_2} - HC_{SO_2})$$

where

$$\frac{1}{K_g} = \frac{1}{k_g} + \frac{H}{\phi k_1}$$

In $CaCO_3$ slurry scrubbing, the equilibrium SO_2 partial pressure of the bulk solution, HC_{SO_2}, is usually negligible compared with that of P_{SO_2}.

If the ratio of the first and second terms, $\phi k_1^°/H k_g$, is much less than 1, SO_2 mass transfer is controlled by liquid film resistance. If it is much greater than 1, gas film resistance is controlling. Typical values of $k_1^°/H k_g$ in $CaCO_3$ slurry scrubbing are 0.05 to 0.20 (Rochelle 1977). Therefore, gas- and liquid-film resistances are equally important with enhancement factors ranging from 5 to 20, whereas gas-film resistance should tend to dominate with enhancement factors greater than 20.

The enhancement factor, ϕ, is a strong function of gas and solution composition (Chang and Rochelle 1980). SO_2 diffusion through the liquid film is enhanced by the following instantaneous reversible reactions, which convert it to bisulfite in the liquid film:

$$SO_2 + H_2O \rightleftarrows H^+ + HSO_3^-$$
$$SO_2 + SO_3^= + H_2O \rightleftarrows 2HSO_3^-$$

The extent to which these reactions occur depends on the bulk solution composition and the partial pressure of SO_2 at the gas-liquid interface. The hydrolysis of SO_2 to H^+ and HSO_3^- is depressed by higher HSO_3^- concentrations and is relatively less important than SO_2 diffusion at high P_{SO_2}. The reaction of SO_2 with sulfite is usually limited by diffusion of $SO_3^=$ to the gas-liquid interface and by supply of $SO_3^=$ in the bulk liquid. Thus, higher concentrations of $SO_3^=$ or sulfite species, such as $CaSO_3°$ or $MgSO_3°$, give greater enhancement factors.

Figure A-1 gives typical enhancement factors as a function of P_{SO_2} at the gas-liquid interface and pH in 0.1 M $CaCl_2$ solution saturated to $CaSO_3$ (Rochelle 1980g).

Effect of SO_2 Gas Concentration

The enhancement factors are a strong function of SO_2 gas concentration.

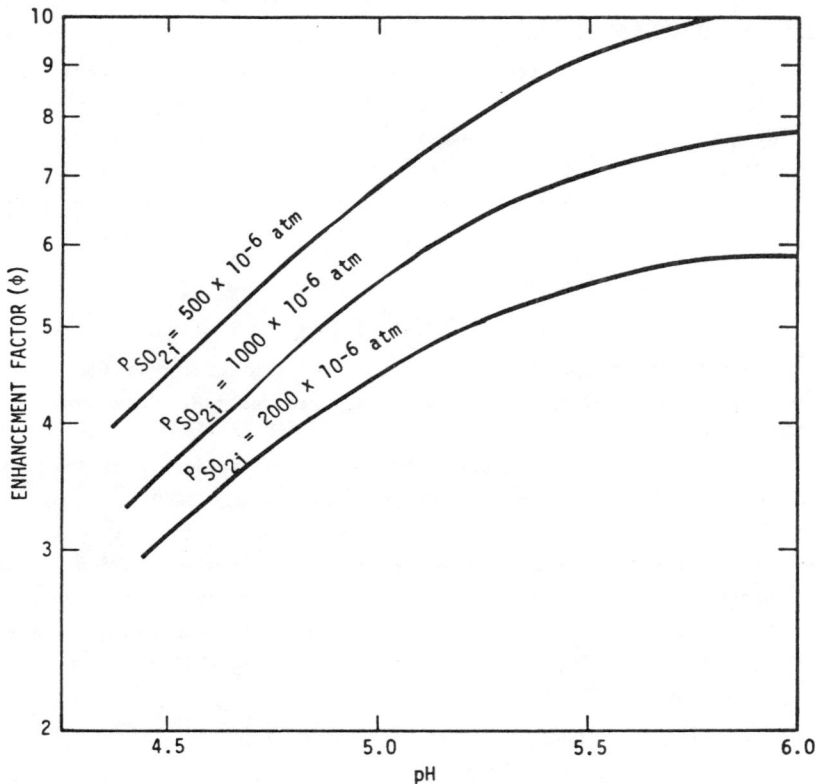

Figure A-1. Effect of pH and P_{SO_2} on SO_2 absorption.

At gas concentrations lower than 100 to 500 ppm SO_2, most $CaCO_3$ slurry scrubbers can be assumed to be limited by gas-film resistance. In this range, the SO_2 flux and the amount of SO_2 to be removed are both proportional to SO_2 gas concentration. As SO_2 gas concentration increases above 500 ppm, the enhancement factor decreases substantially and SO_2 flux does not increase as fast as the amount of SO_2 to be removed. Therefore, the percentage of SO_2 removal decreases at higher SO_2 gas concentrations. As a result, achieving a given percentage of SO_2 removal is usually easier at low gas concentration than at high gas concentration. Because of changing gas concentration across a scrubber, the mass transfer of SO_2 can be limited by liquid-film resistance at the gas inlet and by gas-film resistance at the gas outlet.

Effects of pH and Excess Limestone

The pH level of the scrubber solution is often a direct indicator of concentrations of both HSO_3^- and $SO_3^=$ in the bulk solution and is often correlated with SO_2 removal. Because of the tendency of the solution to be in equilibrium with $CaSO_3$ solids, either by dissolution or by crystallization, the HSO_3^- concentration is roughly correlated with pH by the equilibrium:

$$CaSO_3\,(s) + H^+ \rightleftarrows Ca^{++} + HSO_3^-$$

$$[HSO_3^-] = K\,[H^+]/[Ca^{++}]$$

Hence lower pH at any point in the system always gives higher HSO_3^-, which inhibits the hydrolysis reaction and thereby reduces the enhancement factor and SO_2 removal.

The pH level at the scrubber inlet is also an indicator of the amount of excess limestone; higher pH indicates more excess $CaCO_3$. As solution passes through the scrubber, the drop in pH depends on the extent to which limestone dissolves and replenishes the alkalinity of the solution. With a large excess of $CaCO_3$, high pH and low HSO_3^- are maintained throughout the scrubber, whereas with little excess $CaCO_3$, the pH drops and HSO_3^- increases as SO_2 is absorbed. In the absence of $CaCO_3$, $CaSO_3$ will dissolve with an even greater increase in HSO_3^-:

$$CaSO_3 + SO_2 + H_2O \rightarrow Ca^{++} + 2HSO_3^-$$

In general the $SO_3^=$ concentration is not a function of pH, since it tends to be controlled by the equilibrium:

$$CaSO_3\,(s) \rightleftarrows Ca^{++} + SO_3^=$$

At high pH, however, $CaSO_3$ tends to crystallize in the scrubber, giving somewhat higher $SO_3^=$; at low pH, $CaSO_3$ tends to dissolve, requiring lower $SO_3^=$.

Effect of Alkali Additives

If a soluble alkali salt of hydroxide, carbonate, sulfate, or sulfite is added to a $CaCO_3$ slurry scrubbing system, the alkali will accumulate in solution primarily as the sulfate salt. For example, with Na_2CO_3 addition:

$$Na_2CO_3 + 2HSO_3^- + 2CaSO_4\,(s) \rightarrow 2Na^+ + SO_4^= + 2CaSO_3\,(s) + H_2O + CO_2$$

This reaction results in higher levels of dissolved sulfite because of the equilibrium:

$$CaSO_3(s) + SO_4^= \rightleftarrows CaSO_4(s) + SO_3^=$$

The total sulfite concentration will be proportional to the sulfate concentration and inversely proportional to the relative saturation level with respect to $CaSO_4$ solid (gypsum):

$$[SO_3^=] = K \frac{a_{CaSO_3}}{a_{CaSO_4}} [SO_4^=]$$

The higher level of dissolved sulfite enhances diffusion of SO_2 through the liquid film by reaction with SO_2 to produce bisulfite. It can also reduce the pH drop across the scrubber and minimize the dissolution of $CaCO_3$ or $CaSO_3$ in the scrubber.

The most significant alkali additives are salts of sodium or magnesium. Magnesium generates more sulfite species because of its interactions with sulfite and sulfate, giving the equilibrium:

$$MgSO_4^\circ + CaSO_3 (s) \rightleftarrows CaSO_4 (s) + MgSO_3^\circ$$

The $MgSO_3^\circ$ ion pair diffuses at a rate about 45 percent less than free sulfite ion generated in sodium solution. Therefore, magnesium and sodium additives are probably equivalent in their effects on SO_2 mass transfer. Figure A-2 shows the effect of total dissolved sulfite ($SO_3^=$ + HSO_3^-) on the enhancement factor at pH 4.5 in solutions containing Na_2SO_4 or $MgSO_4$ (Rochelle 1980c).

Calculated concentrations of sulfite and bicarbonate species generated by addition of MgO and Na_2CO_3 are given in Figures A-3 and A-4 (Rochelle 1977). These calculations assume $CaSO_3$ saturation of 1.0, with variable levels of $CaSO_4$ saturation; they also assume that the solutions contain 0.1 M Cl^-. The effect of $CaSO_4$ saturation is significant. In systems where oxidation is less than 15 to 20 percent, gypsum does not crystallize and the $CaSO_4$ saturation can be as low as 20 to 30 percent. With 30 percent gypsum saturation the effectiveness of alkali additives is increased by a factor of 3.

The simplest alkali additives are Na_2CO_3 and MgO. Dolomitic lime can be used to provide MgO; however, the magnesium content of most dolomitic or high-magnesium limestones is not expected to be reactive. Cooling tower blowdown, makeup water, or alkaline fly ash can be a significant source of $MgSO_4$ or Na_2SO_4.

Chloride accumulation in a scrubber system suppresses the desirable effects of alkali additives by permitting their accumulation as chloride salts rather than sulfate salts. The HCl in flue gas is absorbed in dissolved $CaCl_2$ in the absence of alkali additives. The first increment of alkali additive ends up as dissolved chloride by the reaction:

$$2Ca^{++} + 2Cl^- + MgO + 2HSO_3^- \rightarrow Mg^{++} + 2Cl^- + 2CaSO_3(s) + H_2O$$

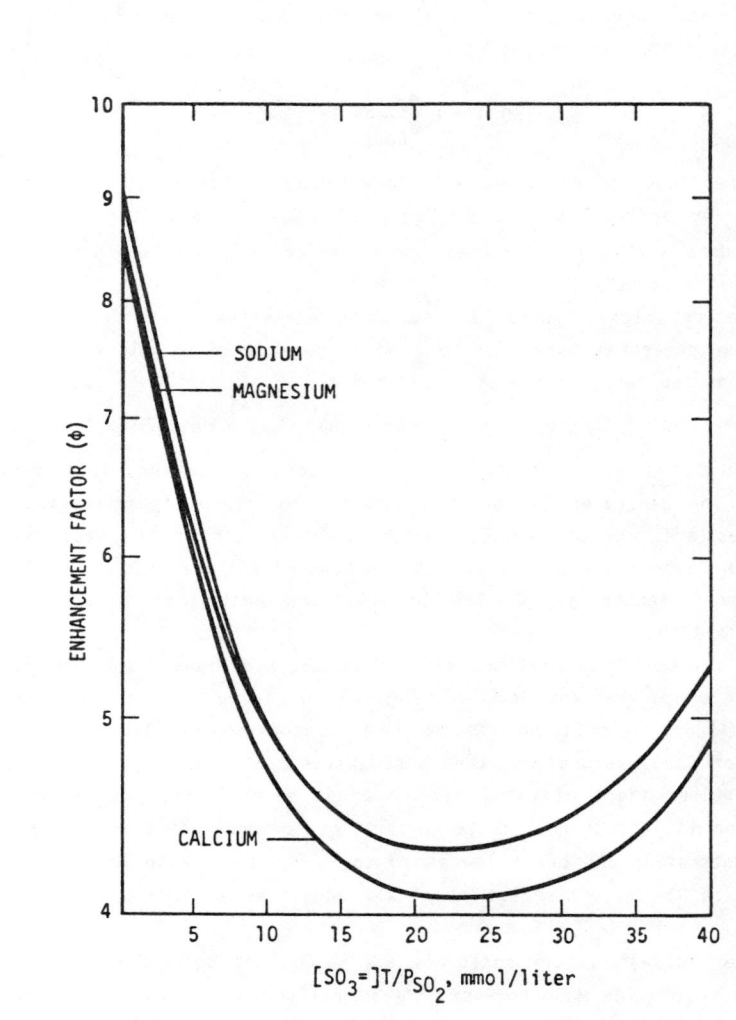

Figure A-2. Effect of total dissolved sulfite concentration on mass transfer enhancement at pH 4.5, 55°C, with 1000 ppm SO_2, 0.3 Molar ionic strength.

Appendix A: Chemistry of Limestone Scrubbing 245

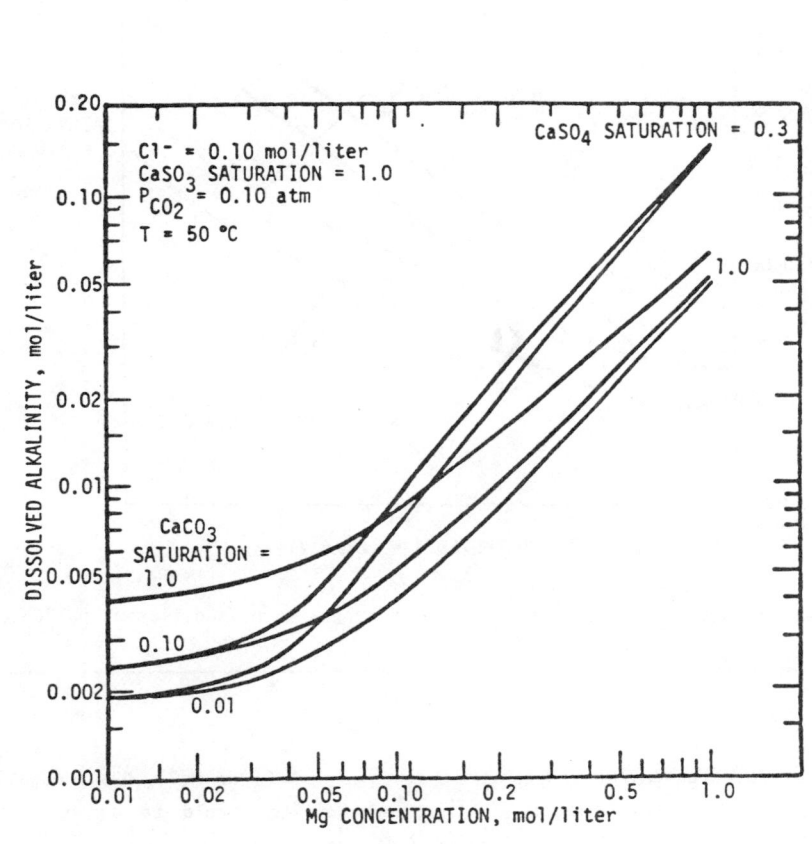

Figure A-3. Dissolved alkalinity generated by addition of MgO.

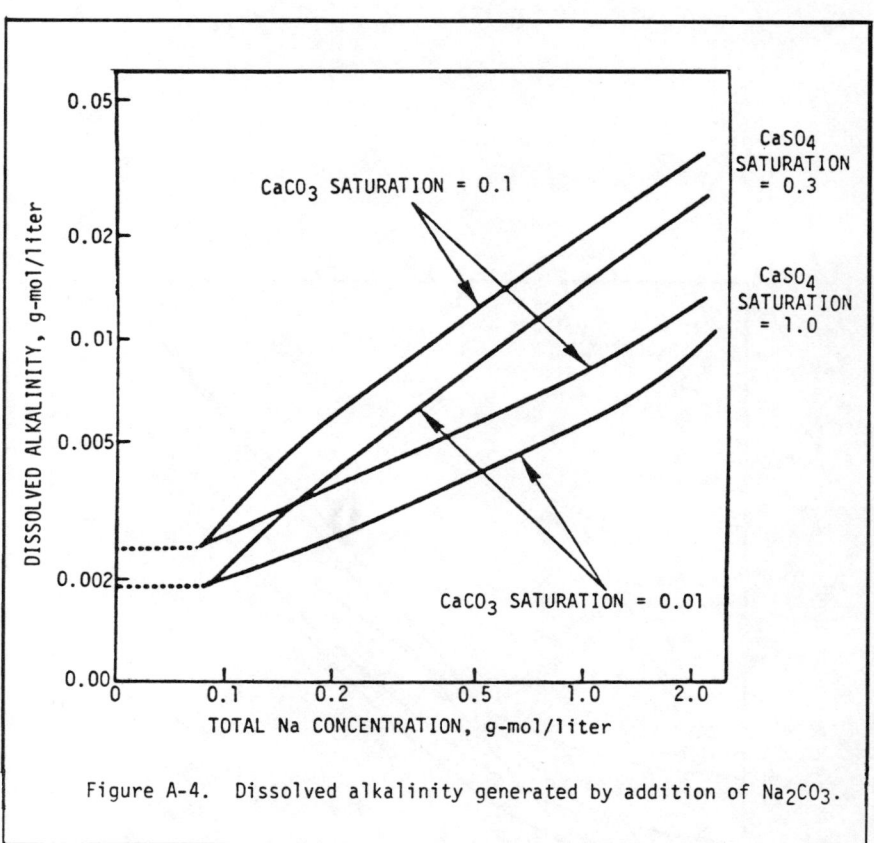

Figure A-4. Dissolved alkalinity generated by addition of Na_2CO_3.

By the same reasoning, makeup water containing NaCl or $MgCl_2$ has a negative effect. The actual amount of dissolved sulfate should be approximately equal to the "liquid goodness factor" (LGF) defined by:

$$[SO_4^=] \sim LGF = [Mg^{++}] + \tfrac{1}{2}[Na^+] - \tfrac{1}{2}[Cl^-]$$

Work at Shawnee has shown that an LGF of 0.2 to 0.4 gmol/liter is sufficient to get significant enhancement of SO_2 absorption (Burbank and Wang 1980a).

Effect of Buffer Additives

Any additive with buffer capacity between the pH of the gas-liquid interface (3 to 4) and the pH of the bulk liquid (4.5 to 5.5) will enhance diffusion of SO_2 through the liquid film by converting it to bisulfite (Rochelle 1977; Chang and Rochelle 1980):

$$SO_2 + A^- + H_2O \rightleftarrows HSO_3^- + HA$$

Such buffer additives also reduce pH drop across the scrubber and minimize dissolution of $CaCO_3$ and $CaSO_3$ in the scrubber. Buffer concentrations as low as 500 to 1000 ppm provide significant enhancement of SO_2 absorption and can also permit satisfactory SO_2 removal at lower pH with improved limestone utilization. Unlike alkali additives, buffer additives are unaffected by chloride accumulation.

Table A-2 gives the concentration of several alternative buffer additives required to get an enhancement factor of 20 in a typical scrubbing solution (Chang and Rochelle 1980). Relative costs of additive makeup have been calculated, on the assumption that makeup rate is proportional to additive concentration. Acetic formic acid appear to be most attractive, but would have problems with volatility. Sulfosuccinic and sulfopropionic are made in situ by adding maleic anhydride or acrylic acid, respectively, to the scrubber loop.

Waste streams of carboxylic acids from the manufacture of adipic acid or cyclohexanone are not listed in Table A-2, but should be more cost-effective than the additives shown there. Figure A-5 shows enhancement factors for several different buffers at typical scrubber conditions (Weems 1981).

Adipic acid has been tested extensively by EPA at Research Triangle Park and Shawnee. Figure A-6 shows the effect of adipic acid on SO_2 removal in the TCA scrubber at the Shawnee test facility (Burbank and Wang 1979). EPA is supporting a demonstration of adipic acid at the Southwest No. 1 station, City of Springfield, Missouri.

Under some conditions organic additives such as adipic acid will oxidize in the scrubber system. Shawnee operation has demonstrated that adipic acid losses are minimized by operating at a scrubber inlet pH less than 5.0 with 10 to 50 ppm of dissolved manganese (Burbank and Wang 1980b; Rochelle 1980f). Under these conditions the adipic acid makeup is equal to that lost with entrained solution in the waste solids.

Effect of Forced Oxidation

Complete oxidation of sulfite by air injection in the EHT or as a result of low SO_2/O_2 ratios in the flue gas will usually improve SO_2 absorption (Borgwardt 1978). Oxidation reduces the bisulfite concentration in the solution and thereby permits increased enhancement of SO_2 diffusion through the liquid film by means of the hydrolysis reaction (Chang and Rochelle 1980):

$$SO_2 + H_2O \rightarrow H^+ + HSO_3^-$$

This is especially true at lower pH levels (4 to 5), where operation with $CaSO_3$ solids would give a high concentration of HSO_3^-.

TABLE A-2. CONCENTRATIONS OF BUFFER ADDITIVES REQUIRED TO ACHIEVE ENHANCEMENT FACTOR OF 20

Basis: $\phi = 20$, 55°C, pH 5.0, $[Ca^{++}] = 0.1$ M, $[SO_2]_i = 0.5$ mmol/liter, $[SO_3^=]_T = 10$ mmol/liter

Acid	Acid, mmol/liter	Cost,[a] $/lb mol	Relative cost
Formic	17.7	12.9	0.51
Acetic	14.3	11.4	0.37
Adipic	7.0	63.5	1.0
Sulfosuccinic	7.8	44.1[b]	0.77
Sulfopropionic	16.1	28.8[c]	1.04
Hydroxypropionic	14.5	28.8[c]	0.94
Phthalic	6.4	58.1[d]	0.84
Succinic	7.3	106.0[e]	1.74
Benzoic	15.8	36.7	1.30
Glycolic	22.4	30.4	1.53
Lactic	24.7	73.8	4.10

[a] Chemical Marketing Reporter, July 2, 1979.
[b] Maleic anhydride
[c] Acrylic acid
[d] Phthalic anhydride
[e] Succinic anhydride

Source: Chang and Rochelle 1980.

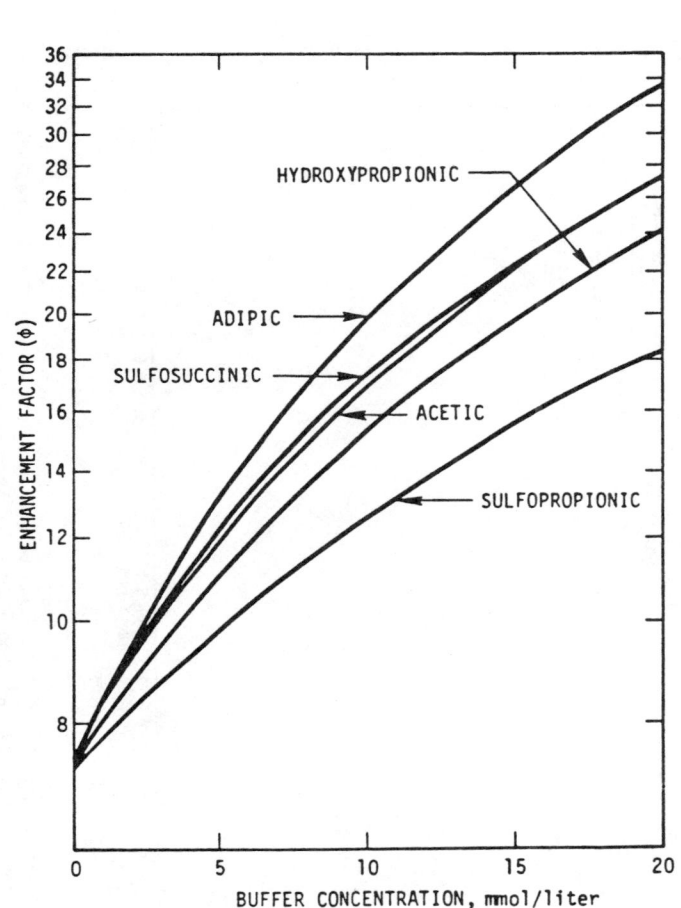

Figure A-5. Effect of organic acids on the enhancement factor, pH 5, 0.3 M $CaCl_2$, 55°C, 1000 ppm SO_2, 3 mM total sulfite.

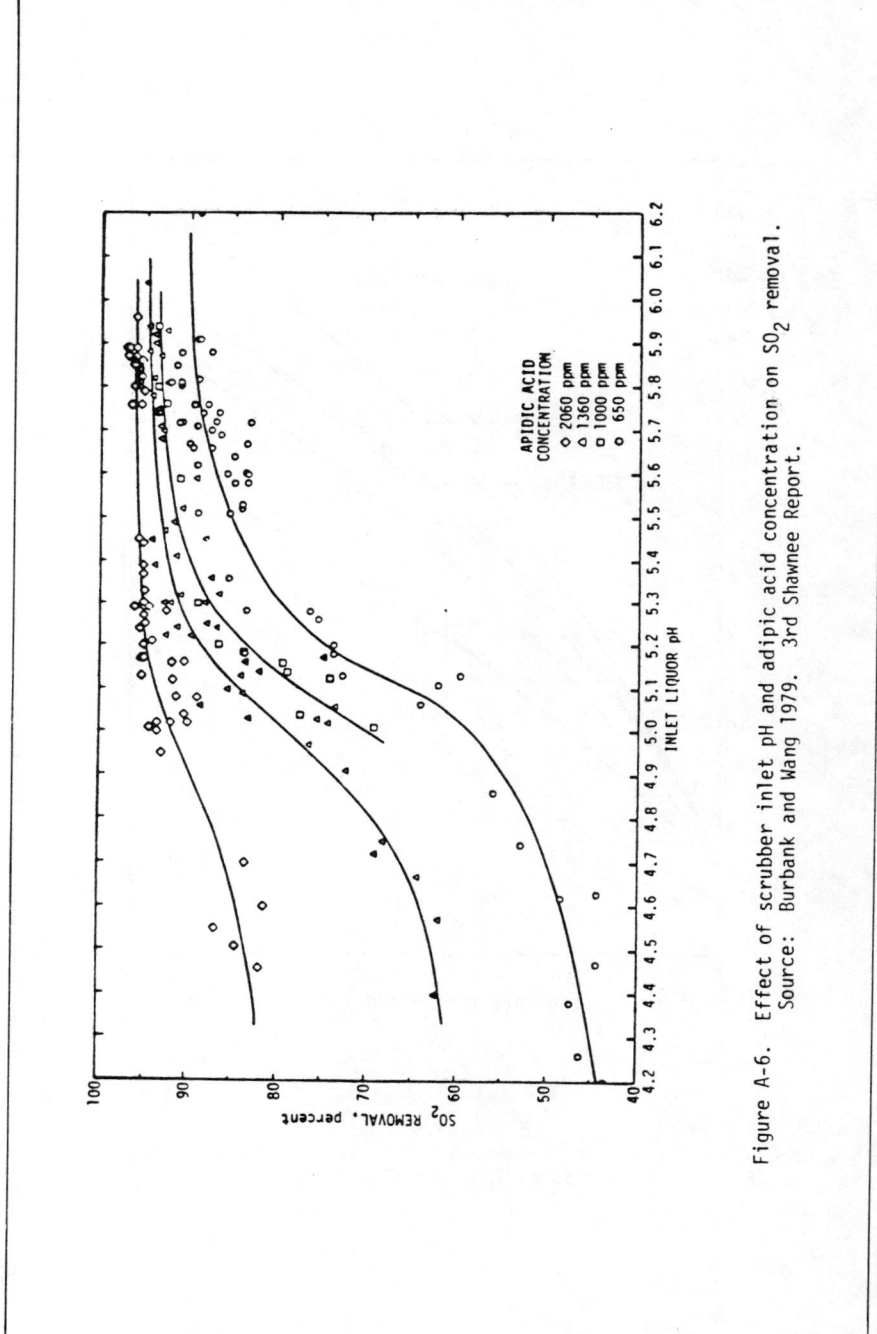

Figure A-6. Effect of scrubber inlet pH and adipic acid concentration on SO_2 removal. Source: Burbank and Wang 1979. 3rd Shawnee Report.

Complete oxidation in the scrubber loop is undesirable with alkali additives because it eliminates dissolved sulfite, which is required to enhance mass transfer. Alkali additives are effective with forced oxidation of a slurry bleed stream. Complete oxidation does not reduce the effectiveness of buffer additives, but with operation above pH 5 it greatly increases oxidative degradation of organic acids (Burbank and Wang 1980b).

LIMESTONE DISSOLUTION

Limestone dissolution occurs in both the scrubber and the EHT. Ideally about half of the limestone would dissolve in the scrubber to maximize SO_2 removal and pH and prevent $CaSO_3$ scaling. In practice the fraction dissolved in the scrubber varies with the amount of unreacted limestone slurry. With a large excess of limestone, practically all of the dissolution will occur in the scrubber. With little excess limestone, most of it will dissolve the the EHT.

Equilibrium

With a large excess of limestone or a large EHT, the solution in the EHT will approach equilibrium with $CaCO_3$ as given by:

$$CaCO_3 \text{ (s)} + 2H^+ \rightleftarrows Ca^{++} + CO_2 + H_2O$$

$$a_{H^+} = K \frac{[Ca^{++}]^{0.5}}{P_{CO_2}^{0.5}}$$

The equilibrium pH depends on the dissolved calcium concentration and the equilibrium partial pressure of CO_2 over the solution. Hence an accumulation of $CaCl_2$ tends to give lower pH, whereas an accumulation of sodium or magnesium sulfate should tend to give higher pH because dissolved calcium is reduced by the equilibrium:

$$CaSO_4 \text{ (s)} \rightleftarrows Ca^{++} + SO_4^=$$

Carbon dioxide generated by limestone dissolution in the EHT or the tank is usually stripped out of the solution by the flue gas in the scrubber. Very little CO_2 should desorb from the EHT unless the CO_2 vapor pressure exceeds 1 atm. Solution entering the EHT is probably saturated with SO_2 at the conditions of the flue gas, about 0.1 atm. As $CaCO_3$ dissolves in the EHT, CO_2 accumulates in the EHT solution. Thus the equilibrium CO_2 partial pressure in the EHT depends on the fraction of limestone dissolved in the EHT, the amount of SO_2 absorbed, and limestone dissolved per pass through the scrubber. Therefore, equilibrium pH out of the EHT tends to be lower with less excess $CaCO_3$ (giving a larger fraction dissolved in the EHT) and with a lower liquid circulation rate or higher SO_2 gas concentration (giving higher make-per-pass). The combined effects of Ca concentration and CO_2 partial pressure can give an equilibrium pH of 5.5 to 6.5 in the EHT.

Mass Transfer

In the scrubber and in the EHT with low excess limestone or short residence time, the solution usually is not in equilibrium with $CaCO_3$. Dissolution of limestone in the system must occur at the same rate as SO_2 absorption. Composition of the system solution will adjust until the rate of dissolution is equal to the rate of SO_2 absorption. Generally a low pH level reduces the rate of SO_2 absorption while increasing the rate of $CaCO_3$ dissolution.

The rate of $CaCO_3$ dissolution is usually controlled by diffusion of acid/base species in the liquid film surrounding a limestone particle (Rochelle 1979; 1980b). In the simplest case, dissolution is controlled by H^+ diffusion from the bulk solution to participate in the reaction:

$$CaCO_3 + H^+ \rightleftarrows Ca^{++} + HCO_3^-$$

Therefore, limestone dissolves about 10 times faster at pH 4.5 than at pH 5.5. In this case the rate of $CaCO_3$ dissolution (R_d, gmol/sec) is given approximately by the mass transfer expression:

$$R_d = k_d \cdot X_{CaCO_3} \cdot V \frac{\pi}{d_{ave}} (10^{-pH} - 10^{-pH^*})$$

where:

k_d = mass transfer coefficient, m/sec

X_{CaCO_3} = volume fraction $CaCO_3$ solids

V = slurry holdup, m^3

d_{ave} = mean particle diameter, m

pH^* = equilibrium pH at the limestone surface

The dissolution rate is a strong function of particle size distribution. The mass transfer coefficient is constant with large particles; it varies inversely with diameter with smaller particles. The total surface area available for mass transfer varies directly with the volume fraction of $CaCO_3$ solids in the slurry and with the specific surface area, which is inversely proportional to particle diameter. Therefore the rate of dissolution always tends to be greater with finer grinds of limestone. With reasonably pure limestones (> 90 percent $CaCO_3$), the reactivity depends only on the particle size distribution and not on the source of the stone (Rochelle 1980a,b). Available stones differ primarily in grindability. Thus a marble will be reactive if it is ground sufficiently fine.

The absolute rate of limestone dissolution varies with liquid holdup. Typical liquid residence time in the scrubber is 1 to 10 seconds, whereas in the EHT it is hundreds or thousands of seconds. The pH, however, is as much as one unit lower in the scrubber, and the EHT frequently operates much closer to the equilibrium pH. Therefore, absolute rates of dissolution are about the same in the scrubber and the EHT.

Effect of Sulfite. In the presence of normal levels of dissolved sulfite, $CaCO_3$ dissolution is 2 to 4 times faster than in solutions with greater amounts of dissolved sulfite (Rochelle 1980e). The mass transfer of acid/base species is enhanced by the diffusion of $SO_3^=/HSO_3^-$ buffer species participating in the reaction:

$$CaCO_3 + HSO_3^- \rightleftarrows Ca^{++} + SO_3^= + HCO_3^-$$

The concentrations of Ca^{++} and $SO_3^=$ are greater at the surface of limestone than in the bulk solution, so there is a tendency for $CaSO_3$ to precipitate at the limestone surface.

At high levels of dissolved $CaSO_3$ in the bulk solution, $CaCO_3$ dissolution can be significantly inhibited or even stopped by formation of a $CaSO_3$ layer on the limestone (Rochelle 1980e). Such high $CaSO_3$ supersaturations typically occur in slurries with no $CaSO_3$ solids but high levels of dissolved sulfite. This condition can be encountered in unsteady-state operation or in steady-state operation where the oxidation rate is high enough to deplete solid sulfite but not to eliminate dissolved sulfite. The condition has occurred frequently at the Shawnee test facility with forced oxidation in the EHT (Burbank and Wang 1980b).

Limestone Utilization. In a simple limestone slurry scrubbing system the limestone utilization is an independent variable that varies directly with the amount of limestone fed to the system. SO_2 removal requires some excess limestone in the scrubber, which usually leads to excess limestone in the solid waste, or utilization less than 100 percent. For a given level of SO_2, limestone utilization can be improved up to a point by using a larger EHT or several EHT's in series (Borgwardt 1975; Burbank and Wang 1980a). Utilization can also be improved by finer grinding of the limestone. With an increase in slurry concentration the concentration of $CaCO_3$ solids in the slurry can be held constant while the fraction of the total solids is reduced, with a resulting increase in limestone utilization. Typically, scrubber systems are designed for use of 5 to 50 percent excess limestone.

Countercurrent and crossflow scrubbers can be useful in improving limestone utilization. With a double-loop countercurrent scrubber, the top scrubber is operated at high pH with poor limestone utilization but good SO_2 removal. Slurry from the top scrubber is fed to the bottom scrubber (gas inlet), which operates at low pH with high limestone utilization. With crossflow scrubbing, the slurry bleed from several parallel scrubbers operating with excess limestone at high-pH is fed to an additional parallel scrubber operating with little excess limestone at low-pH. The reduced SO_2 removal in the low-pH scrubber is more than offset by improved performance of the high-pH scrubbers.

Limestone utilization can also be improved by separating and recycling unreacted limestone from the scrubber bleed. When coarse limestone is used and rather fine $CaSO_3$ crystals are produced, separation can be achieved by a cyclonic separator. This arrangement permits operation of the scrubber with excess limestone even though the system undergoes little loss of unreacted limestone.

Buffer and Alkali Additives. Additives used to enhance SO_2 gas/liquid mass transfer have both good and bad effects on limestone dissolution. In general any additive or scrubber design that improves SO_2 removal at a constant limestone utilization rate can be used to achieve the same SO_2 removal with improved limestone utilization.

Buffer additives such as adipic acid also provide specific enhancement of limestone dissolution. The buffer species contributes to diffusion of acid/base species by the reaction:

$$CaCO_3 + HA \rightleftarrows Ca^{++} + A^- + HCO_3^-$$

Thus diffusion of the buffer acid can substitute for diffusion of H^+. Figure A-7 shows the effect of a simple buffer such as acetic acid on the dissolution rate of $CaCO_3$ (Rochelle 1980d). The effects of adipic acid and other useful buffers are similar.

Alkali additives reduce the level of dissolved calcium and thereby increase the equilibrium pH for limestone dissolution. Increasing the concentrations of the $SO_3^=/HSO_3^-$ buffer (without increasing $CaSO_3$ saturation) may also increase the $CaCO_3$ dissolution rate.

There is reason to believe, however, that Mg additives may have a negative effect on $CaCO_3$ dissolution. There is evidence that Mg^{++} inhibits $CaCO_3$ crystallization and thereby increases $CaSO_3$ saturation (Jones et al. 1976). An increase in $CaSO_3$ saturation is likely in turn to cause $CaSO_3$ blinding of limestone, which will reduce limestone utilization. This negative feature of Mg additives may be offset by the added capability to operate at lower pH and still achieve acceptable SO_2 removal.

Inhibitors. Several substances are known to be potent inhibitors of $CaCO_3$ dissolution. Heavy metals such as Fe and Mg can form insoluble carbonates that adsorb on and blind the $CaCO_3$ surface. Phosphate, polyacrylic acid, and perhaps other polyelectrolytes used in water treatment or

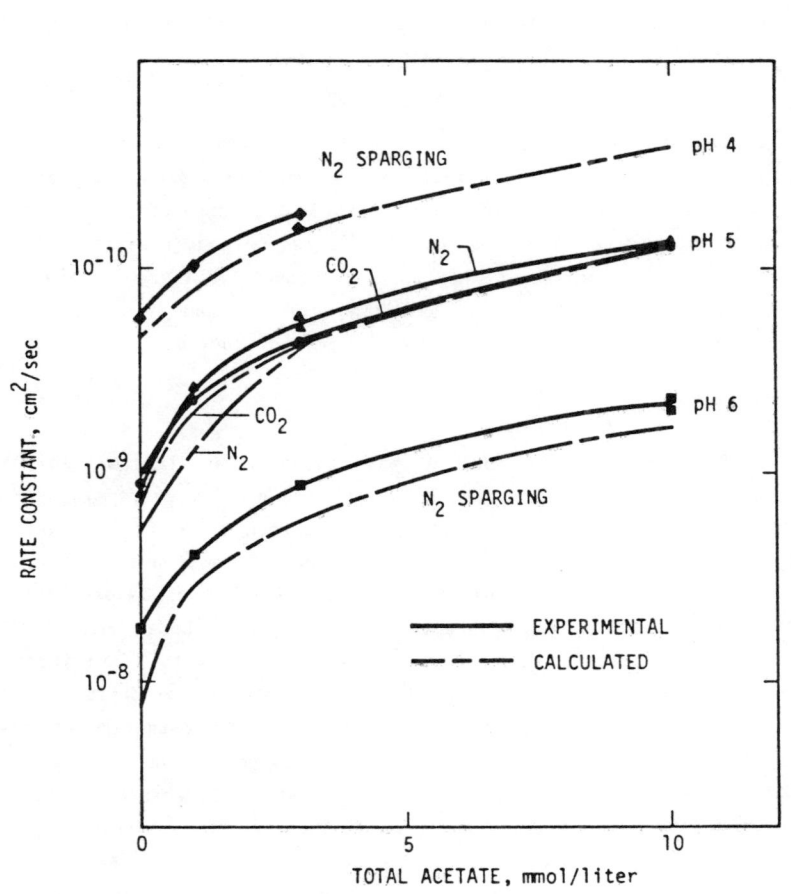

Figure A-7. Dissolution rate of $CaCO_3$ in 0.1M $CaCl_2$, 25°C. Source: Rochelle 1980d.

as flocculating agents inhibit $CaCO_3$ dissolution by surface adsorption (Rochelle 1977).

OXIDATION

Because flue gas contains 3 to 10 percent O_2, SO_2 absorbed as sulfite and bisulfite can be irreversibly oxidized to sulfate:

$$HSO_3^- + \tfrac{1}{2}O_2 \rightarrow SO_4^= + H^+$$

As discussed with regard to $CaSO_3/CaSO_4$ crystallization, it is usually best to oxidize either less than 15 to 20 percent of the absorbed SO_2 or all of it. The most difficult operating regime is in the range of 20 to 50 percent oxidation. To enable operation out of this regime and generation of easily dewatered solids, operators can achieve complete forced oxidation by sparging air in the EHT or by oxidizing the scrubber bleed slurry in a separate sparged tank. Hence, the objectives of controlling oxidation are either to inhibit it or to effect it completely.

Mass Transfer

The reaction rate of $SO_3^=/HSO_3^-$ with dissolved O_2 is reasonably high, especially with dissolved Mn and Fe, which are usually present in scrubber systems (Hudson 1980). The solubility of oxygen in aqueous solutions is very limited. Therefore, the degree of oxidation is frequently limited by the diffusion of O_2 through the liquid film. As soon as O_2 penetrates the liquid film it is consumed by reaction with $SO_3^=/HSO_3^-$. Under this condition, the apparent oxidation rate is unaffected by catalysts, inhibitors, and other variables that would normally affect inherent oxidation kinetics. Rather, the degree of oxidation depends on gas/liquid contacting efficiency, moles of SO_2 removed, and the concentration of oxygen in the flue gas.

The liquid-phase mass transfer coefficient of O_2 absorption is essentially the same as that which applies for physical absorption of SO_2 ($k_l^o a$). Therefore, additional oxidation should be expected if SO_2 removal is improved by adding mass transfer capability in the form of higher liquid flow rate or increased pressure drop, or in some other form. Hence, if oxidation is to be minimized, a scrubber should not be overdesigned for SO_2 removal. The use of alkali or buffer additives could reduce the degree of oxidation by permitting equivalent SO_2 removal with reduced mass transfer capability.

The O_2/SO_2 mole ratio in the gas can directly affect the degree of oxidation. The O_2 absorption rate and oxidation rate are directly proportional to O_2 concentration. If the percent SO_2 removal is constant, the SO_2 absorption rate will be directly proportional to SO_2 inlet gas concentration. Therefore, the percentage of oxidation of the product solids should be directly proportional to the O_2/SO_2 ratio. This factor is evident in the observed trend from 15 to 30 percent oxidation with high-sulfur coal and 50 to 100 percent oxidation with low-sulfur coal.

Reaction Kinetics

With very high concentrations of catalyst and bisulfite it is possible to consume the oxygen even before it gets through the liquid film. Since the effective film thickness is smaller, a very fast chemical reaction will enhance the mass transfer of O_2. The most important catalysts in limestone scrubbing are probably Mn and Fe. As much as 50 ppm Mn has been accumulated in testing of forced oxidation at Shawnee. As little as 0.5 ppm Mn can have a significant effect on the oxidation kinetics (Hudson, 1980). There is also evidence that NO_2 can have a catalytic effect on oxidation (Rosenberg and Grotta 1980). High levels of dissolved $SO_3^=/HSO_3^-$ also appear to enhance oxidation kinetics. Generally higher levels of oxidation are observed in operation at lower pH levels because of the increase in HSO_3^- concentration and in concentrations of metal catalyst, both of which appear to be solubility limited. At Shawnee, oxidation levels with high-sulfur coal have increased from 20 to 30 percent at pH 5.5 to 50 to 60 percent at pH 4.5 (Burbank and Wang 1980b).

In the presence of potent inhibitors or in the absence of catalysts and dissolved sulfite, it is sometimes possible to achieve oxidation rates lower than those predicted from mass transfer theory. Sodium thiosulfate has been identified and tested as a potent inhibitor in lime scrubbing (Holcomb and Luke 1978). At concentrations as low as 25 ppm it essentially stops sulfite oxidation at pH 5.0 (Hudson 1980). This substance should also be an effective inhibitor in limestone scrubbing. Hudson (1980) also showed that glycolic acid was a moderate inhibitor.

Forced Oxidation

Forced oxidation, described in Section 1 among the process options, is considered because it permits more effective sludge dewatering, which reduces the quantity of sludge produced and makes it easier to dispose of. Forced oxidation can be accomplished by sparging air into the EHT or by oxidizing the slurry bleed in a separate sparged reactor. Forced oxidation in the low-pH EHT of a double-loop scrubber is relatively easier because of the sustained low pH. In both bleed stream and double-loop forced oxidation, the dissolution of $CaSO_3$ solids can limit the oxidation rate. In single-loop oxidation, $CaSO_3$ solids are never crystallized and do not need to be dissolved; however, the blinding of limestone can be severe.

Testing of single-loop forced oxidation has shown that complete oxidation can be achieved over the normal pH range, 5.0 to 6.2 (Borgwardt 1978).

CO_2 is stripped by air sparging of the EHT, so that pH values as high as 6 to 6.5 can be achieved with high air stoichiometries. High inlet pH combined with very low dissolved bisulfite gives somewhat better SO_2 removal in the scrubber. Insufficient air stoichiometry or agitation can result in reasonably complete solids oxidation (95 to 98 percent) without oxidation of dissolved sulfite. Because there are no $CaSO_3$ seed crystals, excessive $CaSO_3$ supersaturation can build up in the scrubber and result in blinding of $CaCO_3$. This problem can be avoided by using more air or agitation and by reducing the moles of SO_2 absorbed per pass through the scrubber (increasing the liquid rate) (Burbank and Wang 1980b).

In bleed stream or double-loop oxidation, the pH level usually must be lower than 5.5 to permit rapid dissolution of $CaSO_3$ solids (Head and Wang 1979). Hudson (1980) studied oxidation kinetics of $CaSO_3$ slurries at pH 4.3 to 6.0 and concluded that the oxidation rate declines at higher pH because the reduced solubility of the $CaSO_3$ reduces it dissolution rate. A low pH level in bleed stream oxidation requires either the addition of sulfuric acid to neutralize excess $CaCO_3$ or scrubber-loop operation to achieve very high limestone utilization. High limestone utilization can be achieved with satisfactory SO_2 removal by addition of adipic acid (Burbank and Wang 1980b). Alkali additives permit the use of bleed stream oxidation at pH 6 to 7.5 by increasing the concentration of dissolved sulfite in equilibrium with $CaSO_3$ solids (Head and Wang 1979).

Low-pH Operation. Complete $SO_3^=/HSO_3^-$ oxidation in the scrubber can permit satisfactory SO_2 removal at pH values as low as 3.5 to 4.0. With low-sulfur coals having alkaline fly ash it is sometimes desirable to operate at pH levels in this low range to leach Ca, Mg, and Na alkali from the fly ash and thereby reduce limestone makeup. The combination of low pH, high metals concentrations, and high O_2/SO_2 results in complete oxidation in the scrubber and gives sufficient SO_2 removal. Complete scrubber oxidation at low pH has also been achieved by a special scrubber design in which additional air is sparged in the scrubber vessel (Morasky et al. 1980).

$CaSO_3/CaSO_4$ CRYSTALLIZATION

SO_2 absorbed by $CaCO_3$ slurry scrubbing is removed from the system continuously as $CaSO_3$ and $CaSO_4$ solids. Normally little sulfite or sulfate is lost in solution purged from the system. Therefore $CaSO_4$ must crystallize at the rate of sulfite oxidation and $CaSO_3$ must crystallize at the rate of SO_2 absorption minus oxidation. In a given system the driving forces for crystallization and supersaturation will adjust until these rates are balanced.

The crystallization processes are important because they are directly responsible for the quality of solids, ease of dewatering, and for the level of supersaturation. High levels of supersaturation can lead to scaling and plugging of the scrubber and can inhibit limestone dissolution. Ideally, all crystallization of $CaSO_3$ and $CaSO_4$ should occur in the EHT, but practically, some crystallization in the scrubber is acceptable.

Crystallization Kinetics

The solution driving force for crystallization is defined in terms of relative saturation (RS). For example, the RS for gypsum is given by:

$$RS_{CaSO_4 \cdot 2H_2O} = \frac{a_{Ca^{++}} \cdot a_{SO_4^=}}{K_{SP_{CaSO_4 \cdot 2H_2O}}}$$

The activities of Ca^{++} and $SO_4^=$, $a_{CA^{++}}$ and $a_{SO_4^=}$, vary roughly with concentrations of dissolved calcium. The solubility product, K_{SP}, is derived from solubility data such that RS is equal to 1 when the solution is in equilibrium with gypsum solids. If RS is less than 1, the solids tend to dissolve; if greater than 1, they crystallize.

Figure A-8 shows the general dependence of crystallization or precipitation on RS for both $CaSO_3$ and $CaSO_4$. At low values of RS only crystal growth occurs and little nucleation or formation of new crystals is observed. In this region the crystallization rate is proportional to supersaturation. Above a certain critical level of saturation, 1.3 to 1.4 for gypsum ($CaSO_4$) and 6 to 8 for $CaSO_3$ (Ottmers et al. 1974), nucleation of new crystals occurs at an increasing rate and at higher saturations, and dominates the crystallization. Excessive nucleation results in smaller crystals, which are more difficult to dewater. High saturations giving excessive nucleation also result in crystallization of $CaSO_3$ and $CaSO_4$ on foreign surfaces, i.e., in scaling of scrubber equipment. One objective of EHT design is to minimize the RS of solution going to the scrubber.

At moderate RS levels the crystallization rate is given by:

$$R_c = k_c \beta V_R W_S (RS-1)$$

where:

R_c = net crystallization rate, gmol/sec

k_c = crystallization rate constant, gmol/m²-sec

β = specific surface area of solids, m²/g solids

V_R = volume of slurry, m³

W_S = slurry solids content, grams solids/m³ slurry

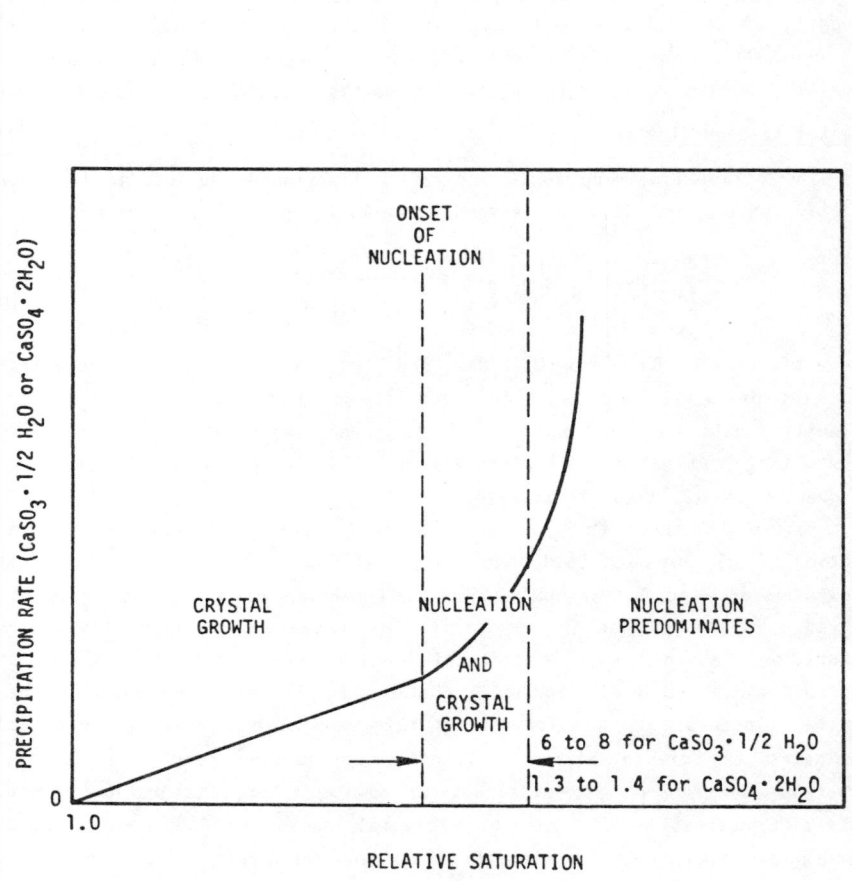

Figure A-8. Precipitation mechanism and rate as a function of relative saturation.

Since R_c is a constant related to the rate of SO_2 absorption and oxidation, an increase in the volume, V_R, of the EHT or of the slurry solids content, W_S, generally reduces RS of solution leaving the hold tank. The specific surface area, β, varies inversely with the mean particle diameter. Since an increase in nucleation gives smaller particles and larger specific surface area, it also increases the crystallization rate. Nucleation is usually a stronger function of saturation than crystal growth, so there can be interacting effects of an increase in RS. In general, operation at larger EHT volume and higher slurry solids content reduces RS and yields larger particles. Typically, EHT residence times of 5 to 30 minutes and solids concentrations of 8 to 15 percent are adequate to desupersaturate scrubber feed solutions.

$CaSO_3$ Scaling

The relative saturation of the sulfite solid product, $CaSO_3 \cdot \frac{1}{2}H_2O$, is strongly dependent on pH because its solubility is dominated by the equilibrium:

$$CaSO_3 + H^+ \rightleftarrows Ca^{++} + HSO_3^-$$

Solution entering the scrubber should be slightly supersaturated to $CaSO_3$. As the solution passes through the scrubber, the HSO_3^- concentration increases because of SO_2 absorption and the Ca^{++} concentration increases because of $CaCO_3$ or $CaSO_3$ dissolution, but the pH will decrease because SO_2 absorption as HSO_3^- adds H^+ to the solution. Therefore the RS of $CaSO_3$ leaving the scrubber depends on the extent to which the pH drop is neutralized by $CaCO_3$. With little $CaCO_3$ dissolution, the RS of $CaSO_3$ at the scrubber exit can be less than 1. If there is a large excess of $CaCO_3$ with resulting $CaCO_3$ dissolution, the RS of $CaSO_3$ will be excessive and will result in $CaSO_3$ crystallization and nucleation in the scrubber. Depending on the concentration of $CaSO_3$ solids and design of the scrubber, these conditions may or may not cause scaling.

Shawnee operation has shown that reliable performance of the mist eliminator requires avoidance of excess $CaCO_3$ (Head 1976). The presence of excess $CaCO_3$ causes $CaSO_3$ to crystallize in the mist eliminator and results in a "sticky" mud deposit. At Shawnee the mist eliminator was kept clean with limestone utilization greater than 85 percent. This result was obtained in operation with Fredonia fine limestone. Less reactive or coarser stones should give reliable operation at lower utilization. One potential problem of double-loop scrubbing is the presence of excessive, unreacted limestone in the high-pH loop.

CaSO₄ Scaling

CaSO₄ can crystallize gypsum, $CaSO_4 \cdot 2H_2O$, and as a hemihydrate in solid solution with $CaSO_3$, $(CaSO_3)_{1-x} \cdot (CaSO_4)_x \cdot \tfrac{1}{2}H_2O$. Relative saturation is usually defined in terms of gypsum. If the sulfate concentration (or oxidation) of the solid solution is less than 15 to 20 mole percent, the gypsum saturation is less than 1.0, so gypsum does not crystallize (Borgwardt 1973; Jones et al. 1976; Setoyami and Takahashi 1978). Under these conditions, crystallization rates are the same as those for $CaSO_3$. Furthermore, since gypsum saturation is less than 1.0, there is usually no problem with gypsum crystallization or scaling in the scrubber.

With oxidation levels greater than 15 to 20 percent, $CaSO_4$ is crystallized both as gypsum and as solid solution containing 15 to 20 percent $CaSO_4$. Gypsum saturation can be estimated at 50°C by the equation (Head 1977):

$$RS = C_{Ca^{++}} \, C_{SO_4^=} \left(\frac{263}{I} + 47 \right)$$

where:

$$I = 3[C_{Ca^{++}} + C_{Mg^{++}}] + C_{SO_4^=}$$

= ionic strength in solution containing only Ca^{++}, Mg^{++}, $SO_4^=$, and Cl^-

C = molarities of Ca^{++}, Mg^{++}, and $SO_4^=$

Solution entering the scrubber should be slightly supersaturated to gypsum. As the solution passes through the scrubber, the sulfate concentration increases because of sulfite oxidation and the calcium concentration increases because of $CaCO_3$ and $CaSO_3$ dissolution. For a given percentage of oxidation, the absolute concentration of changes of $SO_4^=$ and Ca^{++} vary proportionately with the moles of SO_2 absorbed per liter of solution passing through the scrubber (the SO_2 make-per-pass). Therefore, the gypsum saturation leaving the scrubber is greater than that entering, to an extent proportional to the make-per-pass. The make-per-pass varies inversely with the ratio of liquid-to-gas flow rates (L/G) through the scrubber. If the L/G is too low (and the make-per-pass is too large), the gypsum RS leaving the scrubber can exceed the critical level of 1.3 to 1.4 and can lead to severe scaling.

Chloride accumulation of alkali additives can influence gypsum scaling. With the accumulation of $CaCl_2$ in the scrubber inlet solution, the concentration of dissolved sulfate is reduced. Under these conditions a change of Ca^{++} concentration across the scrubber has less effect on the outlet gypsum saturation than a change in sulfate concentration. Therefore, oxidation in the scrubber is more troublesome than solids dissolution. On the other hand, the accumulation of sulfate salts from alkali additives depress the Ca concentration in the scrubber inlet. Therefore oxidation and the change of sulfate concentration are unimportant, and the dissolution of $CaCO_3$ or $CaSO_3$ solids and increase of dissolved calcium are critical to gypsum scaling.

Forced Oxidation

The use of forced oxidation in the scrubber loop can eliminate problems with both $CaSO_3$ scaling and gypsum scaling. If care is taken to oxidize completely both solid and dissolved sulfite and to avoid the use of excessive limestone, crystallization and scaling of $CaSO_3$ are avoided. Gypsum scaling is prevented by the presence of excess gypsum surface area in the scrubber. Increases in Ca^{++} and $SO_4^=$ across the scrubber are depleted by controlled crystallization on the gypsum solids in the scrubber before excessive saturation can develop. Operation of forced oxidation systems has shown no evidence of high saturations of gypsum (Head and Wang 1979).

SUMMARY

The chemical performance of a limestone slurry scrubbing system can be represented by the degree of SO_2 removal and the extent of scale-free operation. A given system must be designed to operate over a specific range of O_2 and SO_2 gas concentrations and with a specific degree of chloride or sulfate accumulation from impurities in the flue gas, fly ash, and makeup water. The important chemical design variables are limestone grind, EHT volume, percentage of slurry solids, and optional process configurations such as forced oxidation and double-loop scrubbing. Other design variables that may be manipulated to control or optimize an operating system include liquid to gas ratio (L/G), limestone utilization (or pH), and concentrations of additives such as soluble alkalis, buffers, and oxidation inhibitors/ catalysts.

SO_2 Removal

SO_2 removal is directly related to limestone utilization, particle size, and solids concentration in the scrubber. Relatively lower utilization, finer grind, and higher solids concentration facilitate SO_2 removal. To a lesser extent, greater volume of the EHT also improves SO_2 removal. High L/G not only increases mass transfer by physical effects but also enhances SO_2 removal by reducing the need for $CaCO_3$ dissolution and by reducing bisulfite concentration in the scrubber. These variables all tend to give higher pH and lower bisulfite concentration in the scrubber, which promote the liquid-phase diffusion of SO_2 as bisulfite through enhancement of the hydrolysis reaction:

$$SO_2 + H_2O \rightleftarrows H^+ + HSO_3^-$$

Alkali and buffer additives improve SO_2 removal without reducing limestone utilization. Alkali additives generate high concentrations of dissolved sulfate, which induce higher sulfite ($SO_3^=$) concentrations by solid/liquid equilibria of the form:

$$CaSO_3 \text{ (s)} + SO_4^= \rightleftarrows CaSO_4 \text{ (s)} + SO_3^=$$

Both sulfite and basic buffer species (A^-) enhance liquid-phase diffusion by reacting with SO_2 to allow its diffusion as bisulfite:

$$A^- + SO_2 + H_2O \rightleftarrows HA + SO_3^-$$

The maximum effect of these additives is achieved when SO_2 absorption is controlled by liquid-film diffusion rather than gas-film diffusion.

Double-loop scrubbing and other process options can lead to lower limestone utilization in the scrubber, and thereby improve SO_2 removal at a given rate of limestone utilization in the system.

Forced oxidation in the scrubber loop improves SO_2 removal by removing dissolved bisulfite from the scrubber feed. This increases the enhancement of mass transfer by the hydrolysis reaction.

The percentage of SO_2 removal is usually greater at lower SO_2 inlet gas concentrations. The enhancement of liquid-film diffusion by the hydrolysis reaction and by reaction with sulfite is greater at lower SO_2 concentration. In the range of 100 to 500 ppm SO_2, SO_2 removal is controlled by gas-film diffusion.

For soluble salts in the flue gas, flyash, makeup water, and alkali additives there is a strong interaction of chloride and sulfate accumulation. With the soluble ions, Na^+, Mg^{++}, and Cl^-, the sulfate accumulation is given by the "liquid goodness factor" (LGF):

$$LGF = Mg^{++} + 2Na^+ - 2Cl^-$$

Therefore in the range of positive LGF, higher chloride levels tend to reduce the accumulation of sulfate in solution and its positive effects on SO_2 removal.

Scale-free Operation

For scale-free operation the EHT must be designed and controlled so that there is no excessive supersaturation of $CaSO_3$ or $CaSO_4$ in the solution returning to the scrubber or in solution leaving the scrubber.

Relatively higher EHT volume and solids concentration reduce supersaturation at the hold tank exit, with a corresponding reduction of supersaturation leaving the scrubber. An increase in solids concentration provides an additional secondary reduction of scrubber supersaturation by means of controlled crystallization in the scrubber.

Low limestone utilization or blinding of finely ground limestone can cause $CaSO_3$ scaling in the scrubber, which in turn gives $CaSO_3$ crystallization by the stoichiometry:

$$CaCO_3(s) + SO_2 \rightarrow CaSO_3(s) + CO_2$$

Moderate levels of $CaCO_3$ dissolution in the scrubber give the acceptable stoichiometry:

$$CaCO_3 + 2SO_2 + H_2O \rightarrow Ca^{++} + 2HSO_3^- + SO_2$$

A high L/G ratio is needed to reduce the increase in gypsum ($CaSO_4 \cdot 2H_2O$) saturation across the scrubber. An increase in L/G ratio reduces the SO_2 make-per-pass and therefore reduces the moles/liter of $CaCO_3$ dissolution and sulfate formation.

Low O_2/SO_2 in the flue gas or the use of a potent oxidation inhibitor (e.g., sodium thiosulfate) can prevent gypsum crystallization and scaling by reducing solids oxidation below 15 to 20 percent. Under these conditions calcium sulfate is crytallized as a solid solution with the $CaSO_3$ solids, and the gypsum saturation can be substantially less than 1.

Forced oxidation in the scrubber loop prevents both $CaSO_3$ and $CaSO_4$ scaling. It eliminates $CaSO_3$ in the solids and solution. The presence of high $CaSO_4$ solids concentrations permits desupersaturation in the scrubber by controlled crystallization on gypsum surfaces.

REFERENCES FOR APPENDIX A

Borgwardt, R. H. 1974. Symposium on Flue Gas Desulfurization, Atlanta, Georgia, November, 1974. EPA-650/2-74-126a. NTIS No. PB-242572.

Borgwardt, R. H. 1975. Increasing Limestone utilization in FGD Scrubbers. Paper presented at the AICHE 68th Annual Symposium, Los Angeles, California, November 16-20, 1975.

Borgwardt, R. H. 1978. Effect of Forced Oxidation on Limestone SO_x Scrubber Performance. In: Proceedings of the Symposium on Flue Gas Desulfurization, Hollywood, Florida. November 1977. Vol. I. EPA-600/7-78-058a. NTIS No. PB-282 090.

Borgwardt, R. H. 1979. Significant EPA/IERL-RTP Pilot Plant Results In Proceedings Industry. Briefing on EPA LIme/Limestone Wet Scrubbing Test Programs, August 1978. EPA-600/7-79-092. NTIS No. PB-296 517.

Burbank, D. A., and S. C. Wang. 1979. Test Results on Adipic Acid-Enhanced Lime/Limestone Scrubbing at the EPA Shawnee Test Facility. Presented at the Industry Briefing on EPA Lime/Limestone Wet Scrubbing Test Program, Raleigh, North Carolina, December 5.

Burbank, D. A., and S. C. Wang. 1980a. EPA Alkali Scrubbing Test Facility: Advanced Program--Final Report (October 1974 to June 1978). EPA 600/7-80-115. NTIS No. PB80-204 241.

Burbank, D. A., and S. C. Wang. 1980b. Test Results on Adipic Acid - Enhanced Limestone Scrubbing at the EPA Shawnee Test Facility - Third Report. Presented at the Symposium on Flue Gas Desulfurization, Houston, Texas, October 28-31, 1980.

Chang, C. S., and G. T. Rochelle. 1980. Effect of Organic Acid Additives on SO_2 Absorption into $CaO/CaCO_3$ Slurries. Proceedings of the Second Conference on Air Quality Management in the Electric Power Industry, University of Texas at Austin, Austin, Texas.

Epstein, M. 1975. EPA Alkali Scrubbing Test Facility: Summary of Testing Through October 1974. EPA-650/2-75-047. NTIS No. PB-244 901.

Head, H. N. 1976. EPA Alkali Scrubbing Test Facility: Advanced Program, Second Progress Report. EPA-600/7-76-008. NTIS No. PB-258 783.

Head, H. N. 1977. EPA Alkali Scrubbing Test Facility: Advanced Program, Third Progress Report. EPA-600/7-77-105. NTIS No. PB-274 544.

Head, H. N., and S. C. Wang. 1979. EPA Alkali Scrubbing Test Facility: Advanced Program, Fourth Progress Report. EPA-600/7-79-244a. NTIS No. PB80-117 906.

Holcomb, L., and K. W. Luke. 1978. Characterization of Carbide Lime to Identify Sulfite Oxidation Inhibitors. EPA-600/7-78-176. NTIS No. PB-286 646.

Hudson, J. L. 1980. Sulfur Dioxide Oxidation in Scrubber Systems, EPA-600/7-80-083. NTIS No. PB80-187 842.

Jones, B. F., P. S. Lowell, and F. B. Meserole. 1976. Experimental and Theoretical Studies of Solid Solution Formation in Lime and Limestone SO_2 Scrubbers. Vol. 1. EPA 600/2-76-273a. NTIS No. PB-264 953.

Lowell, P. S., et al. 1970. A Theoretical Description of the Limestone Injection - Wet Scrubbing Process. Vol. I. NAPCA Report. NTIS No. PB-193 029.

Morasky, T. M., D. P. Burford, and O. W. Hargrove. 1980. Results of the Chiyoda Thoroughbred - 121 Prototype Evaluation. Presented at the EPA Symposium on Flue Gas Desulfurization, Houston, Texas, October 28-31, 1980.

Ottmers, D. M., Jr., et al. 1974. A Theoretical and Experimental Study of the Lime/Limestone Wet Scrubbing Process. EPA-650/2-75-006. NTIS No. PB-243 399.

Rochelle, G. T. 1977. Process Synthesis and Innovation in Flue Gas Desulfurization. EPRI FP-463-SR.

Rochelle, G. T. 1979. Monthly Progress Report for EPA Grant R806251.

Rochelle, G. T. 1980a. Monthly Progress Report (April) for EPA Grant R806251.

Rochelle, G. T. 1980b. Monthly Progress Report (May) for EPA Grant R806251.

Rochelle, G. T. 1980c. Monthly Progress Report (June) for EPA Grant R806743.

Rochelle, G. T. 1980d. Monthly Progress Report (July) for EPA Grant R806251.

Rochelle, G. T. 1980e. Monthly Progress Report (October) for EPA Grant R806251.

Rochelle, G. T. 1980f. Monthly Progress Report (October) for EPA Grant R806743.

Rochelle, G. T. 1980g. Monthly Progress Report (December) for EPA Grant R806743.

Rosenberg, H. S., and H. M. Grotta. 1980. NO_x Influence on Sulfite Oxidation and Scaling in Lime/Limestone FGD Systems. Env. Sci. and Tech., 14(4):470-472.

Setoyamak, K., and S. Takahashi. 1978. Solid Solution of Calcium Sulfite Hemihydrate and Calcium Sulfate. Yogyo-Kyokai-Ski, 86[5].

Weems, W. T. 1981. Enhanced Absorption of Sulfur Dioxide by Sulfite and Other Buffers. M.S. Thesis, University of Texas at Austin.

Appendix B

Operational Factors

The theoretical material in Appendix A, which deals with the important process chemistry parameters, also provides background for understanding the effects of some operational factors of a limestone FGD system. Many chemical and operational factors are interrelated. For example, the determination of liquid-to-gas (L/G) ratio on the basis of liquid phase alkalinity (LPA) in the scrubbing liquor is a chemical procedure, whereas the relation of L/G ratio to liquid/gas contact area is a physical consideration. Similarly, any discussion of scale formation and means of preventing it must consider both chemical and physical aspects.

Because of such interrelationships, this appendix is complementary to Appendix A. It emphasizes on-line operation of a limestone FGD system, and thus many of the references cited pertain to experience gained with operational systems.

LIQUID-TO-GAS RATIO

The proper ratio of slurry flow rate in the scrubber to flue gas flow rate is very important for effective removal of SO_2. The primary effect of increasing liquid circulation flow rate (or a higher L/G ratio for a given gas flow rate) is to increase the rate of mass transfer from gas to liquid, which in turn increases SO_2 removal efficiency.

Effect on SO_2 Removal

According to Corbett et al. (1977), the differential SO_2 absorption rate can be calculated by the following equation:

$$GdY = K_G (adV) P (Y-Y^*) \qquad \text{(Eq. B-1)}$$

where G = molar flow rate of the gas, lb-mol/h

Y = mole fraction of SO_2, lb-mol SO_2/lb-mol flue gas

K_G = over-all mass transfer coefficient, lb-mol/ft^2-h-atm a = gas-liquid interfacial mass transfer area, ft^2/ft^3 slurry

dV = volume of the slurry holdup in a small differential, ft^3

P = total pressure of the system, atm

Y^* = bulk gas-phase mole fraction of SO_2 in equilibrium with the bulk absorbing liquor, lb-mol SO_2/lb-mol gas

If we assume that the equilibrium mole fraction of SO_2 (Y^*) is negligible compared with the actual mole fraction (Y) in the gas phase and that values of the over-all mass transfer coefficient (K_G) and the gas-liquid interfacial area (a) are constant throughout the scrubber, the above equation can be integrated to yield the following (Corbett et al. 1977):

$$\eta = (Y_{in} - Y_{out})/Y_{in} = 1 - \exp[-K_G \, aP \, (V/G)] \qquad \text{(Eq. B-2)}$$

where η = SO_2 removal efficiency, percent

Y_{in} = mole fraction of SO_2 at the inlet of the scrubber, lb-mol SO_2/lb-mol flue gas

Y_{out} = mole fraction of SO_2 at the outlet of the scrubber, lb-mol SO_2/lb-mol flue gas

V = volume of slurry holdup in the scrubber, ft^3

Equation B-2 shows that for a given molar gas flow rate (G), the SO_2 removal efficiency can be increased by increasing the over-all mass transfer coefficient (K_G), the gas-liquid interfacial area (a), the total pressure of the gas phase (P), or the liquid holdup (V). In limestone scrubbing, the pressure is seldom increased and is close to 1 atmosphere. The effect of liquid flow rate on other variables is discussed later.

Corbett et al. (1977) have also shown the following relationship between the overall mass transfer coefficient (K_G) and individual mass transfer coefficients:

$$1/K_G = (1/k_g) + (H/k_\ell) \qquad \text{(Eq. B-3)}$$

where k_g = gas-side mass transfer coefficient, lb-mol/ft^2-h-atm

k_ℓ = liquid-side mass transfer coefficient, ft/h

H = Henry's Law constant, atm-ft^3/lb-mol

Because the individual mass transfer coefficients (k_g and k_ℓ) increase with the gas and liquid flow rates (Wen et al. 1975), an increase in the liquid flow rate increases the overall mass transfer coefficient (K_G).

The gas-liquid interfacial area (a) available for mass transfer depends on the type of the scrubber, but may also be affected by the liquid flow rate. If a venturi or spray scrubber is used and if the droplets are spherical and uniform, the area is given by:

$$a = 6/d \qquad \text{(Eq. B-4)}$$

where a = interfacial area, ft^2/ft^3 slurry

d = diameter of the droplet, ft

For a given number of nozzles, an increase in liquid flow rate produces finer droplets and thereby increases the interfacial area in a venturi or spray scrubber.

If a mobile-bed scrubber is used, the area is usually expressed in square feet per cubic foot of packing. If the packing is a bed of uniform spheres, the area is given by:

$$a' = 6/d' \qquad \text{(Eq. B-5)}$$

where a' = interfacial area, ft^2/ft^3 of packing

d' = diameter of the sphere, ft

In this case, an increase in liquid flow rate does not increase the interfacial area.

The liquid holdup (V) in a scrubber, which affects the degree of SO_2 removal, is given by Equation B-6 for a venturi or spray scrubber and Equation B-7 for a mobile-bed scrubber:

$$V = V_s (1 - E) \qquad \text{(Eq. B-6)}$$

$$V = V_p (1 - E') \qquad \text{(Eq. B-7)}$$

where V_s = volume of the scrubber, ft^3

V_p = volume of the packing, ft^3

E = voidage, ft^3/ft^3 of scrubber volume

E' = voidage, ft^3/ft^3 of packed volume

For a given scrubber, the volume (V_s) is fixed. For a venturi, spray, or mobile-bed scrubber, the voidage decreases with an increase in liquid flow rate, which increases the holdup of liquid in the scrubber (Treybal 1968).

Minimum L/G Ratio and Liquid Phase Alkalinity

The minimum liquid flow rate required for a given amount of SO_2 removal is determined by an overall material balance. Because each mole of SO_2 reacts with 1 mole of alkalinity in the liquid phase,

$$\text{Molar rate of } SO_2 \text{ removal} \leq \text{Molar rate of alkalinity fed to the scrubber} + \text{molar rate of alkalinity formation by dissolution} \quad \text{(Eq. B-8)}$$

If there is no dissolution of solids in the scrubber, gas and liquid flow rates are related as follows:

$$G(Y_{in} - Y_{out}) \leq L[(LPA)_R - (LPA)_S] \quad \text{(Eq. B-9)}$$

where   L = volumetric flow rate of the liquor phase of the recycle slurry, ft^3/h

$(LPA)_R$ = liquid phase alkalinity of the recycle slurry, $lb\text{-}mol/ft^3$

$(LPA)_S$ = liquid phase alkalinity of the spent slurry, $lb\text{-}mol/ft^3$

Equation B-9 can be rearranged to give

$$L/G_{min} = (Y_{in} - Y_{out})/[(LPA)_R - (LPA)_S] \quad \text{(Eq. B-10)}$$

Note that in Equation B-10 the L/G ratio is expressed in cubic feet of liquid per pound-mole of gas because G is a molar flow rate of the flue gas (lb-mol/h). The significance of $(LPA)_R$ and the need for dissolution of $CaCO_3$ in the scrubber are best illustrated by an example. For a high-sulfur coal application, the following values are typical:

Y_{in} = 3 x 10^{-3} lb-mol SO_2/lb-mol flue gas (equivalent to 3000 ppm SO_2)

Y_{out} = 3 x 10^{-4} lb-mol SO_2/lb-mol flue gas (equivalent to 300 ppm SO_2)

$(LPA)_R$ = 1.25 x 10^{-4} lb-mol/ft^3 (equivalent to 2 x 10^{-3} g-mol/liter)

$(LPA)_S$ = 1.25 x 10^{-5} lb-mol/ft^3 (equivalent to 2 x 10^{-4} g-mol/liter)

These values would correspond to a minimum L/G value of 24.0 ft^3/lb-mol or

500 gal/1000 scf. Because this L/G value is economically and technically infeasible, the value of $(LPA)_R$ must be increased by a factor of 6 to 8. This increase can be achieved by the use of alkali or buffer additives. Alternatively, the system must be designed to achieve dissolution of solids in the scrubber. The differential rate of solids dissolution can be calculated as follows:

$$R_d = (k_d)(A_p)(dV)(C^*_{alk} - C_{alk}) \qquad \text{(Eq. B-11)}$$

where
- R_d = rate of solids dissolution, lb-mol/h
- k_d = mass transfer coefficient for solid dissolution, ft/h
- A_p = surface area of the particles, ft²/ft³ of slurry
- dV = differential of the volume of the slurry holdup, ft³
- C^*_{alk} = solubility or equilibrium concentration of the alkaline species, lb-mol/ft³
- C_{alk} = actual concentration of the alkaline species, lb-mol/ft³

For simplicity, let us assume that the average rate throughout the scrubber is given by

$$\bar{R}_d = (k_d)(A_p)(dV)(\bar{C}^*_{alk} - \bar{C}_{alk}) \qquad \text{(Eq. B-12)}$$

where the bars indicate an average value across the scrubber. Combining Equations B-8, B-9, and B-12, we obtain

$$(k_d)(A_p)(dV)(\bar{C}^*_{alk} - \bar{C}_{alk}) \geq G(Y_{in} - Y_{out}) \qquad \text{(Eq. B-13)}$$

or

$$(L/G)_{min} = \frac{Y_{in} - Y_{out}}{(LPA)_R - (LPA)_s + k_d A_p d(\frac{V}{L})(\bar{C}^*_{alk} - \bar{C}_{alk})} \qquad \text{(Eq. B-14)}$$

Thus, the minimum L/G ratio needed to produce a given SO_2 gradient $(Y_{in} - Y_{out})$ can be decreased by increasing the alkalinity of the recycle slurry, the dissolution rate constant (k_d), the surface area of particles (A_p), and the liquid holdup (V) or the residence time of the slurry in the scrubber (V/L). To bring the minimum (L/G) value between 3 and 4 ft³/lb-mol (62.5 to 83.3 gal/1000 scf) for the LPA values used above would require that the dissolution term (k_d) in Equation B-14 be between 7.8 and 5.6 x 10⁻⁴ lb-mol/ft³ (12.5 and 8.9 x 10⁻³ g-mol/liter).

In the foregoing example, some typical LPA values are used to illustrate that limestone scrubbing without additives requires most of the alkalinity to be provided by dissolution of limestone in the scrubber. The effect of SO_2 make-per-pass, as described in Appendix A, can reduce this dissolution requirement. Certain minimum values of LPA are obtainable in both the recycle and the spent slurry, based on driving force considerations. With a countercurrent scrubber, these minimum values correspond to equilibrium SO_2 mole fractions of less than 5 times the actual SO_2 mole fraction in the flue gas at the scrubber outlet (Y_{out}) and inlet (Y_{in}). For a specific case, these minimum values of $(LPA)_R$ and $(LPA)_S$ should be used in Equation B-14 to determine the minimum L/G ratio.

Actual L/G Ratio

In a tray or mobile-bed scrubber, the maximum permissible liquid and gas flow rates are determined by the flooding characteristics of the scrubber (see the later discussion of gas velocity in this appendix). Another consideration in determining the actual L/G ratio is the control of scaling and plugging (also discussed later). The actual L/G design ratio should be higher than the minimum L/G ratio.

GAS/LIQUID DISTRIBUTION

In limestone scrubbing, proper distribution of gases and liquids is critical for maintaining the design SO_2 removal efficiency. Poor distribution will reduce both the time of gas-liquid contact and the effective interfacial mass transfer area. Plexiglas models have been used to verify and support the gas/liquid distribution design of commercially available scrubbers. This model information should be considered in the design of a full-scale unit.

Gas Distribution

Uniformity of gas distribution across the scrubber should be a primary design consideration in the selection of a scrubber. Analysis of Plexiglass models have shown that even gas distribution can be maintained by the use of aids such as ladder vanes or perforated trays. Spray tower scrubbers require no aids because the energy expended by the sprays against the rising flue gas is sufficient to redistribute the gas uniformly. Enough spray nozzles must be used to cover the scrubber cross-sectional area with a spray pattern providing considerable overlap and uniform dense spray zones through which the gas must pass (Saleem 1980).

Gas Inlet Design

Gas inlet design should also be optimized in the scrubber because of its effect on gas distribution. Long uninterrupted sections of inlet ductwork should be provided to minimize any disturbances caused by bends. If long runs of ductwork cannot be used and if the direction of the gas flow changes at the scrubber inlet, the use of turning vanes is recommended.

Additionally, the gas inlet design must keep liquid from entering and drying on the hot surface and thus prevent the creation of a wet/dry interface that will allow buildup of deposits. These deposits can seriously interfere with gas flow. Diversion plates should be installed around the inlet duct opening to prevent liquid entry. Spray nozzles in the vicinity of the duct must be carefully angled to avoid spraying into the duct (Saleem 1980).

Liquid Distribution

In a venturi scrubber, the gas pressure drop across the variable throat atomizes the liquid drops into finer droplets, and control of the pressure drop can be effected by positioning the variable throat. The degree of atomization is thus limited by the allowable power consumption of the fan.

In a spray scrubber, the droplet size is controlled primarily by the nozzle type, size of the nozzle opening, and pressure drop across the nozzle. Typical nozzle pressure drop is 10 to 20 $lb/in.^2$ Although finer droplets are desirable because they increase the interfacial area, the lower limit on the droplet size is set by the allowable power consumption of the pumps and the entrainment of the droplets by the gas. The spray nozzles should produce droplets with an average diameter of roughly 2500 microns (Saleem 1980). The angle of the spray pattern is another important variable. The angle should be large enough to effect maximum gas-liquid contact, but small enough to minimize the coalescence of droplets from two adjacent nozzles and the impingement of droplets on the walls. In a vertical spray scrubber, it is essential that spray headers be situated at various levels to prevent channeling (segregation of the gas and liquid flows) caused by coalescence of the droplets as they flow downwards.

In a mobile-bed or packed-bed scrubber, fine droplets of slurry are not necessary. At least five points of liquid introduction should be provided per square foot of tower cross section (Treybal 1968). In a mobile-bed scrubber, movement of the packing on each stage ensures proper liquid redistribution. In the only packed-bed scrubber used commercially (Research-Cottrell/Munters) the packing has built-in liquid redistribution. In a tray scrubber, the design of inlet nozzles, downcomers, and trays is critical to proper liquid distribution.

Appendix B: Operational Factors   275

GAS VELOCITY AND PRESSURE DROP

For a given superficial molar gas flow rate (G), the cross-sectional area of the scrubber is determined on the basis of an operating gas velocity, which may be expressed for purposes of calculation in molar (lb-mol/h-ft^2), mass (lb/h-ft^2), or volumetric (ft^3/h-ft^2) units. Designing for operation at the highest possible gas velocity minimizes the cross section and capital cost of the scrubber. The upper limit on the gas velocity is set by the flooding potential (or the pressure drop) and entrainment potential, which are discussed below.

Flooding Potential

In a mobile-bed or packed-bed scrubber, the pressure drop of gas through the system is influenced by the gas and liquid flow rates in a manner shown in Figure B-1 (Treybal 1968). In the region below Line A of the figure, the pressure drop increases with gas velocity at a given liquid

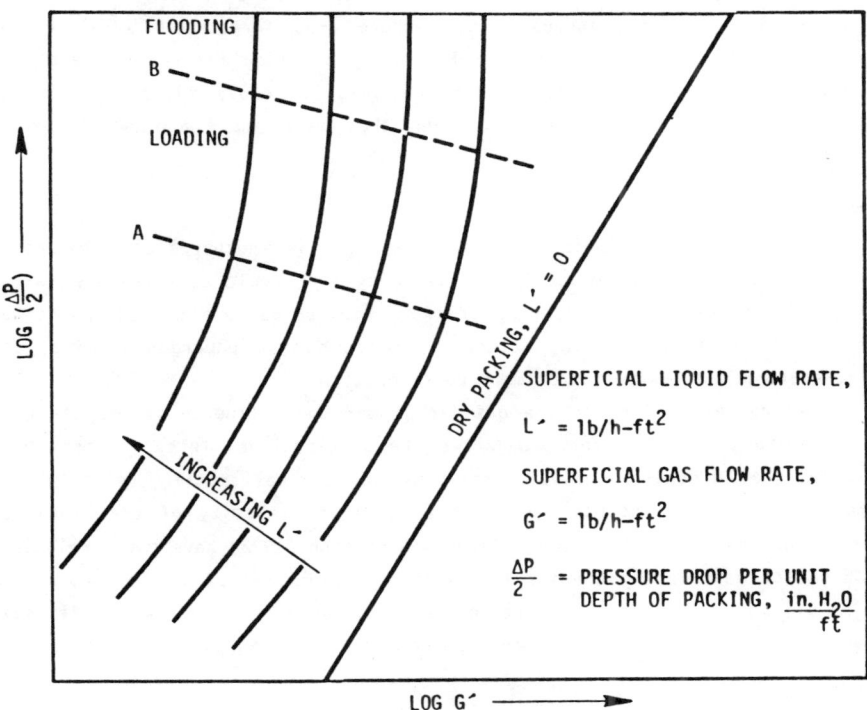

Figure B-1. Effect of gas and liquid flow rates on gas pressure drop in a packed-bed scrubber (Treybal 1968).

flow rate. The liquid holdup is reasonably constant with changing gas velocity, although it increases with liquid flow rate. In the region between Lines A and B, the liquid holdup increases rapidly with gas flow rate, the free area for gas flow becomes smaller, and the pressure drop rises rapidly. This condition is known as loading. As the gas rate is increased to Line B at a fixed liquid rate, there is a change from a liquid-dispersed to a gas-dispersed state (inversion). A layer of liquid may appear at the top of the packing, and entrainment of liquid by the effluent gas may increase rapidly. This condition is known as flooding. At a given gas flow, flooding will occur if the liquid flow rate is increased beyond a maximum value (see Figure B-1). It is not practical to operate a scrubber in a flooded condition; most are operated in the lower part of the loading zone.

In a tray scrubber, as the gas velocity increases at a fixed liquid flow rate, the gas is dispersed thoroughly into the liquid, which in turn is agitated into a froth. This action provides a large interfacial surface area. At high gas velocities, however, the entrainment of liquid droplets above the upper tray increases, and the absorption efficiency is reduced. In addition, a rapid increase in pressure drop forces the level of the liquid in the downcomer to rise. Ultimately the liquid level may reach the tray above. Further increases in gas velocity cause flooding when the liquid fills the entire space between the trays and the flow of gas is erratic.

Entrainment Potential

Entrainment of liquid droplets by the gas is another factor in determining the maximum permissible gas velocity and should be considered along with the potential for flooding of tray, mobile-bed, and packed-bed scrubbers. In venturi or spray scrubbers, which do not undergo flooding, entrainment potential is the primary consideration.

A droplet falling into a gas flow under the influence of gravity will accelerate until drag force balances the gravitational force. After that, it falls at a constant velocity known as the terminal settling velocity. If the gas velocity exceeds the terminal settling velocity of the droplets, heavy entrainment will occur. Perry and Chilton (1973) have given relationships for calculating the terminal settling velocities of liquid droplets in gas streams. The terminal settling velocity decreases with the droplet size and thus increases the entrainment potential at a given gas velocity.

Pressure Drop

The operating gas velocity is determined from the maximum gas velocity, which is dependent on the aforementioned factors. The gas pressure drop through the scrubber depends on the gas velocity and the type of scrubber internals. The total gas pressure drop consists of losses in the bends and losses by contraction and expansion in the inlet and outlet ductwork, scrubber, mist eliminator, reheater, and stack. The pressure drop in the scrubber and the mist eliminator usually constitutes the major portion of the total pressure drop.

As described in Section 3 (see discussion of process control), changes in pressure drop across the scrubber, mist eliminators, or reheater can be a symptom of internal plugging. Normal pressure drop ranges from 5 in. H_2O (in spray towers) to 15 in. H_2O (in mobile-bed scrubbers).

TURNDOWN CAPABILITY

In a limestone FGD system, it is important that the design level of SO_2 removal provide a safety margin when the flow of flue gas is reduced because of reduced boiler load. The ratio of maximum to minimum gas flow that a scrubber can handle without reducing SO_2 removal or causing unstable operation is called turndown capability.

Equation B-2 indicates that any reduction in the gas flow rate (G) tends to increase the SO_2 removal efficiency. A reduction of gas flow, however, decreases the overall mass transfer coefficient (K_G) by decreasing the individual coefficients (k_g, k_ℓ). A decrease in the gas flow also reduces the interfacial area (a) by decreasing the gas dispersion (tray scrubber), liquid agitation (mobile-bed and packed-bed scrubbers), or the pressure drop (venturi or rod-deck scrubbers). Thus, the effect of reduced gas flow rate on SO_2 removal depends primarily on the type of scrubber. In a spray scrubber, the interfacial area is not dependent on the gas flow rate or the pressure drop. Thus, the SO_2 removal efficiency increases with a reduction in gas flow.

Turndown capability is affected by some mechanical limitations. At a given liquid flow rate in a tray scrubber, the liquid may start to drip through the tray openings as the gas flow is reduced. This phenomenon, known as weeping, is caused by reduced gas pressure. At a given liquid flow rate in a mobile-bed or packed-bed scrubber, reduction of gas flow may lead to channeling.

In general, the turndown capability of a spray scrubber (3 to 4) is superior to those of tray scrubbers (2 to 3), mobile-bed scrubbers (about 2), and venturi scrubbers (less than 2).

An FGD system incorporating parallel scrubber modules renders good overall turndown capability, regardless of the type of scrubber, provided that the minimum load is limited to the minimum capacity of a single module. Such an arrangement allows a stepwise turndown capability, but requires good control and distribution of flue gas through the parallel modules. Gas flow distribution problems must be carefully considered in design of a new power-plant, when the flue gas is to be distributed to parallel modules by use of dampers in a common duct rather than by use of an individual fan for each module. Many operators believe that use of individual fans provides better control of gas distribution.

PREVENTION OF SCALE FORMATION

In limestone FGD systems the major operating problem is formation of calcium sulfite and calcium sulfate scale. The causes of scale formation are discussed in Appendix A; the effects of scale formation and methods of controlling it are considered here. Every effort must be made to prevent or control scaling because it can lead to plugging by accumulation of scale or other solids such as fly ash and recycle slurry solids, which in turn can necessitate a scrubber shutdown. When screens, piping, nozzles, packing material, mist eliminator blades, or liquid distribution internals become plugged with scale, the pressure differential increases and flow rate capacities are reduced. Scale formation can also occur in instruments and sensor lines such as pH sample taps, pressure differential sensors, level indicators, pressure gauges, and gas sampling taps. When this type of scaling is severe, the system cannot be reliably controlled.

Plugging can result from sudden scale formation during an upset condition or from scale buildup over a long interval. Calcium sulfite scale is soft, whereas calcium sulfate scale is very hard. The soft scale hardens, however, when scrubbers are shut down for more than a few hours and accumulations are not immediately washed away; under those conditions the soft sulfite scale begins to oxidize and forms the much harder sulfate scale. Scale accelerates the effect of corrosion either by concentrating electrochemical attack beneath a layer of scale deposited on a metallic surface or by damaging protective coatings when a chunk of scale is dislodged. Even stainless steel can be severely damaged by stress-corrosion attack and pitting underneath scale deposits, especially if the slurry contains a high concentration of chloride in solution.

Scale formation also can significantly influence gas flow distribution, especially in the mist eliminator area, where uniform distribution is critical for preventing high local velocities and subsequent carryover of solids and liquids.

Calcium Sulfite Scaling

In Appendix A, the scaling phenomenon was explained in terms of relative saturation (RS). The critical RS value at which nucleation of a species begins to occur should not be exceeded because nucleation would lead to uncontrolled precipitation or scaling. The rate of precipitation of a species is as follows (Corbett et al. 1977):

$$R_p = (K_p)(\beta)(V_R)(W_S)(RS-1) \qquad (Eq. \ B-15)$$

where R_p = net precipitation rate, lb-mol/h

K_p = precipitation rate constant, lb-mol/ft^2-h

β = specific surface area of solids, ft^2/lb of solids

V_R = volume of slurry in the reaction tank, ft^3

W_S = slurry solids content, lb solids/ft^3 slurry

RS = relative saturation (unitless)

The molar precipitation rate (R_p) is fixed for a system, and is equal to the molar rate of SO_2 removal. Thus, the RS value can be minimized by increasing the slurry solids content (W_S) or the recycle tank volume (V_R). By expressing precipitation rate in the form of reaction kinetics, Borgwardt (1975) has shown that the volume required for a given amount of precipitation is considerably less with a series of mixed tanks than with one mixed tank. Basically, an FGD system should be operated in such a way as to confine calcium sulfite precipitation to the scrubber effluent hold tank. This is achieved by controlling pH to keep the limestone feed stoichiometry below 1.4.

Control of pH. Wen et al. (1975) report that calcium sulfite scale can be minimized by keeping pH at the scrubber inlet below 6.2. They found that at a pH of 6 or less the rate of scale formation was only 5 percent of that at pH greater than 6.2. The factors determining the pH of slurry in the hold tank are SO_2 content of the flue gas, residence time in the tank, L/G ratio, and stoichiometric ratio. Achieving a high rate of SO_2 removal from combustion of high-sulfur coals requires a relatively longer residence time and higher L/G and stoichiometric ratios to maintain a given pH level.

Operating at reduced pH can lead to reduced SO_2 removal, especially with high inlet SO_2 concentrations. When this occurs, a part of the limestone slurry may be fed to the top of the scrubber. The horizontal spray scrubber is especially suitable for this arrangement.

The soft calcium sulfite scale dissolves as the pH is reduced, because of its increased solubility. Thus, when formation of calcium sulfite scale is noticed, the pH of the recycle liquor should be reduced for a short time by reducing limestone feed rate.

Limestone Utilization. Appendix A discusses the effect of stoichiometric ratio on calcium sulfite scaling. Operations at the EPA Shawnee Test Facility have shown that maintaining a low stoichiometric ratio or high limestone utilization can aid in keeping the mist eliminator free of sulfite scale (Head et al. 1977). Where limestone utilization is greater than 85 percent, an intermittent bottom wash keeps the mist eliminator free of soft (calcium sulfite or carbonate) scale deposits. Where utilization is less than 85 percent, a continuous bottom wash is needed to limit the accumulation of soft solids to less than 10 percent of the open area.

Within economic and design constraints, the system should be operated at the highest possible L/G ratio. A high liquid flow rate would reduce the concentration and thus the RS of the calcium sulfite. It also reduces stagnation of the slurry and thus reduces plugging.

Calcium Sulfate Scaling

Two basic modes have been used to prevent calcium sulfate scaling in limestone FGD systems: coprecipitation and control of supersaturation (Devitt, Laseke, and Kaplan 1980). These techniques are discussed below.

When a system is operated so that the maximum oxidation level in the slurry is less than 16 percent, the scrubbing liquor remains subsaturated with calicum sulfate (gypsum), which is removed from the system as a coprecipitate or solid solution with calcium sulfite. In this case, the discussion of calcium sulfite scale control is applicable.

When the oxidation level exceeds 16 percent, the liquor is supersaturated with gypsum (RS > 1). In this case, a part of calcium sulfate must be removed from the system as gypsum. The critical level of gypsum RS is 1.3 to 1.4. Thus, the supersaturation must be kept below 1.3 to prevent gypsum scaling.

A means of controlling calcium sulfate supersaturation is to circulate a minimum amount of calcium sulfate seed crystals, which act as nucleation sites that enhance the homogenous precipitation of calcium sulfate. Wen et al. (1975) found that circulation of 1 percent gypsum seed crystals reduced the rate of scaling to about 40 percent of that in saturated solutions without gypsum solids. Concentrations of gypsum greater than 1 percent did not reduce the scaling rate further. Concentration of the seed crystals should be maintained by keeping the solids content of the slurry not less than 8 percent. It should be higher if fly ash is also present. Concentration of the seed crystals can also be maintained at a stable level, with relative saturation maintained at about 1.1, by operating in a forced oxidation mode. An EPA pilot-plant study (Borgwardt 1978) indicates that when the forced oxidation was transferred from the scrubber to an external oxidizer tank, gypsum scaling was minimized, operating stability was improved, and the need for constant monitoring to maintain supersaturation at less than 130 to 140 percent was eliminated. Later EPA testing has shown that forced oxidation by sparging air directly into the effluent hold tank causes carbon dioxide to be stripped from the solid slurry; carbon dioxide stripping enhances limestone dissolution and nullifies any adverse effects on SO_2 removal.

The RS of gypsum in the recycle liquor, similar to that of calcium sulfite, can be reduced by providing a hold tank large enough to allow a minimum of eight (8) minutes residence time for complete precipitation of gypsum in the tank.

Mechanical Considerations

In spite of efforts to prevent scaling, some scale formation can occur in an operating limestone FGD system. The following mechanical considerations, therefore, should be addressed in the design phase so as to facilitate occasional cleaning of the FGD system.

Depending on the type of scrubber, the scale deposits occur at various locations in the scrubber internals. Manholes should be installed at each stage of the scrubber for easy access by maintenance personnel. View plates can allow observation of deposits or mechanical problems when the system is shut down. Also, the scrubber effluent hold tank should be equipped with side doors to allow entry of maintenance personnel for removal of solids deposits during shutdown periods.

It is essential to prevent plugging of pump suction lines and spray nozzles. Thus, strainers should be installed in pump suction lines, and the spray nozzles and headers should be checked during regular maintenance.

The transformation of soft sulfite deposits into hard sulfate scale has been observed in the wet/dry regions of separated flow, (including the venturi section of a presaturator or an adjustable venturi) and in the sudden cross-sectional expansion at the entrance to the scrubber. In addition to the chemical transformation of soft deposits to hard scales, deposition of slurry solids and fly ash can occur in the wet/dry regions. The design therefore should incorporate into the quiescent zone some means of washing away deposits with intermittent sprays or blowing them away with sootblowers while they are still soft.

CHLORIDE CONTROL

Chlorides enter an FGD system from chlorine in the coal and in the makeup water. Chlorine in the makeup water is generally a significant source of chloride only if cooling tower blowdown or other wastewater is used for scrubber makeup. In any case, most chlorides come from chlorine in the coal. When the coal is combusted, the chlorine is converted to hydrogen chloride (HCl) gas. The hydrogen chloride in the flue gas is captured by the scrubber and at steady state leaves the system as chloride ions in the interstitial water of the waste sludge. When the chlorine content of the coal is high and is bound to organic compounds, so that no significant reduction is achieved by washing, the steady-state chloride ion concentrations in the scrubbing liquor and the dewatered sludge are significantly high. High concentrations of chloride ions in the scrubbing liquor adversely affect the process chemistry and promote corrosion and subsequent failure of construction materials (see Section 3).

Chloride ions are among the effluents from the FGD system that must be discharged in an environmentally acceptable manner. The RCRA provisions limit the trace element and major anion content of leachate from a disposal site to 100 times the drinking water criteria specified under the National Interim Primary Drinking Water Regulations. The water-recycle requirements are critical for FGD systems in water-short areas, where salinites and river volume control the withdrawals as well as the discharges (Dascher and Lepper 1977).

In view of these constraints, options should be considered to reduce high concentrations of chloride ion in scrubbing liquor by external means. Various methods are available for extraction of chlorides from the scrubbing liquor. Vapor-compression (V-C) evaporation is presented as an example. A slipstream from the quencher loop (if separate from the scrubber loop) or a thickener overflow stream can be treated for chloride ion removal. The treated water may then be reused as FGD system makeup water (Borgwardt 1980) or as makeup to boiler feed water demineralizers. A chloride removal system can effectively reduce the total dissolved solids (TDS) level from 30,000 ppm to less than 50 ppm (Weimer 1977). Such a system, however, generates a concentrated chloride brine (about 300,000 ppm or 30 percent), which must be disposed of.

Simply stated, V-C evaporation concentrates the decanted feed liquor and returns the condensate as purified water. The concentrate of dissolved salts from the evaporator bottom is sent to disposal.

The methodology involves initial pH adjustment, heat exchange to recover the heat from the product water, vacuum deaeration to remove noncondensible gases, and evaporation of water from the feed liquor. The ratio of the concentration of TDS achieved in the concentrate to that in the feed liquor is approximately 14, if one assumes a TDS level of 30,000 ppm in the feed liquor.

A V-C evaporation system handling 150 gpm of feed liquor has been successfully operated to purify wastewater from the Wellman-Lord FGD system at the San Juan Generating Station, which is jointly owned by the Public Service Co. of New Mexico and the Tucson Gas and Electric Co. (Dascher and Lepper 1977). This chloride treatment system is simpler and requires a lower capital investment than other chloride control options, although operating costs are somewhat higher because of higher energy consumption.

ENERGY DEMAND

A limestone FGD system consumes energy to force the flue gas through the system, to grind the limestone, to handle various liquid/solid streams, and sometimes to reheat the exiting gas. On the gas side, a flue gas fan (forced draft or induced draft) uses energy to offset the gas pressure drop produced by various portions of the FGD system, such as the ductwork, presaturator, scrubber, and mist eliminator. On the liquid side, energy is expended to recirculate the slurry to the scrubber, to pump water and slurry streams to the various parts of the system, and to prepare the effluent sludge for final disposal.

There is an increasing emphasis on dewatering and stabilization of FGD wastes for managed landfill disposal and a corresponding decrease in ponding of the wastes. Formerly, a pond was expected to fulfill three functions: clarification, dewatering, and temporary or final sludge storage. The increased emphasis on disposal in landfill and structural fills, and on attaining closed-loop operation, has stimulated the use of clarifiers, vacuum filters, and centrifuges, all of which consume energy.

Reheating the saturated flue gas consumes more energy than any other part of the scrubber system. The reheat may be required for several reasons. One is to provide buoyancy to the flue gas and thus reduce the nearby ground-level concentrations of pollutants. Another reason for reheat is to prevent condensation of acid-containing, wet, cool gas from the absorber in the induced draft fan, exit duct, or stack; such condensation can accelerate corrosion of downstream equipment. Further, reheat minimizes the settling of mist droplets as localized fallout and the formation in cold weather of a heavy steam plume with resultant high opacity.

Most of the energy consumed by the FGD system is attributed to reheat, the flue gas fan, and slurry recirculation pumps. The energy used by the fan depends upon the pressure drop across the scrubber, which varies according to the type of scrubber. Spray tower scrubbers, for example, require lower fan power because of low pressure drop; however, a high slurry recirculation rate and the nozzle pressure drop required for efficient SO_2 removal add to the pump head and power requirements.

A TVA study (McGlamery, Tarkington, and Tomlinson 1979) has estimated

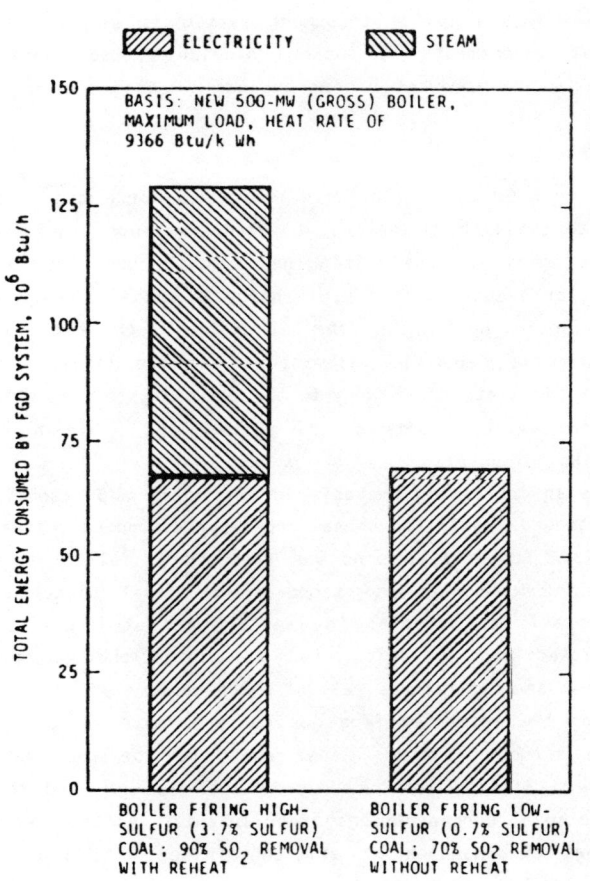

a. As steam and electricity used by the system.

Figure B-2. Total energy consumption by typical limestone FGD systems.

(continued)

Appendix B: Operational Factors 285

Figure B-2. (continued)

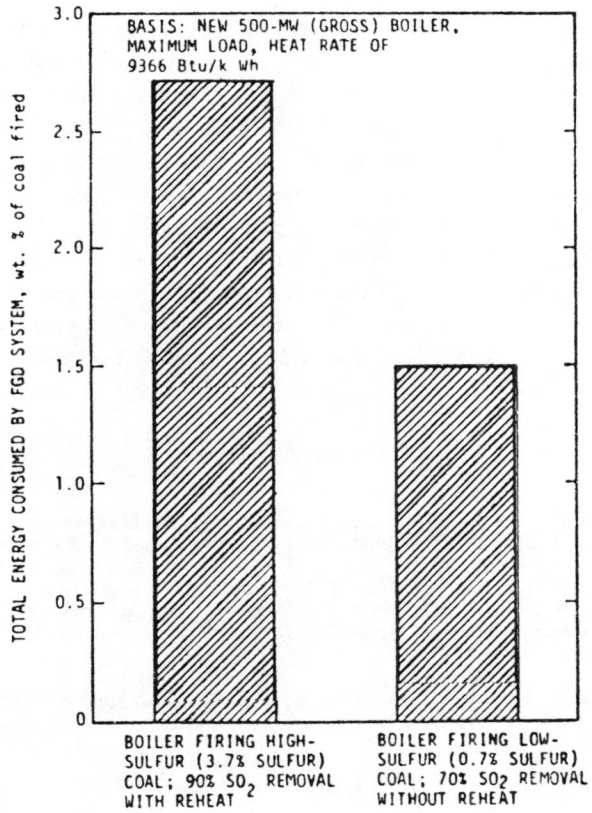

b. As portion of total coal fired in the boiler.

that total energy consumption of a typical limestone FGD system, including reheat, on a 500-MW boiler with a gross heat rate of 9000 Btu/kWh for generation of electricity is 3.3 percent of the input energy. The computations were based on 3.5 percent sulfur in the coal and an allowable emission of 1.2 lb SO_2/million Btu heat input. The system employs a mobile-bed scrubber and a settling pond for sludge disposal. Flue gas reheat of 50°F is provided. The indirect in-line reheat steam consumption is 1.8 percent of the input energy.

The total energy consumption of typical limestone FGD systems on a 500-MW boiler firing high-sulfur coal with reheat and firing low-sulfur coal without reheat is shown in Figure B-2. The values were calculated from energy requirements for the base case as described in Section 2. The electricity consumption is higher for the low-sulfur coal because the flue gas flow rate is higher than with the high-sulfur coal. The total electricity consumption, therefore, is determined primarily by the gas volume (the fan) and is influenced only marginally by total SO_2 removal. Thus, excess combustion air and total air inleakage can have a significant effect on energy consumption in the FGD system. The total energy consumptions with the high- and low-sulfur coals are 2.7 and 1.5 percent of the total input energy, respectively.

REFERENCES FOR APPENDIX B

Borgwardt, R. H. 1975. Increasing Limestone Utilization in FGD Scrubbers. Presented at the 68th AIChe Annual Meeting, Los Angeles, California.

Borgwardt, R. H. 1978. Effect of Forced Oxidation on Limestone/SO_x Scrubber Performance. In: Proceedings of the Symposium on Flue Gas Desulfurization, Hollywood, Florida, November 1977. Vol. I EPA-600/7-78-058a. NTIS No. PB-282 090.

Borgwardt, R. H. 1980. Combined Flue Gas Desulfurization and Water Treatment in Coal-Fired Power Plants. Environmental Science and Technology, 14(3):294-298.

Corbett, W. E., et al. 1977. A Summary of the Effects of Important Chemical Variables Upon the Performance of Lime/Limestone Wet Scrubbing Systems. Electric Power Research Institute, Palo Alto, California. EPRI FP-639.

Dascher, R. E. and R. Lepper. 1977. Meeting Water-Recycle Requirements at a Western Zero-Discharge Plant. Power, 121(8):23-28.

Devitt, T. W., B. A. Laseke, and N. Kaplan. 1980. Utility Flue Gas Desulfurization in the U.S. Chem. Eng. Prog., 76(5):45-57.

Head, H. N., et al. 1977. EPA Alkali Scrubbing Test Facility: Advanced Program, Third Progress Report. EPA-600/7-77-105. NTIS No. PB-274 544.

McGlamery, G. G., T. W. Tarkington, and S. V. Tomlinson. 1979. Economics and Energy Requirements of Sulfur Oxides Control Processes, TVA. In: Proceedings of the Symposium on Flue Gas Desulfurization, in Las Vegas, Nevada, March 1979. Vol. I. EPA-600/7-79-167a. NTIS No. PB 80-133 168.

Perry, R. H., and G. H. Chilton, eds. 1973. Chemical Engineers' Handbook. 5th ed. McGraw-Hill Book Co., New York.

Saleem, Abdus. 1980. Spray Tower: The Workhorse of Flue-Gas Desulfurization. Power, 124(10):73-77.

Treybal, R. E. 1968. Mass Transfer Operations. 2d ed. McGraw-Hill Book Co., New York.

Wen, C. Y. and L. S. Fan. 1975. Absorption of SO_2 in Spray Column and Turbulent Contacting Absorbers. EPA-600/2-75-023. NTIS No. PB-247 334.

Weimer, L. D. 1977. Effective Control of Secondary Water Pollution From Flue Gas Desulfurization Systems. EPA-600/7-77-106.

BIBLIOGRAPHY FOR APPENDIX B

Head, H. N. EPA Alkali Scrubbing Test Facility: Advanced Program, Second Progress Report. EPA-600/7-76-008. NTIS No. PB-258 783.

Head, H. N. EPA Alkali Scrubbing Test Facility: Advanced Program, Third Progress Report. EPA-600/7-77-105. NTIS No. PB-274 544.

Head, N. N. and S. C. Wang. EPA Alkali Scrubbing Test Facility: Advanced Program, Fourth Progress Report. Vol. 1, Basic Report; and Vol. 2, Appendices. EPA-600/7-79-244a and EPA-600/7-79-244b. NTIS Nos. PB 80-117906 and PB 80-117914.

Jones, D. G., A. V. Slack, and K. S. Campbell. 1978. Lime/Limestone Scrubber Operation and Control Study. Electric Power Research Institute, Palo Alto, California. RP 630-2.

Slack, A. V. 1978. Lime-Limestone Scrubbing: Design Considerations. Chem. Eng. Prog., February 1978. 74(2):71-75.

Uchida, S., C. Y. Wen, and W. J. McMichael. 1976. Role of Holding Tank in Lime and Limestone Slurry Sulfur Dioxide Scrubbing. Ind. Eng. Chem., Proc. Des. Dev., 15(1):18-27.

Wen, C. Y., and C. S. Chang. 1978. Absorption of SO_2 in Lime and Limestone Slurry: Pressure Drop Effect on Turbulent Contact Absorber Performance. Environmental Science and Technology, 12(6):703-707.

Appendix C
Computer Programs

Two computerized systems are available for use by the utility Project Manager during selection and evaluation of a proposed limestone scrubbing system. A third computer program has resulted in the development of simplified equations with which to calculate relative gypsum saturation levels that will reduce scaling potential during system operation. This appendix describes these three programs and their use.

The Lime/Limestone Economics Computer Model was sponsored by the EPA and developed by the TVA and Bechtel National, Inc. This program can assist the Project Manager in selecting the most cost-effective options for a scrubbing system, in developing initial plans, and in evaluating trade-offs. It is also useful for determining the size of individual components and the effect of SO_2 removal efficiency on energy demand.

The FGD Information System (FGDIS), also sponsored by the EPA, was developed by PEDCo Environmental, Inc. This program can supply the Project Manager with needed experience information and can serve as a tool for initial design costing and feasibility screening. The Project Manager can use the FGDIS to determine predominant scrubber design configurations, average L/G ratios, prevailing types of reheat systems, situations requiring no reheat, and materials of construction used in specific components. The design data can be coupled with performance information to develop a design-versus-reliability analysis. Additionally, the FGDIS enables the Project Manager to identify problem areas and to alter proposed design features to prevent these problems.

The Bechtel-Modified Radian Equilibrium Program is based on EPA/TVA experience at Shawnee Station. This program has been simplified into two convenient equations, which can be used to maintain the relative gypsum saturation level of an operating limestone scrubbing system below 130 to 140 percent. Thus, these equations constitute an important tool in reducing the scrubbing unit's scaling potential.

LIME/LIMESTONE ECONOMICS COMPUTER MODEL

Background

Under EPA sponsorship, TVA and Bechtel developed the Lime/Limestone Economics Computer Model, which can be used to estimate the capital investment and annual lifetime revenue requirements for a full-scale lime/limestone FGD system. The process and material balance relationships used in the model are based on data obtained from the EPA test facility at the TVA Shawnee station in Paducah, Kentucky.

The test facility is incorporated into the flue gas ductwork of Boiler 10 at the Shawnee plant. The facility originally consisted of three parallel scrubber units, each capable of processing about 30,000 acfm of flue gas. The original scrubbers were a venturi/spray tower combination unit, a turbulent contact scrubber, and a marble-bed scrubber. The marble-bed scrubber was shut down in 1973 and converted to a cocurrent scrubber in 1978. Tests conducted on venturi/spray towers and turbulent contact scrubbers provided the basic data for the computer model.

Development of the FGD computer model began in 1975. The current model is a combination of two separate models: one predicting process parameters, equipment sizes, and capital costs, and the other projecting the annual and lifetime revenue requirements based on the parameter, size, and cost predictions.

Bechtel developed the material balance relationships, flow rates, and stream compositions. On the basis of Bechtel's input, the TVA determined the equipment size relationships, developed cost bases, and performed the system analysis. These efforts yielded a model capable of estimating capital investment costs. The TVA then developed procedures for using this output to project the annual and lifetime revenue requirements. Recent enhancements of the model allow automatic calculation of revised NSPS emission limits based on coal composition; also included are an option for partial scrubbing with bypass and revised TVA/EPA premises for comparative economic evaluations of emission control processes.

The TVA has published several papers (McGlamery et al. 1975; Stephenson and Torstrick 1978; Torstrick 1976; and Torstrick, Henson, and Tomlinson 1978) relating to the development of the model and its use. Stephenson and Torstrick (1979) provide user's guidelines in a concise form. An updated user's manual describing recent developments is to be available early in 1981. Additionally the TVA staff will assist a user either by running the program with data or by providing a copy of the program. Further information may be obtained from Mr. Robert L. Torstrick, Tennessee Valley Authority, Energy Design Operations, 501 Chemical Engineering Building (CEB), Muscle Shoals, Alabama 35660 [telephone (205) 386-2814].

Using the Lime/Limestone Economics Computer Model, TVA staff members have evaluated the economics of flue gas desulfurization and of byproduct handling and waste disposal. The evaluation report (McGlamery et al. 1980) summarizes capital investments and annual revenue requirements based on a three-phase analysis of FGD processes and of sludge disposal practices. The report includes projections of the potential 1985 market for FGD byproduct sulfur/sulfuric acid. The authors suggest new premises that will be used to evaluate FGD processes in the early 1980's and indicate some effects of these premises on assessments of limestone scrubbing economics. The report also presents the results of a recent evaluation of limestone scrubbing in a spray tower; forced oxidation, use of adipic acid, and gypsum disposal by stacking are considered.

An updated version of the model was recently published by EPA (Anders and Torstrick 1981). This version supplements Torstrick's 1979 report and extends the number of scrubber options that can be evaluated. It includes spray tower and venturi/spray tower scrubbers, forced oxidation systems, systems with scrubber loop additives (MgO or adipic acid), revised design and economic premises, and other changes reflecting process improvements and variations.

Model Description

The TVA/Bechtel model is intended for use by Project Managers during the preliminary engineering phase of an FGD project to analyze the process and system options and their effects on system costs. The model can generate a preliminary conceptual design and cost package for a lime or limestone FGD system. Since the costs estimated by the model are based on conceptual input parameters, the cost values predicted by the model must be considered as preliminary. These values should be useful, however, in weighing the impact of various process parameters on final costs.

The model calculates the flow rates and compositions of various streams in the FGD system; sizes of the major equipment items are then determined from the flow rates. The costs of materials and installation labor are determined for the equipment items, and capital investment is calculated by adding indirect costs to the installed costs of the system equipment.

Annual operation and maintenance costs are based upon the annual consumption of raw materials and utilities. Annual capital charges are added to these costs to arrive at the total annual cost of the system. The lifetime annual revenue requirements for the system are generated from data on the input capacity and projected useful life of the system.

The input to the program consists of the major parameters of the FGD system. The input parameters needed to run the program are listed in Table C-1. The ranges of major input parameters used in development of the model

TABLE C-1.   REQUIRED INPUT PARAMETERS, TVA/BECHTEL PROGRAM

Plant and Site Data

    Plant rating
    Annual rainfall
    Seepage rate
    Annual evaporation
    Land area available for disposal pond
    Expected pond capacity
    Plant remaining life
    Plant operating pattern for the remaining life

Boiler and Fuel Data

    Boiler heat rate
    Excess air
    Temperature of flue gas to scrubber
    Temperature of flue gas to stack
    Heating value of coal
    Coal composition (C, H, O, N, S, Cl, ash, and H_2O)

Process Parameters

    Temperature of reheater steam
    Heat of vaporization of reheater steam
    Scrubber L/G ratio
    Superficial gas velocity in the scrubber
    Face velocity of gas through reheater
    Amount of SO_2 to be removed
    Effluent hold tank residence time
    Value for stoichiometry
    Soluble MgO in limestone
    Soluble MgO added to the system
    Insolubles in limestone
    Moisture content of limestone
    Soluble CaO in particulates
    Soluble MgO in particulates
    Percent solids in scrubber slurry
    Percent solids in discharged sludge
    Percent solids in clarifier underflow
    Percent oxidation of sulfite in scrubber
    Percent solids in filter cake
    Filtration rate
    Limestone hardness index
    Size of ground limestone
    Entrainment level as percentage of wet gas

(continued)

TABLE C-1 (continued)

System Options
Solids settling rate in clarifier
Number of turbulent contact stages in the scrubber
Number of turbulent contact scrubber grids
Height of spheres in each stage
Number of spare limestone preparation units
Number of operating scrubber modules
Number of spare scrubber modules
Sludge disposal option
Depth of finished pond
Maximum excavation depth
Distance between pond and scrubber area
Pond lining option
Cost Parameters
Maintenance cost factor expressed as percent of capital investment exclusive of pond
Pond maintenance factor
Plant overheads
Administrative research and services rate
Annual capital charge basis
Insurance and interim replacement cost factor
Limestone unit rate
Operating labor unit rate
Supervision labor unit rate
Steam rate
Process water rate
Electricity rate
Chemical engineering material cost index
Chemical engineering labor cost index
Capital investment base year
Revenue requirement base year
Operating profile
Chemical engineering plant index

are shown in Table C-2. Values beyond the indicated ranges are not necessarily invalid, but the potential for error is greater when these ranges are exceeded.

The user can select program options by following the input instructions outlined in the user's manual. The model options are summarized in Table C-3.

TABLE C-2. BASIS FOR MAJOR VARIABLES, TVA/BECHTEL PROGRAM

Variable	Basis/range
Type of plant	New
Plant size, MW	100-1300
Fuel sulfur, %	2-5
Scrubber gas velocity, ft/s	8-12.5
L/G ratio, gal/1000 acf	28-75
Effluent hold tank residence time, min	2-25
Number of operating scrubber modules	1-10
Number of spare scrubber modules	0-10
Sulfur converted to SO_2, %	0-100
System pressure drop	3 in. H_2O maximum per turbulent contact scrubber stage
Base year for capital investment	Midpoint of project duration
Base year for revenue requirement	First year of FGD operation

TABLE C-3. MODEL OPTIONS, TVA/BECHTEL PROGRAM

Process Options

    Limestone FGD
    Lime FGD

System Options

    Number of scrubbers
    Redundancy
    Sludge processing
    Disposal pond
    Pond liner
    Pond capacity
    Operating profile
    Fixation/stabilization
    Landfill
    Additions in 1981

Cost Factor Options

    User capability to select percentages for individual indirect cost components

Output Options

    User capability to select any or all available output reports

TABLE C-4. INPUT DATA REPORT

```
BASE CASE EXAMPLE 500 MW

                    *** INPUTS ***

BOILER CHARACTERISTICS
------ ---------------
MEGAWATTS = 500.
BOILER HEAT RATE = 9000. BTU/KWH
EXCESS AIR = 33. PERCENT, INCLUDING LEAKAGE
HOT GAS TEMPERATURE = 300. DEG F
COAL ANALYSIS, WT % AS FIRED :

  C      H     O     N     S     CL    ASH    H2O
 57.56  4.14  7.00  1.29  3.12  0.10  16.00  10.76
SULFUR OVERHEAD = 95.0 PERCENT
ASH OVERHEAD = 80.0 PERCENT
HEATING VALUE OF COAL = 10500. BTU/LB

                         EFFICIENCY,   EMISSION,
   FLYASH REMOVAL             %        LBS/M BTU
   --------------        -----------   ---------
   UPSTREAM OF SCRUBBER       98.5        0.18
   WITHIN SCRUBBER            50.0        0.09

ALKALI
------
LIMESTONE :

       CACO3     =  97.15 WT % DRY BASIS
       SOLUBLE MGO =  0.0
       INERTS    =   2.85
       MOISTURE CONTENT =  5.00 LB H2O/100 LBS DRY LIMESTONE
       LIMESTONE HARDNESS WORK INDEX FACTOR = 10.00
       LIMESTONE DEGREE OF GRIND FACTOR = 1.35

FLY ASH :

       SOLUBLE CAO =  0.0  WT %
       SOLUBLE MGO =  0.0
       INERTS    = 100.00


RAW MATERIAL HANDLING AREA
--- -------- -------- ----

NUMBER OF REDUNDANT ALKALI PREPARATION UNITS =    1
```

(continued)

TABLE C-4 (continued)

```
SCRUBBER SYSTEM VARIABLES

NUMBER OF OPERATING SCRUBBING TRAINS =    4
NUMBER OF REDUNDANT SCRUBBING TRAINS =    1
NUMBER OF BEDS =   3
NUMBER OF GRIDS =   4
HEIGHT OF SPHERES PER BED =  5.0 INCHES
LIQUID-TO-GAS RATIO =  95. GAL/1000 ACF
SCRUBBER GAS VELOCITY =  12.5 FT/SEC
SO2 REMOVAL =   85. PERCENT
STOICHIOMETRY RATIO      TO BE CALCULATED
ENTRAINMENT LEVEL =  0.10 WT %
EMT RESIDENCE TIME =  12.0 MIN
SO2 OXIDIZED IN SYSTEM =  30.0 PERCENT
SOLIDS IN RECIRCULATED SLURRY =  15.0 WT %

SOLIDS DISPOSAL SYSTEM

COST OF LAND =  3500.00 DOLLARS/ACRE
SOLIDS IN SYSTEM SLUDGE DISCHARGE =  40.0 WT %
MAXIMUM POND AREA =   500. ACRES
MAXIMUM EXCAVATION =  25.00 FT
DISTANCE TO POND =  5280. FT
POND LINED WITH 12.0 INCHES CLAY

STEAM REHEATER (IN-LINE)

SATURATED STEAM TEMPERATURE =  470. DEG F
HEAT OF VAPORIZATION OF STEAM =  791. BTU/LB
OUTLET FLUE GAS TEMPERATURE =  175. DEG F
SUPERFICIAL GAS VELOCITY (FACE VELOCITY) =  25.0 FT/SEC

     IT     SR      SROLD
      1    1.41     1.50
      2    1.41     1.41
```

TABLE C-5. PROCESS PARAMETER REPORTS

FLUE GAS TO STACK

	MOLE PERCENT	LB-MOLE/HR	LB/HR
CO2	11.675	0.2089E+05	0.9193E+06
SO2	0.033	0.5943E+02	0.3807E+04
O2	4.472	0.8000E+04	0.2560E+06
N2	68.865	0.1232E+06	0.3452E+07
H2O	14.955	0.2676E+05	0.4820E+06

SPECIFIED SO2 REMOVAL EFFICIENCY = 85.0 %

CALCULATED SO2 EMISSION = 0.85 POUNDS PER MILLION BTU

CALCULATED SO2 CONCENTRATION IN STACK GAS = 332. PPM

FLYASH EMISSION = 0.09 LBS/MILLION BTU
              = 0.042 GRAINS/SCF (WET)   OR   411. LB/HR

STACK GAS FLOW RATE = .1130E+07 SCFM (60 DEG F, 1 ATM)
                    = .1380E+07 ACFM (175. DEG F, 1 ATM)

STEAM REHEATER (IN-LINE)

SUPERFICIAL GAS VELOCITY (FACE VELOCITY) = 25.0 FT/SEC

SQUARE PIPE PITCH = 2 TIMES ACTUAL PIPE O.D.

SATURATED STEAM TEMPERATURE = 470. DEG F

OUTLET FLUE GAS TEMPERATURE = 175. DEG F

REQUIRED HEAT INPUT TO REHEATER = 0.6882E+08 BTU/HR

STEAM CONSUMPTION = 0.9163E+05 LBS/HR

OUTSIDE PIPE DIAMETER, IN.	PRESSURE DROP, IN. H2O	HEAT TRANSFER COEFFICIENT, BTU/HR FT2 DEG F
1.00	0.76	0.2082E+02

	REHEATER OUTSIDE PIPE AREA, SQ FT PER TRAIN	NUMBER OF PIPES PER BANK PER TRAIN	NUMBER OF BANKS (ROWS) PER TRAIN
INCONEL	0.1284E+04	87	3
CORTEN	0.1313E+04	87	4
TOTAL	0.2597E+04	87	7

(continued)

TABLE C-5 (continued)

```
WATER BALANCE INPUTS
-----  -------  ------

    RAINFALL(IN/YEAR)                35.
    POND SEEPAGE(CM/SEC)*10**8       10.
    POND EVAPORATION(IN/YEAR)        30.


WATER BALANCE OUTPUTS
-----  -------  -------

WATER AVAILABLE
    RAINFALL                        562. GPM         281059. LB/HR
    ALKALI                            5. GPM           2453. LB/HR
         TOTAL                      567. GPM         283512. LB/HR


WATER REQUIRED
    HUMIDIFICATION                  428. GPM         216145. LB/HR
    ENTRAINMENT                      10. GPM           5102. LB/HR
    DISPOSAL WATER                  187. GPM          93329. LB/HR
    HYDRATION WATER                  11. GPM           5724. LB/HR
    CLARIFIER EVAPORATION             0. GPM              0. LB/HR
    POND EVAPORATION                522. GPM         260805. LB/HR
    SEEPAGE                          20. GPM           9970. LB/HR

    TOTAL WATER REQUIRED           1179. GPM         589075. LB/HR

    NET WATER REQUIRED              611. GPM         305563. LB/HR
```

SCRUBBER SYSTEM
-------- ------

TOTAL NUMBER OF SCRUBBING TRAINS (OPERATING+REDUNDANT) = 5

SO2 REMOVAL = 85.0 PERCENT

PARTICULATE REMOVAL IN SCRUBBER SYSTEM = 50.0 PERCENT

TCA PRESSURE DROP ACROSS 3 BEDS = 8.6 IN. H2O

TOTAL SYSTEM PRESSURE DROP = 14.8 IN. H2O

OVERRIDE TOTAL SYSTEM PRESSURE DROP = 20.0 IN. H2O

SPECIFIED   LIQUID-TO-GAS-RATIO   = 55. GAL/1000 ACF

LIMESTONE ADDITION = 0.4906E+05 LB/HR DRY LIMESTONE

CALCULATED LIMESTONE STOICHIOMETRY = 1.41 MOLE CACO3 ADDED AS LIMESTONE
                                          PER MOLE SO2 ABSORBED

SOLUBLE CAO FROM FLY ASH = 0.0 MOLE PER MOLE SO2 ABSORBED

TOTAL SOLUBLE MGO   = 0.0 MOLE PER MOLE SO2 ABSORBED

TOTAL STOICHIOMETRY   = 1.41 MOLE SOLUBLE (CA+MG)
                                 PER MOLE SO2 ABSORBED

(continued)

TABLE C-5 (continued)

SCRUBBER INLET LIQUOR PH = 5.64

MAKE UP WATER = 611. GPM

CROSS-SECTIONAL AREA PER SCRUBBER = 425. SQ FT

SYSTEM SLUDGE DISCHARGE

SPECIES	LB-MOLE/HR	LB/HR	SOLID COMP, WT %	LIQUID COMP, PPM
CASO3 .1/2 H2O	0.2356E+03	0.3042E+05	48.45	
CASO4 .2H2O	0.9996E+02	0.1720E+05	27.41	
CACO3	0.1333E+03	0.1334E+05	21.26	
INSOLUBLES	----------	0.1810E+04	2.88	
H2O	0.5180E+04	0.9333E+05		
CA++	0.7219E+01	0.2893E+03		3073.
MG++	0.0	0.0		0.
SO3--	0.1539E+00	0.1233E+02		131.
SO4--	0.1066E+01	0.1024E+03		1088.
CL-	0.1199E+02	0.4250E+03		4513.

TOTAL DISCHARGE FLOW RATE = 0.1569E+06 LB/HR
= 237. GPM

TOTAL DISSOLVED SOLIDS IN DISCHARGE LIQUID = 8805. PPM

DISCHARGE LIQUID PH = 7.32

SCRUBBER SLURRY BLEED

SPECIES	LB-MOLE/HR	LB/HR
CASO3 .1/2 H2O	0.2356E+03	0.3042E+05
CASO4 .2H2O	0.9996E+02	0.1720E+05
CACO3	0.1333E+03	0.1334E+05
INSOLUBLES	----------	0.1810E+04
H2O	0.1957E+05	0.3526E+06
CA++	0.2727E+02	0.1093E+04
MG++	0.0	0.0
SO3--	0.5816E+00	0.4656E+02
SO4--	0.4028E+01	0.3870E+03
CL-	0.4529E+02	0.1605E+04

TOTAL FLOW RATE = 0.4185E+06 LB/HR
= 760. GPM

TOTAL SUPERNATE RETURN

SPECIES	LB-MOLE/HR	LB/HR
H2O	0.1496E+05	0.2694E+06
CA++	0.2084E+02	0.8353E+03
MG++	0.0	0.0
SO3--	0.4445E+00	0.3558E+02
SO4--	0.3078E+01	0.2957E+03
CL-	0.3461E+02	0.1227E+04

TOTAL FLOW RATE = 0.2718E+06 LB/HR
= 544. GPM

SUPERNATE TO WET BALL MILL

SPECIES	LB-MOLE/HR	LB/HR
H2O	0.1689E+04	0.3044E+05

TABLE C-5 (continued)

	LB-MOLE/HR	LB/HR
CA++	0.2354E+01	0.9436E+02
MG++	0.0	0.0
SO3--	0.5021E-01	0.4020E+01
SO4--	0.3478E+00	0.3341E+02
CL-	0.3910E+01	0.1386E+03

TOTAL FLOW RATE = 0.3071E+05 LB/HR
= 61. GPM

LIMESTONE SLURRY FEED

SPECIES	LB-MOLE/HR	LB/HR
CACO3	0.4761E+03	0.4766E+05
SOLUBLE MGO	0.0	0.0
INSOLUBLES	----------	0.1398E+04
H2O	0.1799E+04	0.3242E+05
CA++	0.2508E+01	0.1005E+03
MG++	0.0	0.0
SO3--	0.5348E-01	0.4281E+01
SO4--	0.3704E+00	0.3555E+02
CL-	0.4164E+01	0.1476E+03

TOTAL FLOW RATE = 0.8177E+05 LB/HR
= 103. GPM

SUPERNATE RETURN TO SCRUBBER OR EHT

SPECIES	LB-MOLE/HR	LB/HR
H2O	0.1327E+05	0.2390E+06
CA++	0.1849E+02	0.7409E+03
MG++	0.0	0.0
SO3--	0.3942E+00	0.3156E+02
SO4--	0.2731E+01	0.2623E+03
CL-	0.3070E+02	0.1088E+04

TOTAL FLOW RATE = 0.2411E+06 LB/HR
= 482. GPM

RECYCLE SLURRY TO SCRUBBER

SPECIES	LB-MOLE/HR	LB/HR
CASO3 .1/2 H2O	0.2172E+05	0.2805E+07
CASO4 .2H2O	0.9218E+04	0.1586E+07
CACO3	0.1229E+05	0.1230E+07
INSOLUBLES	----------	0.1669E+06
H2O	0.1805E+07	0.3251E+08
CA++	0.2513E+04	0.1008E+06
MG++	0.0	0.0
SO3--	0.5363E+02	0.4294E+04
SO4--	0.3715E+03	0.3568E+05
CL-	0.4176E+04	0.1480E+06

TOTAL FLOW RATE = 0.3859E+08 LB/HR
= 70087. GPM

FLUE GAS COOLING SLURRY

SPECIES	LB-MOLE/HR	LB/HR
CASO3 .1/2 H2O	0.1580E+04	0.2040E+06
CASO4 .2H2O	0.6704E+03	0.1154E+06
CACO3	0.8940E+03	0.8949E+05
INSOLUBLES	----------	0.1214E+05
H2O	0.1312E+06	0.2365E+07
CA++	0.1829E+03	0.7330E+04
MG++	0.0	0.0
SO3--	0.3900E+01	0.3123E+03
SO4--	0.2702E+02	0.2595E+04
CL-	0.3037E+03	0.1077E+05

TOTAL FLOW RATE = 0.2806E+07 LB/HR
= 5097. GPM

TABLE C-6. POND SIZE AND COST PARAMETERS

POND DESIGN

OPTIMIZED TO MINIMIZE TOTAL COST PLUS OVERHEAD

POND DIMENSIONS

DEPTH OF POND	21.34	FT
DEPTH OF EXCAVATION	9.14	FT
LENGTH OF PERIMETER	14803.	FT
LENGTH OF DIVIDER	2703.	FT
AREA OF BOTTOM	1366.	THOUSAND YD2
AREA OF INSIDE WALLS	149.	THOUSAND YD2
AREA OF OUTSIDE WALLS	113.	THOUSAND YD2
AREA OF POND	1505.	THOUSAND YD2
AREA OF POND SITE	1677.	THOUSAND YD2
AREA OF POND SITE	346.	ACRES
VOLUME OF EXCAVATION	1526.	THOUSAND YD3
VOLUME OF SLUDGE TO BE DISPOSED OVER LIFE OF PLANT	10269.	THOUSAND YD3
	6365.	ACRE FT

POND COSTS (THOUSANDS OF DOLLARS)

	LABOR	MATERIAL	TOTAL
CLEARING LAND	917.		917.
EXCAVATION	2846.		2846.
DIKE CONSTRUCTION	1034.		1034.
LINING(12. IN. CLAY)	1263.		1263.
SODDING DIKE WALLS	65.	53.	118.
ROAD CONSTRUCTION	9.	19.	27.
POND CONSTRUCTION	5734.	72.	5806.
LAND COST			1213.
POND SITE			7019.
OVERHEAD			3948.
TOTAL			10967.

TABLE C-7. EQUIPMENT SIZES AND COST REPORT

RAW MATERIAL HANDLING AND PREPARATION

INCLUDING 2 OPERATING AND 1 SPARE PREPARATION UNITS

ITEM	DESCRIPTION	NO.	MATERIAL	LABOR
CAR SHAKER AND HOIST	20HP SHAKER 7.5HP HOIST	1	28582.	1866.
CAR PULLER	25HP PULLER, 5HP RETURN	1	49345.	1866.
UNLOADING HOPPER	16FT DIA, 10FT STRAIGHT SIDE HT, CS	1	4180.	7711.
UNLOADING VIBRATING FEEDER	3.5HP	1	12134.	1866.
UNLOADING BELT CONVEYOR	20FT HORIZONTAL, 5HP	1	17927.	0.
UNLOADING INCLINE BELT CONVEYOR	310FT, 50HP	1	60670.	24875.
UNLOADING PIT DUST COLLECTOR	POLYPROPYLENE BAGTYPE, 2200 CFM, 7.5HP	1	5258.	12438.
UNLOADING PIT SUMP PUMP	60GPM, 70FT HEAD, 5HP	1	3371.	746.
STORAGE BELT CONVEYOR	200FT, 5HP	1	57974.	16169.
STORAGE CONVEYOR TRIPPER	30FPM, 1HP	1	13482.	2488.
MOBILE EQUIPMENT	SCRAPPER TRACTOR	1	136171.	0.
RECLAIM HOPPER	7FT WIDE, 4.25FT HT, 2FT WIDE BOTTOM, CS	2	1079.	1741.
RECLAIM VIBRATING FEEDER	3.5HP	2	24268.	3731.
RECLAIM BELT CONVEYOR	200FT, 5HP	1	40447.	8706.
RECLAIM INCLINE BELT CONVEYOR	193FT, 40HP	1	37750.	13930.
RECLAIM PIT DUST COLLECTOR	POLYPROPYLENE BAG TYPE	1	5258.	12438.
RECLAIM PIT SUMP PUMP	60GPM, 70FT HEAD, 5HP	1	3371.	746.
RECLAIM BUCKET ELEVATOR	90FT HIGH, 75HP	1	80894.	1617.
FEED BELT CONVEYOR	60.FT HORIZONTAL 7.5HP	1	20223.	1368.
FEED CONVEYOR TRIPPER	30 FPM, 1HP	1	13482.	2488.

(continued)

TABLE C-7 (continued)

ITEM	DESCRIPTION	NO.	MATERIAL	LABOR
FEED BIN	13FT DIA, 21FT STRAIGHT SIDE HT, COVERED, CS	3	16179.	29851.
BIN WEIGH FEEDER	16FT PULLEY CENTERS, 2HP	3	34603.	3731.
GYRATORY CRUSHERS	75HP	3	189765.	13672.
BALL MILL DUST COLLECTORS	POLYPROPYLENE BAG TYPE 2200 CFM, 7.5HP	3	19774.	37313.
BALL MILL	12.3TPH, 166.HP	3	475996.	43912.
MILLS PRODUCT TANK	5500 GAL 10FT DIA, 10FT HT, FLAKEGLASS LINED CS	3	14561.	22761.
MILLS PRODUCT TANK AGITATOR	10HP	3	24673.	1119.
MILLS PRODUCT TANK SLURRY PUMP	52.GPM, 60FT HEAD, 2.HP, 2 OPERATING AND 1 SPARES	3	7810.	1493.
SLURRY FEED TANK	54506.GAL, 21.0FT DIA, 21.0FT HT, FLAKEGLASS-LINED CS	1	12189.	26122.
SLURRY FEED TANK AGITATOR	47.HP	1	34432.	2541.
SLURRY FEED TANK PUMPS	26.GPM, 60 FT HEAD, 1.HP, 4 OPERATING AND 4 SPARE	8	20063.	3980.
TOTAL EQUIPMENT COST			1451506.	305284.

SCRUBBING

INCLUDING 4 OPERATING AND 1 SPARE SCRUBBING TRAINS

ITEM	DESCRIPTION	NO.	MATERIAL	LABOR
MECHANICAL ASH COLLECTOR	33% PARTICULATE REMOVAL	1	424515.	78325.
F.D. FANS	20.0IN H2O, WITH 1615. HP MOTOR AND DRIVE	5	1873839.	113586.
SHELL			812283.	
RUBBER LINING			1199962.	
MIST ELIMINATOR			368717.	
SLURRY HEADER AND NOZZLES			313679.	
GRIDS			471794.	
SPHERES			175683.	
TOTAL TCA SCRUBBER COSTS		5	3342115.	278548.
REHEATERS		5	1046932.	43280.
SOOTBLOWERS		60	404468.	298505.
EFFLUENT HOLD TANK	231287.GAL, 34.0FT DIA, 34.0FT HT, FLAKEGLASS-LINED CS	5	181225.	354208.
EFFLUENT HOLD TANK AGITATOR	63.HP	5	345851.	127622.

(continued)

TABLE C-7 (continued)

ITEM	DESCRIPTION	NO.	MATERIAL	LABOR
COOLING SPRAY PUMPS	1274.GPM 100FT HEAD, 59.HP, 4 OPERATING AND 6 SPARE	10	117520.	17352.
ABSORBER RECYCLE PUMPS	8761.GPM, 100FT HEAD, 406.HP, 8 OPERATING AND 7 SPARE	15	658657.	51730.
MAKEUP WATER PUMPS	2549.GPM, 200.FT HEAD, 215.HP, 1 OPERATING AND 1 SPARE	2	19790.	1826.
TOTAL EQUIPMENT COST			8414911.	1364980.

WASTE DISPOSAL

ITEM	DESCRIPTION	NO.	MATERIAL	LABOR
ABSORBER BLEED RECEIVING TANK	57760.GAL, 17.0FT DIA, 34.0FT HT, FLAKGLASS-LINED CS	1	14979.	31228.
ABSORBER BLEED TANK AGITATOR	36.HP	1	20467.	1511.
POND FEED SLURRY PUMPS	760.GPM, 130.FT HEAD 66.HP, 1 OPERATING AND 1 SPARE	2	14772.	2887.
POND SUPERNATE PUMPS	544.GPM, 192.FT HEAD, 44.HP, 1 OPERATING AND 1 SPARE	2	8512.	785.
TOTAL EQUIPMENT COST			58331.	36411.

TABLE C-8. CAPITAL INVESTMENT REPORT

LIMESTONE SLURRY PROCESS -- BASIS: 500 MW UNIT, 1980 STARTUP
PROJECTED CAPITAL INVESTMENT REQUIREMENTS - BASE CASE EXAMPLE 500 MW

CASE 001

INVESTMENT, THOUSANDS OF 1979 DOLLARS

	RAW MATERIAL HANDLING AND PREPARATION	SCRUBBING	WASTE DISPOSAL	TOTAL	DISTRIBUTION PERCENT OF DIRECT INVESTMENT
EQUIPMENT					
MATERIAL	1452.	7990.	58.	9500.	30.1
LABOR	305.	1287.	36.	1628.	5.2
PIPING					
MATERIAL	232.	2529.	936.	3698.	11.7
LABOR	93.	741.	341.	1176.	3.7
DUCTWORK					
MATERIAL	0.	1982.	0.	1982.	6.3
LABOR	0.	1349.	0.	1349.	4.3
FOUNDATIONS					
MATERIAL	126.	92.	12.	229.	0.7
LABOR	525.	276.	36.	826.	2.7
POND CONSTRUCTION	0.	0.	5800.	5800.	18.4
STRUCTURAL					
MATERIAL	270.	171.	1.	442.	1.4
LABOR	100.	300.	6.	486.	1.5
ELECTRICAL					
MATERIAL	172.	542.	100.	814.	2.6
LABOR	342.	815.	228.	1405.	4.4
INSTRUMENTATION					
MATERIAL	105.	763.	8.	856.	2.7
LABOR	76.	122.	8.	165.	0.5
BUILDINGS					
MATERIAL	36.	0.	0.	36.	0.1
LABOR	57.	0.	0.	57.	0.2
SERVICES AND MISCELLANEOUS	130.	640.	258.	1032.	3.3
SUBTOTAL DIRECT INVESTMENT	3969.	19725.	7826.	31520.	100.0
ENGINEERING DESIGN AND SUPERVISION	357.	1775.	704.	2837.	9.0
CONSTRUCTION EXPENSES	635.	3156.	1252.	5043.	16.0
CONTRACTOR FEES	198.	986.	391.	1576.	5.0
CONTINGENCY	397.	1972.	783.	3152.	10.0
SUBTOTAL FIXED INVESTMENT	5557.	27615.	10956.	44128.	140.0
ALLOWANCE FOR STARTUP AND MODIFICATIONS	445.	2209.	877.	3530.	11.2
INTEREST DURING CONSTRUCTION	667.	3314.	1315.	5295.	16.8
SUBTOTAL CAPITAL INVESTMENT	6669.	33138.	13148.	52956.	168.0
LAND	7.	3.	1221.	1231.	3.9
WORKING CAPITAL	150.	744.	295.	1189.	3.8
TOTAL CAPITAL INVESTMENT	6826.	33884.	14664.	55374.	175.7

TABLE C-9. FIRST YEAR ANNUAL REVENUE REQUIREMENT REPORT

LIMESTONE SLURRY PROCESS -- ASSIST 500 MW UNIT, 1980 STARTUP

PROJECTED REVENUE REQUIREMENTS - BASE CASE EXAMPLE 500 MW   CASE 001

```
                    DISPLAY SHEET FOR YEAR   1
                    ANNUAL OPERATION KW-HR/KW =  4512

           31.39 TONS PER HOUR                           DRY             SLUDGE
                    TOTAL FIXED INVESTMENT    55373000                    TOTAL
                                                                         ANNUAL
                                ANNUAL QUANTITY        UNIT COSTS         COSTS
DIRECT COSTS

   Raw Material
      LIMESTONE          110.7 K TONS          8.00/TON                   885400
      LIME                 0.0 K TONS         40.00/TON                        0
                                                                         -------
      SUBTOTAL RAW MATERIAL                                                885400

   Conversion Costs
      OPERATING LABOR AND
        SUPERVISION      20860.0 MAN-HR       12.00/MAN-HR                250400
      UTILITIES
        STEAM           413440.0 K LB          1.70/K LB                  702900
        PROCESS WATER   165510.0 K GAL         0.12/K GAL                  19900
        ELECTRICITY    38819120.0 KWH          0.030/KWH                 1164600
      MAINTENANCE
        LABOR AND MATERIAL                                               1714400
      ANALYSES            2420.0 HR           17.00/HR                     41200
                                                                         -------
      SUBTOTAL CONVERSION COSTS                                          3893400

      SUBTOTAL DIRECT COSTS                                              4778800

INDIRECT COSTS
   DEPRECIATION                                                          1795100
   COST OF CAPITAL AND TAXES, 17.20% OF UNDEPRECIATED INVESTMENT         9324300
   INSURANCE & INTERIM REPLACEMENTS, 1.17% OF TOTAL CAPITAL INVESTMENT    647800
   OVERHEAD
     PLANT, 50.0% OF CONVERSION COSTS LESS UTILITIES,
     ADMINISTRATIVE, RESEARCH, AND SERVICE,                              1003000
     10.0% OF OPERATING LABOR AND SUPERVISION                              25000
                                                                         -------
     SUBTOTAL INDIRECT COSTS                                            12905300

   TOTAL ANNUAL REVENUE REQUIREMENT                                     17784100

   EQUIVALENT UNIT REVENUE REQUIREMENT, MILLS/KWH                           7.87
-----------------------------------------------------------------------------
HEAT RATE  9000. BTU/KWH   •   HEAT VALUE OF COAL   10500 BTU/LB   •   CCAL RATE  906000 TONS/YR
```

TABLE C-10. LIFETIME ANNUAL REVENUE REQUIREMENT REPORT

LIMESTONE SLURRY PROCESS -- BASIS: 500 MW UNIT, 1980 STARTUP

PROJECTED LIFETIME REVENUE REQUIREMENTS - BASE CASE EXAMPLE 500 MW

CASE 001

TOTAL CAPITAL INVESTMENT: $ 55376000

YEARS AFTER OPERA-TION, POWER UNIT START /KW	ANNUAL KW-HR	POWER UNIT HEAT REQUIREMENT, MILLION BTU /YEAR	POWER UNIT FUEL CONSUMPTION, TONS COAL /YEAR	SULFUR REMOVED BY POLLUTION CONTROL PROCESS, TONS/YEAR	BYPRODUCT RATE: EQUIVALENT TONS/YEAR DRY SLUDGE	SLUDGE FIXATION FEE $/TON DRY SLUDGE	ADJUSTED GROSS ANNUAL REVENUE REQUIREMENT EXCLUDING SLUDGE FIXATION COST, $/YEAR	TOTAL ANNUAL SLUDGE FIXATION COST, $/YEAR	NET ANNUAL INCREASE IN TOTAL REVENUE REQUIREMENT, $	CUMULATIVE NET INCREASE IN TOTAL REVENUE REQUIREMENT, $
1	4512	20306000	966900	26300	141600	0.0	17764100	0	17764100	17764100
2	4643	20893300	994900	25000	143700	0.0	17573100	0	17573100	35317200
3	4775	21487500	1023200	25800	149900	0.0	17402800	0	17402800	52719800
4	4906	22077300	1051300	26300	154000	0.0	17230000	0	17230000	69950000
5	5037	22668500	1079400	27200	158100	0.0	17057800	0	17057800	87008000
6	5169	23260000	1107600	27900	162200	0.0	16885500	0	16885500	103893500
7	5300	23850000	1135700	28400	166300	0.0	16711800	0	16711800	120605300
8	5432	24444000	1164000	29300	170500	0.0	16538800	0	16538800	137144000
9	5563	25033500	1192100	30000	174600	0.0	16364100	0	16364100	153508100
10	5695	25623000	1220200	30700	178700	0.0	16189100	0	16189100	169697200
11	5695	25627500	1220200	30700	178700	0.0	16080300	0	16080300	185853500
12	5695	25627500	1220200	30700	178700	0.0	15562700	0	15562700	201416200
13	5695	25627500	1220200	30700	178700	0.0	15257700	0	15257700	216673900
14	5695	25627500	1220200	30700	178700	0.0	14775300	0	14775300	231449200
15	5695	25627500	1220200	30700	178700	0.0	14213100	0	14213100	245662300
16	5337	24016300	1165500	29000	168800	0.0	13573000	0	13573000	259235300
17	5379	24205500	1152900	28200	163900	0.0	13293800	0	13293800	272529100
18	5221	23494500	1118800	27300	159900	0.0	12834100	0	12834100	285363200
19	5064	22788400	1085100	26500	154000	0.0	12337000	0	12337000	297700200
20	4906	22077200	1051300	25800	150100	0.0	12125600	0	12125600	309825800
21	4748	21366300	1017600	25000	145100	0.0	11810400	0	11810400	321636200
22	4591	22659500	983800	24000	141100	0.0	11468500	0	11468500	333104700
23	4433	19948500	950000	23900	137100	0.0	11084700	0	11084700	344189400
24	4275	19237500	916100	23100	134200	0.0	10519700	0	10519700	354709100
25	4118	18531300	882600	22200	128300	0.0	10053100	0	10053100	364762200
26	3960	17820000	848600	21400	123300	0.0	9588200	0	9588200	374350400
27	3802	17109000	814700	20500	119300	0.0	9120000	0	9120000	383470400
28	3645	16402500	781100	19700	114400	0.0	8652700	0	8652700	392123100
29	3487	15691500	747200	18800	109400	0.0	8182500	0	8182500	400305600
30	3329	14980500	713300	18000	104300	0.0	7711300	0	7711300	410016900

TOT 160001 657004500 31286100 787700 4982000 410736000 0 410736000

REVENUE REQUIREMENT DISCOUNTED AT 11.5% TO UP SULFUR REMOVED
LEVELIZED INCREASE IN UNIT REVENUE REQUIREMENT EQUIVALENT TO DISCOUNTED REQUIREMENT OVER LIFE OF POWER UNIT 132962000

LIFETIME AVERAGE INCREASE IN UNIT REVENUE REQUIREMENT
- DOLLARS PER TON OF COAL BURNED   13.13   13.13
- MILLS PER KILOWATT-HOUR   3.63   3.63
- CENTS PER MILLION BTU HEAT INPUT   62.32   62.32
- DOLLARS PER TON OF SULFUR REMOVED   521.44   521.44
- DOLLARS PER TON OF COAL BURNED   14.80
- MILLS PER KILOWATT-HOUR   6.34
- CENTS PER MILLION BTU HEAT INPUT   70.47
- DOLLARS PER TON OF SULFUR REMOVED   587.98

The program output consists of several reports:

1. Input data
2. Process parameters
3. Pond size parameter and costs
4. System equipment sizes and costs
5. Capital investment
6. First-year annual revenue requirement
7. Lifetime annual revenue requirement

Samples of these reports are presented in Tables C-4 through C-10.

Model Usefulness

The TVA/Bechtel computer model can be highly useful to the utility Project Manager in evaluating various limestone FGD scenarios. It can significantly reduce the time and effort expended in the planning stages to analyze the effects of various operating variables on process parameters and final costs. For example, an economic evaluation of sludge disposal options can be performed very rapidly and will delineate the effect of each sludge disposal option on final costs.

The project engineer can apply the model by analyzing the effects of various input variables in greater detail. The process parameter reports indicate the effects of input parameters on system flow rates and on sizes of individual equipment items. For example, the model's output display of the effects of various L/G ratios on system performance will aid in selection of the proper L/G ratio for optimum efficiency of SO_2 removal.

FLUE GAS DESULFURIZATION INFORMATION SYSTEM

Background

Since July 1974, PEDCo Environmental, Inc., has conducted for the EPA a program called the Utility Flue Gas Desulfurization Survey. The object of the program has been to provide the EPA with information and technical assistance concerning U.S. utility FGD technology. The program has recently been expanded to include Japanese utility FGD applications and domestic particulate scrubbers. The major product of this program is a quarterly status report, which is distributed worldwide to recipients who are directly or indirectly involved in development of FGD technology.

The quarterly survey reports cover all aspects of operational and planned FGD systems and the units to which they are (or will be) applied. The plant data range from simple identification of unit and location to more specific information regarding such matters as applicable environmental regulations, boiler type and flow rates, and average fuel analysis. Information on emission controls identifies the primary means of particulate and SO_2 control. Wherever possible the report presents information on component designs and operating parameters, waste disposal strategy, reheater data, and other technical data, as well as both the reported and adjusted capital costs and annual revenue requirements.

The major feature of the report is the identification and description of operational FGD systems. The performance of these systems is described in depth, with emphasis upon mechanical reliability, removal efficiency, problems encountered, and solutions implemented. The importance of an accurate description of the operational FGD systems is twofold: (1) it provides the owner/operator utilities, system suppliers, and design engineering firms with information upon which to base present and future design strategies, and (2) it provides the EPA staff with information needed to develop enforcement strategies and to plan and implement research, development, and demonstration projects.

At first the summary report was prepared and published in a semi-automated manner. A series of computer files were developed and manually updated. This method of information update and retrieval was adequate at the time; only 19 FGD systems were in operation at the end of 1974, and the total committed capacity was less than 40,000 MW. The report has become progressively more comprehensive and voluminous as the number of committed FGD systems continues to grow. Figure C-1 illustrates by year the rapid increase in planned and operational systems. This growth was caused primarily by the promulgation and enforcement of more stringent SO_2 emission standards. As the number of operational FGD systems increased and more performance data became available, the need for a more efficient system for information storage and retrieval became apparent.

In mid-1976 a fully automated information storage and retrieval system replaced the manually updated files and significantly improved the efficiency of report preparation. From 1976 to 1978 the number of operating systems increased to 46, and the total committed capacity rose to approximately 63,000 MW. A corresponding sharp rise in the amounts of available data made it evident that the report should be produced more rapidly to reflect current technology. Preparing meaningful analyses of the data became increasingly difficult, and therefore a decision was made to convert to a data-base system.

System Description

A data-base is a collection of information files consisting of data elements linked in a logical manner. The data elements are stored in groups, or blocks, of similar data that can be repeated as the need arises.

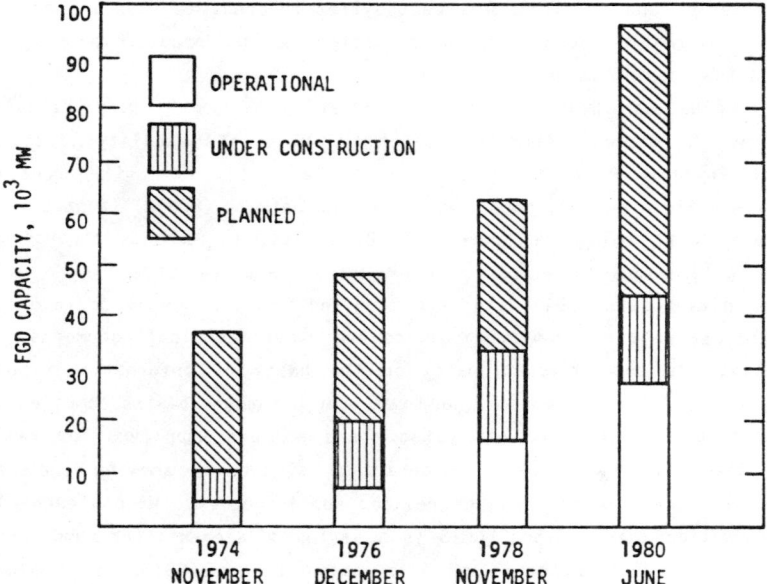

Figure C-1. Growth in the number of FGD systems which are planned, under construction and operating.

With this flexibility such a system can easily accommodate the ever-increasing amounts of data.

An important feature of the data-base system is that it provides users with on-line access to the data files, with comprehensive data manipulation capabilities. The data files are updated continuously to ensure that the information is as complete and current as possible. Interested users can access the data files in the interim between published reports; they can thus examine and analyze the more specific information no longer included in the quarterly report. This data-base system responds to standard and predictable information needs and to unexpected, ad hoc requests arising from unique situations.

The Flue Gas Desulfurization Information System (FGDIS) files are stored at the National Computer Center (NCC), an EPA facility located in Research Triangle Park, North Carolina. The system is easily accessible through a nationwide telephone communications network (COMNET) that currently offers local telephone numbers in 21 cities, as well as through WATS service to locations for which local numbers are not available.*

The data in the FGDIS files are acquired from FGD system operators and suppliers, as well as from publications and other technical documents. The data files are updated continually on the basis of information supplied monthly by FGD system operators, and computer printouts of the compiled data are submitted periodically to system operators and suppliers for review.

Figure C-2 illustrates the data-base structure, showing the major categories of information and the related subcategories. Within each block (or information area) of the FGDIS is a series of elements (or components). Table C-11 presents a sample of information in the FGDIS. Each element represents a single piece of data stored within that block that is consistent with the particular information area in which that piece of data resides. Each element consists of an element number (for identification purposes) and an element name, which identifies the type of data that is stored (or can be stored) within the element. Because of continuing changes in FGD technology, the FGD structure diagram and definition are subject to change.

Access to the FGDIS files and data elements is provided through a user-oriented language, by which information in the data files can be examined without the necessity of prior programming experience. The data can be tabulated in a manner consistent with the specific information needs of the user, and functions are provided for statistical analyses of the numerical data. A detailed description of the use of the FGDIS and the

---

* Information about access to the FGDIS can be obtained from Mr. Walter L. Finch, Product Manager, National Technical Information Service, 5285 Port Royal Road, Springfield, Virginia 22161 [telephone (703) 487-4807].

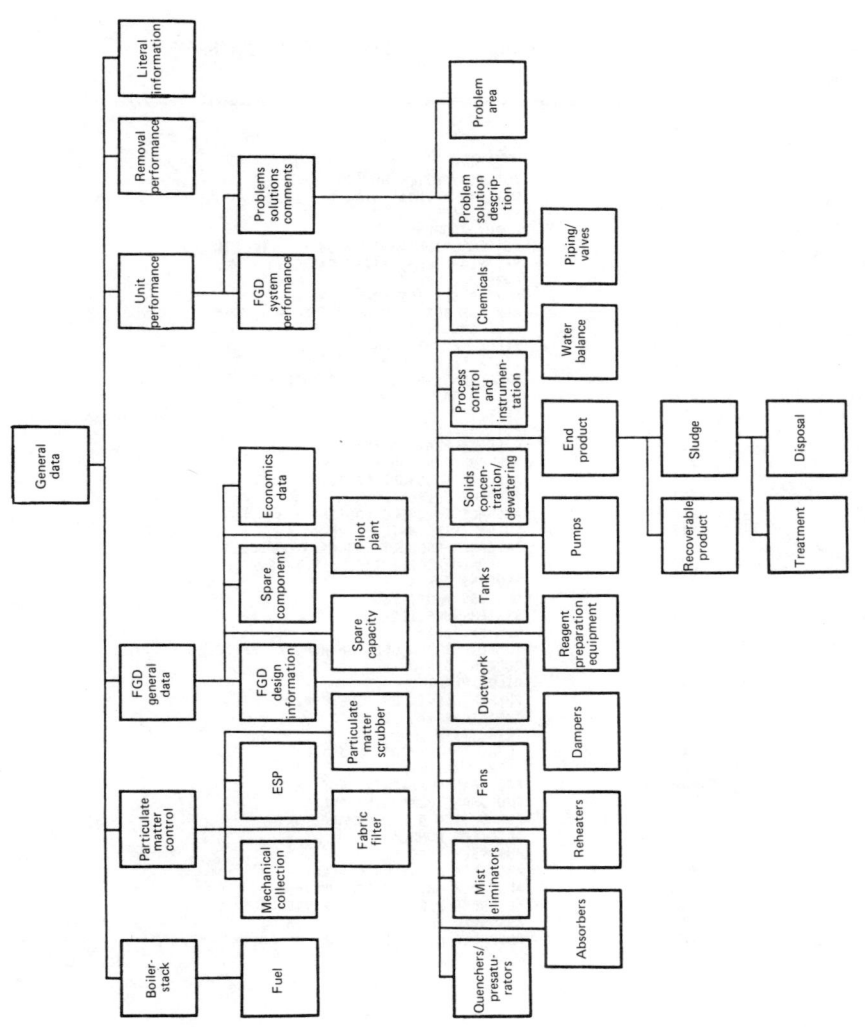

Figure C-2. Computerized data base structure, FGDIS program.

TABLE C-11. MAJOR DATA FIELDS, FGDIS PROGRAM

Field No.	Description
1	IDENTIFICATION NUMBER
3	UTILITY - INDUSTRIAL
6	PLANT NAME
8	PLANT ADDRESS
17	SO_2 EMISSION LIMITATION - LB/MM BTU
18	NET PLANT GENERATING CAPACITY - MW
102	FURNACE TYPE
103	BOILER SERVICE LOAD
105	MAXIMUM BOILER FLUE GAS FLOW - ACFM
106	FLUE GAS TEMPERATURE - F
107	STACK HEIGHT - FT
109	STACK FLUE LINER
112	STACK GAS INLET TEMPERATURE - F
200	FUEL DATA
201	FUEL TYPE
202	FUEL GRADE
207	AVERAGE HEAT CONTENT - BTU/LB
209	AVERAGE ASH CONTENT - %
211	AVERAGE MOISTURE CONTENT - %
213	AVERAGE SULFUR CONTENT - %
215	AVERAGE CHLORIDE CONTENT - %
217	FUEL FIRING RATE - TPM
301	SALEABLE PRODUCT/THROWAWAY PRODUCT
302	GENERAL PROCESS TYPE - WET/DRY/ETC
303	PROCESS TYPE
304	PROCESS ADDITIVES
305	SYSTEM SUPPLIER
309	NEW/RETROFIT
311	TOTAL UNIT SO_2 DESIGN REMOVAL EFFICIENCY
312	CURRENT STATUS
400	PILOT PLANT
410	CURRENT LEVEL OF DEVELOPMENT
506	TOTAL CAPITAL COST - $
507	TOTAL $/KW
512	TOTAL ANNUAL COST - $
513	TOTAL MILLS/KWH
600	FGD SPARE CAPACITY INDEX - %
700	FGD SPARE COMPONENT INDEX
800	FGD DESIGN & PERFORMANCE DATA
901	ABSORBER NUMBER
902	ABSORBER TYPE
916	ABSORBER GAS FLOW - ACFM
917	ABSORBER GAS TEMPERATURE - F
919	ABSORBER L/G RATIO - GAL/1000ACF
920	ABSORBER PRESSURE DROP - IN H_2O
921	ABSORBER SUPERFICIAL GAS VELOCITY - FT/SEC

(continued)

TABLE C-11 (continued)

Field No.	Description
932	ABSORBER NUMBER OF SPARES
1000	CENTRIFUGE
1100	FANS
1101	FAN NUMBER
1104	FAN TYPE
1107	FAN LOCATION
1110	FAN GAS CAPACITY - ACFM
1111	FAN GAS TEMPERATURE - F
1200	VACUUM FILTER
1300	MIST ELIMINATOR
1301	MIST ELIMINATOR NUMBER
1304	MIST ELIMINATOR TYPE
1314	MIST ELIMINATOR SUPERFICIAL GAS VELOCITY - FT/SEC
1315	MIST ELIMINATOR PRESSURE DROP - IN H_2O
1400	PROCESS CHEMISTRY CONTROL
1500	PUMPS
1600	TANKS
1700	REHEATER
1701	REHEATER NUMBER
1704	REHEATER TYPE
1712	REHEATER TEMPERATURE BOOST - F
1713	REHEATER ENERGY REQUIRED
1800	THICKENER
1900	DUCTWORK
2000	WATER BALANCE
2001	WATER LOOP TYPE (CLOSED/OPEN)
2100	REAGENT PREPARATION EQUIPMENT
2200	END PRODUCT
2400	SLUDGE
2500	TREATMENT
2501	TREATMENT TYPE
2600	DISPOSAL
2602	DISPOSAL TYPE
2603	DISPOSAL LOCATION
2607	DISPOSAL CAPACITY - ACRE-FT
2700	PARTICULATE CONTROL
2800	FABRIC FILTER
2900	ESP
3100	PARTICULATE SCRUBBER
3200	LITERAL
3300	EMISSION CONTROL SYSTEM REMOVAL PERFORMANCE
3307	SO_2 INLET CONCENTRATION
3309	SO_2 OUTLET CONCENTRATION
3312	% SO_2 REMOVAL
3313	SO_2 ANALYSIS METHOD
3400	SYSTEM DEPENDABILITY PERFORMANCE
3402	PERIOD - HR
3403	BOILER - HR
3405	UNIT AVAILABILITY - %
3447	SYSTEM AVAILABILITY - %
3448	SYSTEM OPERABILITY - %
3449	SYSTEM RELIABILITY - %
3450	SYSTEM UTILIZATION - %

individual commands available for data manipulation is provided in the user's manual (PEDCo Environmental, Inc. 1979); the data base management system is described more fully in the system reference manual (MRI Systems Corporation 1974).

System Usefulness

The extensive design and performance data stored within the FGDIS and the virtually unlimited data manipulation capabilities of System 2000, a general data base management system, make the FGDIS a very useful design tool based on actual and current FGD system information. The designer can determine, for example, the predominant limestone scrubber design configurations, average L/G ratios, or prevailing types of reheat systems. He can couple the design data with performance information to develop an analysis of design features versus reliability. In addition, the designer can identify and examine the actual problems associated with a specific component design or application and thus possibly avoid the problems encountered at current installations. The design information can also be used in conjunction with the reported and adjusted cost data for an economic evaluation of various systems. Such an evaluation could be based not only on process type, but also on such parameters as fuel sulfur content, geographical location, and boiler size.

Although the number of ways of compiling FGDIS information is virtually limitless, the following examples provide a sample of the output resulting from commands of the following four general categories:

    1.   Data listing
    2.   Data tally
    3.   Statistical analysis
    4.   Tabular report generation

Data Listing.

```
---
>PRINT FUEL DATA WHERE COMPANY NAME EQ TENNESSEE VALLEY AUTHORITY AND
---
>PLANT NAME EQ WIDOWS CREEK AND UNIT NUMBER EQ 8:
      FUEL TYPE= COAL
      AVERAGE HEAT CONTENT - BTU/LB= 10000
      AVERAGE ASH CONTENT - %= 25.00
      AVERAGE MOISTURE CONTENT - %= 10.00
      AVERAGE SULFUR CONTENT - %= 3.70
      FUEL FIRING RATE - TPH= 231
---
```

The data listing commands available for use with the FGDIS provide a simple means of retrieving data from the system in a sequential list. This example shows output from a request for a listing of fuel data about TVA's Widows Creek unit 8. Through the use of commands similar to the one above, any portion of the FGDIS can be examined; the data listed will be only those that meet the qualifications specified in the "Where" clause.

Data Tally.

The tally commands available for use with the FGDIS provide a means of obtaining statistical information about unique values of elements stored in the system. In this example, a tally of the element "Plant EPA Region" was requested. The output shows two frequencies for 1 (indicating the FGDIS contains information for two units located in EPA Region 1), three frequencies for 2, and so on. The output also indicates that the nine EPA regions in the system include a total of 232 plants for which data are recorded in the FGDIS.

Statistical Analysis.

```
>PRINT AVG C19 WHERE C303 EQ LIMESTONE:
AVG GROSS UNIT GENERATING CAPACITY - MW= 474.291
```

System functions can be used in conjunction with the FGDIS to perform the following statistical analyses of numerical data: average, standard deviation, summation, maximum value, minimum value, and number of occurrences. In the above example, the request was for the average value of "C19" (average gross unit generating capacity) for systems that have a process type "C303" (limestone systems); the output appears below the request as 474.291 MW. The system functions can be helpful in determining such things as maximum or minimum pressure drops for a specific component, or average L/G ratio for a specific process type.

Tabular Report Generation.

	NUMBER AND TOTAL MW OF FGD SYSTEMS AND PARTICLE SCRUBBERS		
STATUS	UNITS	TCC-MW	ESC-MW
OPERATIONAL	73	27155	24766

UNDER CONSTRUCTION	39	17855	16854
PLANNING			
CONTRACT AWARDED	29	13769	12919
LETTER OF INTENT	7	5590	5590
REQUESTING-EVALUATING BIDS	15	8424	8424
CONSIDERING ONLY	40	24200	23980
SUBTOTAL	203	96993	92533
PARTICLE SCRUBBERS	15	2998	2077
SUBTOTAL-DOMESTIC	218	99991	94610
JAPANESE FGD	6	1455	1455
TOTAL	224	101446	96065

A feature designated "Report Writer" enables the user to define and generate formatted reports of data contained in the system. In the above example, a report program was defined by a set of commands to generate a tabular listing of the number, total controlled capacity (TCC), and equivalent scrubbed capacity (ESC) of domestic and foreign FGD systems and domestic particle scrubbers by status category. The TCC is the sum of the gross generating capacities of units brought into compliance with SO_2 regulations by FGD systems, and the ESC is the product of the TCC times the average fraction of flue gas scrubbed. The report writer capability is helpful when a tabular listing of FGDIS data is required.

BECHTEL-MODIFIED RADIAN EQUILIBRIUM PROGRAM

Background

In 1970, Radian Corporation developed a computer program under sponsorship of an EPA predecessor agency. This program provided a theoretical description of the limestone injection, wet scrubbing process for removal of SO_2 from power plant flue gas (Lowell et al. 1970). The Radian computer program calculated numerous chemical equilibria for the aqueous lime/limestone scrubbing system.

The original information and guidelines of the Radian program were modified by Bechtel Corporation under contract to the EPA as part of the ongoing research and development effort at the Shawnee test facility. This Bechtel-Modified Radian Equilibrium Program is used extensively to support the Shawnee effort (Burbank and Wang 1980).

Examples of the program output are presented in progress reports prepared by the Bechtel Corporation about the Shawnee work. Especially useful is the "Third Progress Report" (Head et al. 1977), which describes the input requirements and output capabilities of the program. The report also presents parametric plots and nomographs based on a semitheoretical mathematical model that predicts SO_2 removal by limestone wet scrubbing as a function of operating variables. The equations used to construct these plots and nomographs were based on and verified by Shawnee data; the results compared favorably with results predicted by the Bechtel-Modified Radian Equilibrium Program.

Recent work sponsored by EPA involves the use of adipic acid as an effective additive in limestone scrubbing systems (Head et al. 1979). This work has resulted in updating and expansion of the equilibrium program to include a version applicable to use of adipic acid. This version gives detailed chemical process information for use of adipic acid to achieve high SO_2 removal efficiency (Burbank et al. 1980).

Information on how to use this program can be obtained by contacting Mr. Robert H. Borgwardt, U.S. EPA, Industrial Environmental Research Laboratory, Research Triangle Park, North Carolina 27711.

Model Usefulness

The Bechtel-Modified Radian Equilibrium Program is useful for evaluating limestone scrubber performance, especially for monitoring scaling potential by calculation of the gypsum relative saturation levels. The monitoring of gypsum is important because the Shawnee testing has shown that scaling usually occurs when the saturation level exceeds 130 percent. Other user benefits are the SO_2 removal equations, parametric plots, and nomographs presented in the "Third Progress Report" cited earlier.

Simplified equations for calculation of gypsum saturation in limestone wet scrubbing liquors at 25° and 50°C were fitted to the predictions of the modified Radian program by use of liquor data from the Shawnee long-term wet scrubbing reliability tests. Results obtained with these equations differed little from those generated by the Radian program. The equations are accurate for concentrations of total dissolved magnesium and chloride ions up to 15,000 ppm.

Persons not having access to the modified Radian program can use these simplified equations for simple, accurate, and convenient prediction of gypsum saturation levels:

Fraction $CaSO_4$ saturation at 25°C = $(Ca)(SO_4)[(300/I)+76]$

Fraction $CaSO_4$ saturation at 50°C = $(Ca)(SO_4)[(263/I)+47]$

where I = ionic strength of the liquor, g-mol/liter = $3[(Ca)+(Mg)]+(SO_4)$

(Ca) = measured concentration of total dissolved calcium, g-mol/liter

(Mg) = measured concentration of total dissolved magnesium, g-mol/liter

(SO_4) = measured concentration of total dissolved sulfate, g-mol/liter

The calculation of ionic strength assumes that the liquor contains only calcium, magnesium, sulfate, and chloride ions in solution. The ionic balance is as follows:

$$(Cl) = 2\,[(Ca)+(Mg)-(SO_4)]$$

Therefore,

$$I = 1/2\,\Sigma M_i Z_i^2$$
$$= 1/2\,[4(Ca)+(Mg)+4(SO_4)+(Cl)]$$
$$= 3\,[(Ca)+(Mg)]+SO_4$$

where M_i = molarity of component i

Z_i = unit charge of component i

Concentrations of potassium and sodium ions have been small at the Shawnee test facility and thus are not included in the ionic strength equation. Preliminary evaluation of liquor from other limestone and lime wet scrubbing systems indicates that substantial amounts of dissolved sodium can be accounted for by adding the concentration of dissolved sodium (in gram-moles per liter) to the ionic strength, I.

REFERENCES FOR APPENDIX C

Anders, W. L., and R. L. Torstrick. 1981 (in press). Computerized Shawnee Lime/Limestone Scrubbing Model: User's Manual. EPA 600/8-81-081.

Burbank, D. A., and S. C. Wang. 1980. EPA Alkali Scrubbing Test Facility: Advanced Program Final Report (October 1974 to June 1978). EPA-600/7-80-115. NTIS No. PB 80-204241.

Burbank, D. A., et al. 1980. Test Results on Adipic Acid-Enhanced Limestone Scrubbing at the EPA Shawnee Test Facility--Third Report. Presented at the Symposium on Flue Gas Desulfurization, Houston, Texas, October 28-31, 1980.

Head, H. N. 1977. EPA Alkali Scrubbing Test Facility: Advanced Program, Third Progress Report. EPA-600/7-77-105. NTIS No. PB-274 544.

Head, H. N., et al. 1979. Recent Results From EPA's Lime/Limestone Scrubbing Programs--Adipic Acid as a Scrubber Additive. In: Proceedings: Symposium on Flue Gas Desulfurization, Las Vegas, Nevada, March 1979. Vol. 1. EPA-600/7-79-167a. NTIS No. PB80-133168.

Lowell, P. S., et al. 1970. A Theoretical Description of the Limestone Injection - Wet Scrubbing Process. Vol. 1, Final Report for the National Air Pollution Control Administration under Contract No. CPA-22-69-138. NTIS No. PB-193 029.

McGlamery, G. G., et al. 1975. Detailed Cost Estimates for Advanced Effluent Desulfurization Processes. EPA-600/2-75-006. NTIS No. PB-242 541.

McGlamery, G. G., et al. 1980. FGD Economics in 1980. Presented at the Symposium on Flue Gas Desulfurization, Houston, Texas, October 28-31, 1980.

MRI Systems Corporation. 1974. System 2000, Reference Manual. Austin, Texas.

PEDCo Environmental, Inc. 1979. Flue Gas Desulfurization Information System Data Base User's Manual. Cincinnati, Ohio.

Stephenson, C. D., and R. L. Torstrick. 1978. Current Status of Development of the Shawnee Lime/Limestone Computer Program. In: Proceedings, Industry Briefing on EPA Lime/Limestone Wet Scrubbing Test Programs, August 1978. EPA-600/7-79-092. NTIS No. PB-296 517.

Stephenson, C. D., and R. L. Torstrick. 1979. Shawnee Lime/Limestone Scrubbing Computerized Design/Cost-Estimate Model Users Manual. EPA-600/7-79-210. NTIS No. PB80-123037.

Torstrick, R. L. 1976. Shawnee Limestone/Lime Scrubbing Process Computerized Design Cost Estimates Program: Summary Description Report. Presented at the Industry Briefing Conference, Raleigh, North Carolina, October 19-21, 1976.

Torstrick, R. L., L. J. Henson, and S. V. Tomlinson. 1978. Economic Evaluation Techniques, Results, and Computer Modeling for Flue Gas Desulfurization. In: Proceedings of the Symposium on Flue Gas Desulfurization, Hollywood, Florida, November 1977. Volume I. EPA-600/7-78-058a. NTIS No. PB-282 090.

Appendix D

Innovations in Limestone Scrubbing

This appendix discusses new types of scrubbers and new process modifications that are commercially available for use with limestone scrubbing or that have strong commercial potential. It also includes a preliminary economic evaluation of current and future limestone systems. The information presented is not intended to be exhaustive; further details are given in the references indicated.

NEW TYPES OF SCRUBBERS

Jet bubbling and cocurrent scrubbers are two innovative kinds of scrubbers that can be used with limestone scrubbing systems. Additionally, a charged particulate separator (CPS) can be added to a conventional scrubber. This subsection examines these two types of scrubbers and the CPS option.

Jet Bubbling Scrubber

Chiyoda Chemical Engineering and Construction Company, Ltd., has built and operated a limestone scrubbing system that produces gypsum as a byproduct. The central feature of this Chiyoda Thoroughbred 121 system is a jet bubbling reactor (JBR), in which SO_2 scrubbing, forced oxidation, neutralization, and crystallization all occur.

As shown in Figure D-1, flue gas enters a relatively shallow liquid layer through vertical spargers. The velocity of the flue gas causes it to entrain the surrounding liquid, so that a jet bubbling or froth layer is created. The resulting gas-liquid interface is large and enhances SO_2 removal.

Liquid in the JBR is moderately agitated by air bubbling and mechanical stirring. Oxidation air is introduced into the reactor by air spargers at rates 200 to 300 percent of stoichiometric requirements (Laseke 1979). Excess oxidation air, adequate residence time, and sufficient suspended solids promote gypsum crystal growth. Gypsum settles to the bottom of the JBR, and a bleed stream of gypsum is continuously drawn off.

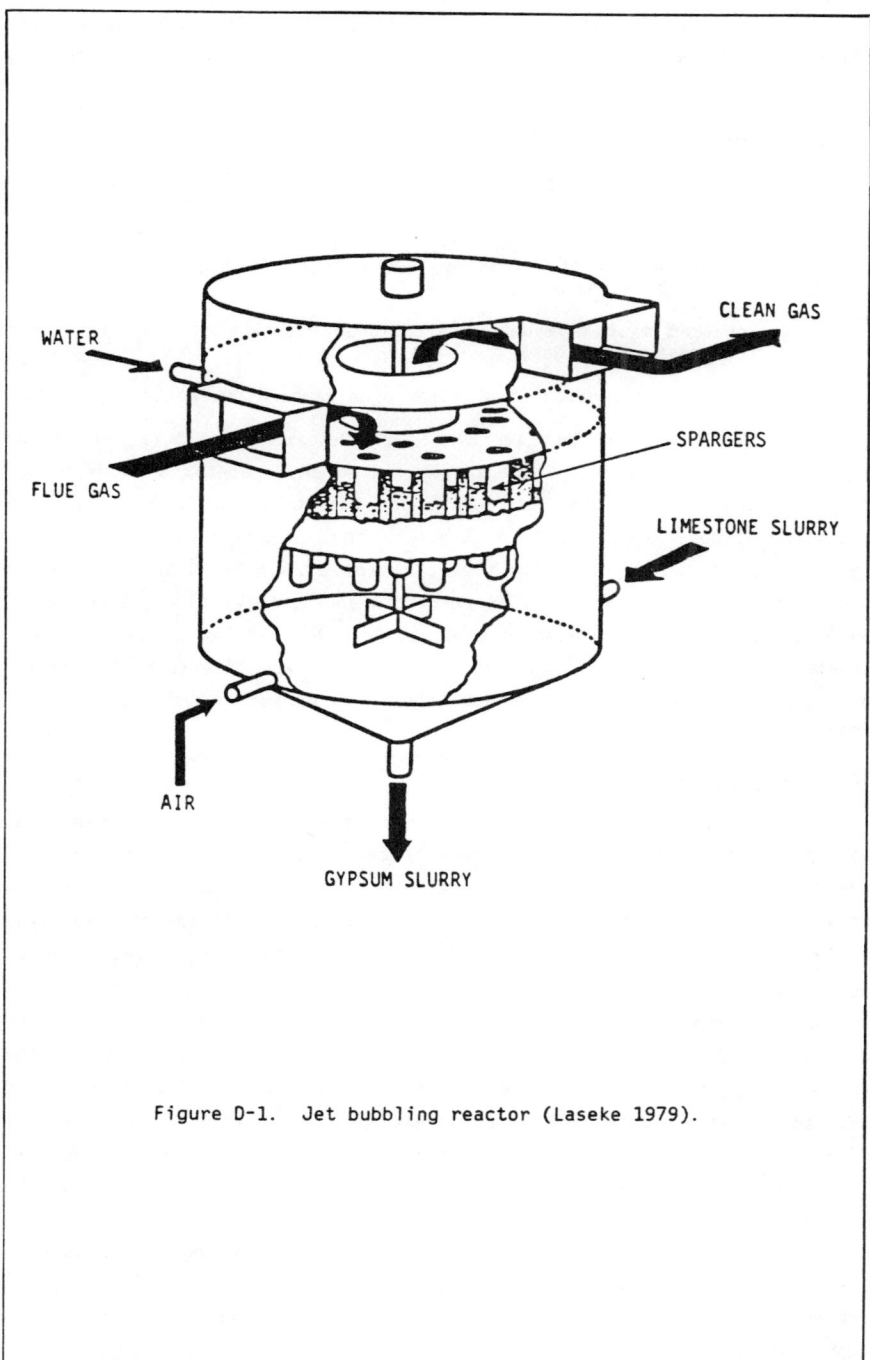

Figure D-1. Jet bubbling reactor (Laseke 1979).

Compared with conventional limestone scrubbers, a jet bubbling scrubber offers several design and operating features that can improve operability and reduce costs:

- No large slurry recirculation pumps

- No nozzles or screens

- High limestone utilization

- Reduced effect of limestone type and grind on operation (because of low operating pH)

- Enhanced mist eliminator performance (because of low slurry entrainment in the gas)

- Minimal scale deposition over a wide range of operating conditions

With funding from the Electric Power Research Institute (EPRI) and Southern Company Services, Chiyoda constructed and operated a 20-MW prototype jet bubbling scrubber at Gulf Power Company's Scholz Station. The prototype scrubber was tested for 9 months and was shown to function reliably at various conditions with flue gas from a coal-fired boiler. Sulfur dioxide removal efficiency was 95 percent when the inlet flue gas contained 3500 ppm of SO_2, and gypsum produced by the JBR settled quickly and was dewatered easily (Radian Corporation 1980). Average limestone utilization exceeded 98 percent, and scale deposition was minimal (Radian Corporation 1980).

Cocurrent Scrubber

A cocurrent scrubber offers several advantages over a conventional countercurrent scrubber. The equipment configuration better fits ductwork and fan arrangements in most powerplants. Because flue gas enters the scrubber at a high elevation and leaves near ground level, the mist eliminator and reheat system, which tend to require the most maintenance, can be near ground level. The scrubber system fans can be on the ground, and ductwork to the stack can usually be shorter and less complex. Also, the change in flue gas direction at the bottom of a cocurrent scrubber and the vertical orientation of the mist eliminator enhance liquid separation and drainage. Another benefit of a cocurrent scrubber is that flooding is less likely to occur. Because of the enhanced liquid separation and the reduced tendency to flood, gas velocity is increased and scrubber size can be smaller.

With funding from EPRI, TVA tested cocurrent scrubbing at the Colbert 1-MW pilot plant. These tests provided design data used to modify a 10-MW prototype unit at the Shawnee test facility for cocurrent scrubbing. From August 1978 to July 1979, TVA tested the 10-MW prototype scrubber. Sulfur dioxide removal efficiency during initial tests of the prototype consistently exceeded 90 percent at inlet flue gas SO_2 concentrations ranging from 1500 to 3000 ppm (Jackson 1980).

In tests conducted at the Shawnee test facility from August 1979 to July 1980, the cocurrent scrubber was altered for forced oxidation with a single effluent hold tank (Figure D-2) and with multiple tanks (Figure D-3). During these more recent tests, TVA identified operating conditions that consistently removed more than 90 percent of SO_2 from the flue gas and oxidized more than 95 percent of calcium sulfite in the scrubber slurry to gypsum (Jackson 1980).

Charged Particulate Separator

A recent commercial development is the use of a CPS after a conventional limestone scrubber (e.g., a venturi-spray tower combination). The CPS has been designed as a wet electrostatic precipitator and, in combination with a conventional scrubber, offers several advantages over other scrubbing systems. If a dry precollector (e.g., an electrostatic precipitator) is used ahead of an SO_2 scrubber, fly ash is collected and removed. A scrubber with a CPS, however, allows the use of alkaline material in the fly ash for SO_2 scrubbing.

A scrubbing system with a dry precollector followed by a wet scrubber usually requires two waste disposal systems: one for dry waste, the other for sludge. In contrast, a scrubber with a CPS allows discharge of waste into a common disposal system, which has fewer components to operate and maintain.

In a conventional venturi-spray tower combination, the pressure drop across the venturi unit is often set at 15 to 20 in. H_2O to remove most particulates. When a CPS is used after a venturi-spray tower combination, however, the venturi serves only as a precollecting device and operates at a pressure drop of 3 to 5 in. H_2O (Martin, Malki, and Graves 1979). This smaller pressure drop can permit a significant reduction in operating costs.

A scrubber with a CPS can remove SO_2 to any required emission level regardless of the sulfur content of the coal. Further, limestone comsumption by a scrubber with a CPS is only slightly greater than the theoretical requirement, whereas consumption by a conventional limestone scrubbing system is substantially greater.

Combustion Engineering has conducted pilot tests at the Sherburne Station of Northern States Power Company to develop the CPS option with a venturi-spray tower combination. According to Martin, Malki, and Graves (1979), this pilot system (known as the Two Stage Plus) not only can meet the particulate and SO_2 standards set in June 1979, but also costs less than

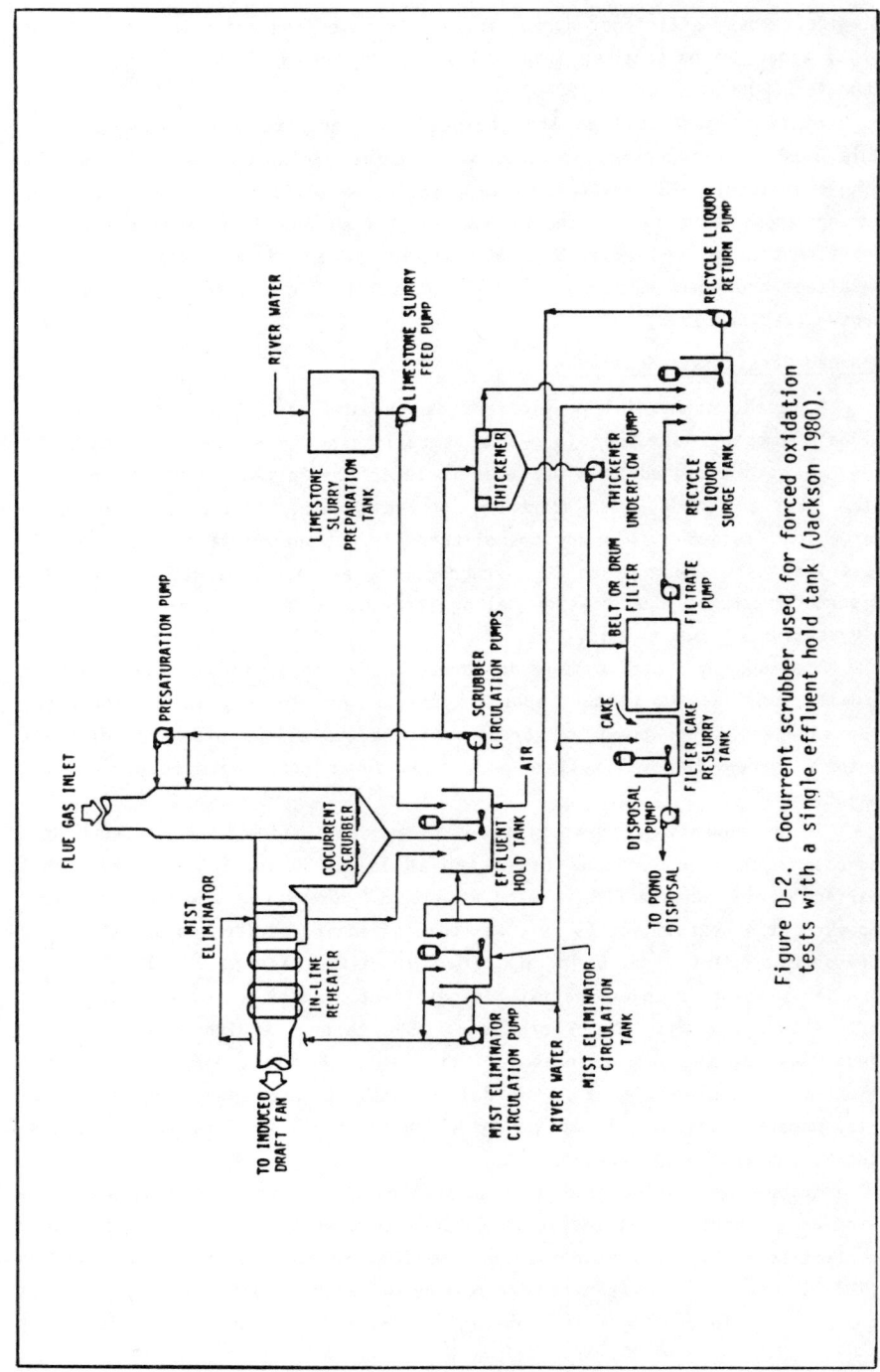

Figure D-2. Cocurrent scrubber used for forced oxidation tests with a single effluent hold tank (Jackson 1980).

Appendix D: Innovations in Limestone Scrubbing 325

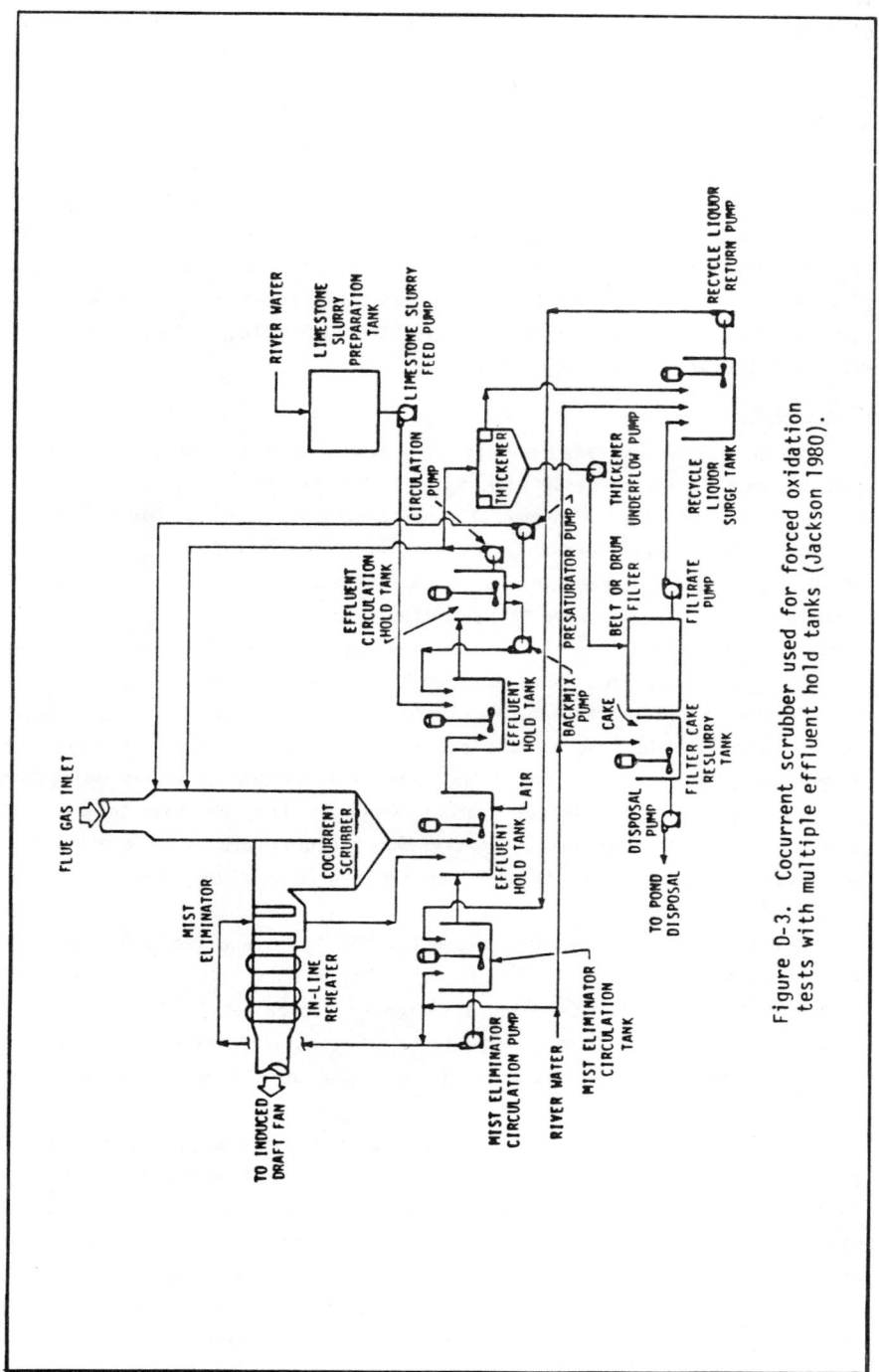

Figure D-3. Cocurrent scrubber used for forced oxidation tests with multiple effluent hold tanks (Jackson 1980).

most wet scrubbers. Also, Peabody Process Systems has developed a CPS option for use with a conventional scrubbing system.

NEW PROCESS MODIFICATIONS

Conventional limestone scrubbing can be modified in several ways. For example, limestone can be used with aluminum sulfate as in the Dowa process, or it can serve as a regenerant in dual-alkali systems. Alterations in the type and grind of limestone can significantly affect SO_2 removal efficiency. Also, forced oxidation, gypsum stacking, adipic acid addition, and magnesium addition can improve a conventional limestone scrubbing system. This subsection discusses these process modifications.

Dowa Process

The Dowa process (Figure D-4) is a dual-alkali SO_2 scrubbing process that uses a solution of basic aluminum sulfate $[Al_2(SO_4)_3]$ to absorb SO_2 and a slurry of limestone to regenerate the absorbent. The Dowa Mining Company of Tokyo, Japan, developed this process, which the Air Correction Division of UOP, Inc., will market in the United States. In Japan, the process is commercially used with smelters, sulfuric acid plants, and utility and industrial oil-fired boilers. At the Shawnee test facility, the Dowa process has been tested with flue gas from a coal-fired boiler.

Several potential advantages of the Dowa process prompted the Shawnee tests. For example, use of a clear solution (rather than a slurry) for absorption eliminates erosion of equipment and buildup of slurry solids on internals of the mist eliminator and scrubber. Also, the Dowa process requires lower limestone stoichiometry and produces byproduct gypsum with dewatering characteristics better than those of unoxidized limestone scrubbing sludge.

A disadvantage of the Dowa process is that it requires more equipment and is more complex than conventional, single-loop, limestone scrubbing. Further, the pH of the scrubbing solution is approximately 3, whereas that of limestone slurry is 5 to 6. The lower pH requires materials of construction that are more acid resistant (e.g., 316L or 317L stainless steel or lined carbon steel).

Initial Shawnee tests of the Dowa process with a packed mobile-bed scrubber (known as a Turbulent Contact Absorber and supplied by the Air Correction Division of Universal Oil Products) showed a maximum SO_2 removal efficiency of 85 to 90 percent (Jackson, Dene, and Smith 1980). Because of problems with gas flow distribution, rigid packing was used in later factorial absorption tests, in which the operating conditions were identified for consistent achievement of an SO_2 removal efficiency greater than 90

Appendix D: Innovations in Limestone Scrubbing 327

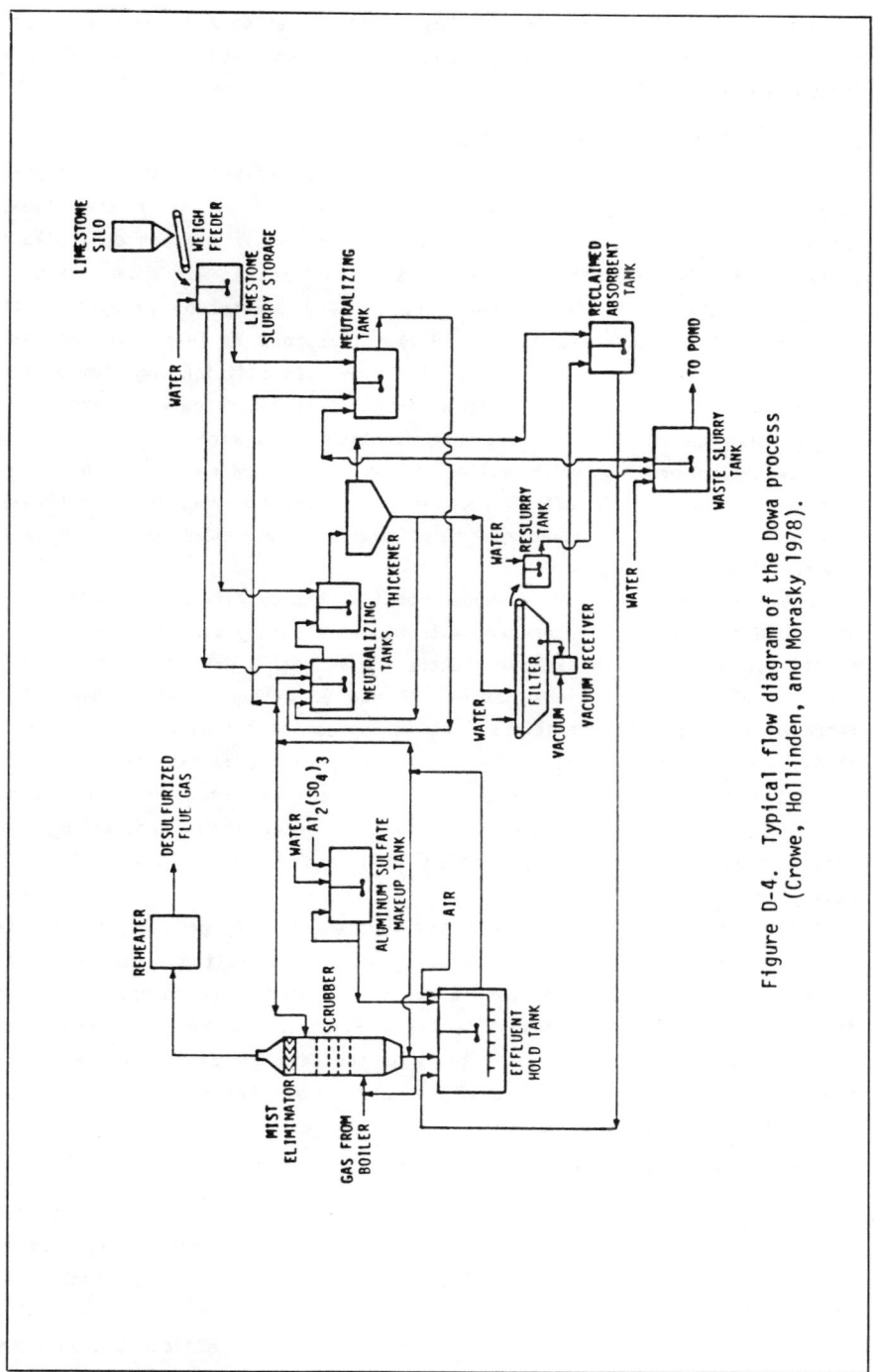

Figure D-4. Typical flow diagram of the Dowa process (Crowe, Hollinden, and Morasky 1978).

percent (Jackson, Dene, and Smith 1980). Neutralization and gypsum dewatering were generally satisfactory, and no problems with reliability were encountered.

Limestone Regeneration in Dual-Alkali Systems

Limestone can be used as a regenerant in dual-alkali systems. Available near most industrial sites, limestone is less expensive than lime, which typically is used for absorbent regeneration in present dual-alkali systems. Studies indicate, however, that impurities in limestone, particularly magnesium, can impair the settling and dewatering properties of byproduct solids (LaMantia et al. 1977). Limestone is less reactive than lime and thus requires a longer reaction time. In addition, calcium utilization rates tend to be lower when absorbent is regenerated with limestone.

Raising temperatures to increase reaction rates might reduce settling and dewatering problems with solids. Also, magnesium can be precipitated from a slipstream. To effect such improvements, however, would probably eliminate the economic advantage of using limestone rather than lime as a reagent (LaMantia 1977).

Oberholtzer et al. (1977) suggest that within certain constraints limestone can be used as a regenerant without inordinately complicating a dual-alkali system. Tests of a four-reactor system with a 2-hour residence time yielded calcium utilizations between 78 and 92 percent over a reasonable range of solution concentrations. As is typical of actual operating conditions, the solution was not heated above 122°F. Fredonia limestone containing 1.1 percent magesium as magnesium carbonate was used, and no efforts were made to control magnesium solubility. Under these conditions, the system produced solids that settled well and could be vacuum-filtered easily.

To increase the use of dual-alkali systems for SO_2 control, the EPA is sponsoring a program in which limestone will serve as the regenerant. In Japan, limestone has already been successfully used to regenerate absorbent in dual-alkali systems. Louisville Gas & Electric Company will attempt to duplicate the Japanese success in tests at the Cane Run Station and will use the results of these tests to design future commercial dual-alkali systems with limestone regeneration.

Effect of Limestone Type and Grind

The type and grind of limestone used in a single-loop, packed mobile-bed scrubber can significantly affect SO_2 removal efficiency (Borgwardt et al. 1979). Tests of three types of high-calcium limestone at the Industrial Environmental Research Laboratory (IERL), Research Triangle Park (RTP), North Carolina, showed that Fredonia limestone was more efficient than Stone

Man limestone, which in turn was more efficient than Georgia marble. The ranking was the same for fine and coarse grinds. Borgwardt et al. (1979) report that at a stoichiometric ratio of 1.5, SO_2 removal efficiencies were 88 percent with fine Fredonia limestone, 74 percent with fine Stone Man limestone, and 70 percent with fine Georgia marble.

The IERL-RTP tests also indicated that fine grinding enhances SO_2 removal and that sludge quality is affected by the type of limestone used. According to Borgwardt et al. (1979), the limestone feed rate needed to maintain a given SO_2 removal efficiency with coarse grinds (70 percent -200 mesh) was roughly 50 percent greater than that needed to maintain the same efficiency with fine grinds (84 percent -325 mesh). Also, the settling rate and filterability of scrubber slurry increased as SO_2 reactivity of the limestone decreased.

Under current EPA sponsorship, Dr. Gary T. Rochelle has developed additional data showing that limestone type and grind can enhance SO_2 removal. The reader is referred to references in Appendix A for details of Dr. Rochelle's work.

Forced Oxidation

Tests in 10-MW prototype units at the Shawnee test facility have demonstrated that forced oxidation of waste sludge material into calcium sulfate (gypsum) reduces its volume and increases its settling speed by an order of magnitude (Head, Wang, and Keen 1977). Further, oxidized calcium sulfate (gypsum) can be filtered to more than 80 weight percent solids and handled like moist soil, whereas unoxidized calcium sulfite byproduct can be filtered to only 50 to 60 weight percent solids and displays thixotropic properties (Head, Wang, and Keen 1977).

Forced oxidation in a single scrubber loop was successfully demonstrated with limestone slurry in the packed mobile-bed scrubber at the Shawnee test facility (Figure D-5). Contact between the air and slurry was achieved in the scrubber loop by pumping slurry from a small downcomer hold tank through an air eductor to a larger oxidation tank, where limestone was added. Slurry from the oxidation tank was returned to the scrubber. Typically, the eductor pH was 5.15, the oxidation tank pH was 5.5, and the air stoichiometry was 2.5 pound-atoms of oxygen per pound-mole of SO_2 removed. Effective oxidation required maintenance of air/slurry contact in the discharge plume from the eductor (Head, Wang, and Keen 1977).

Tests at the IERL-RTP pilot plant also indicate the feasibility of oxidizing calcium sulfite slurry within the scrubbing loop of a single-stage limestone scrubber at normal operating pH (Borwardt 1977). Forced oxidation improved the dewatering properties of sludge and did not adversely affect

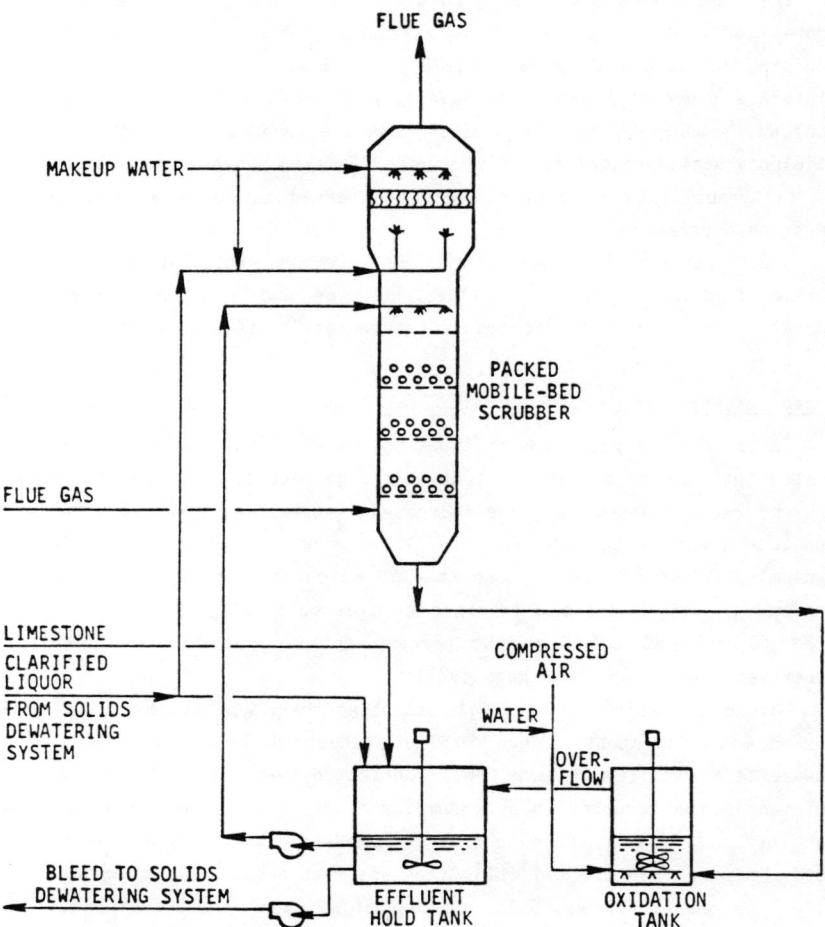

Figure D-5. Packed mobile-bed scrubber used for forced oxidation tests (Burbank and Wang 1980).

SO_2 removal efficiency, limestone utilization, and scrubber feed supersaturation. Good performance can be anticipated with 98 percent oxidation and at chloride concentrations at least as great as 20,000 ppm (Borgwardt 1977).

Tests at a 140-MW unit have shown that forced oxidation of limestone scrubber sludge to gypsum is a viable technique for water disposal. At air stoichiometries of 1.75 to 2.0 pound-atoms of oxygen per pound-mole of SO_2 removed, approximately 95 percent of the slurry was oxidized (Massey et al. 1980). As a result of these tests, TVA will use forced oxidation as the preferred method for disposing of scrubber sludge from Widows Creek Units 7 and 8 and is designing scrubber trains for Paradise Units 1 and 2 with a forced oxidation option to produce a calcium sulfate waste product (Massey et al. 1980).

Many commercial scrubber suppliers now offer forced oxidation as an effective means of reducing disposal problems associated with handling waste sludge materials. The reader is referred to Appendix F for information on commercial systems using forced oxidation.

In Japan and Germany, forced oxidation systems convert waste sludge into gypsum pure enough for wallboard manufacture. Research-Cottrell has designed such a forced oxidation system for Tampa Electric Company's Big Bend Unit 4. This is expected to be the first of many U.S. systems that will produce salable byproduct gypsum.

Gypsum Stacking

The U.S. phosphate fertilizer industry has used gypsum stacking to dispose of waste gypsum for more than 20 years. Typically, gypsum stacks are structurally stable stockpiles that cover 50 to 300 acres and can reach heights of 150 feet (Morasky et al. 1980). Unoxidized calcium sulfite sludges from conventional limestone scrubbing have not been stacked because of their poor handing and dewatering characteristics. Forced oxidation, however, can be used to produce calcium sulfate (gypsum) instead of calcium sulfite waste. This gypsum dewaters easily and can support a considerable load (Radian Corporation 1980).

Construction of an FGD gypsum stack (Figure D-6) is fairly simple. Slurry is fed to an area bounded by a starter dike and having a decant pipe or pond for removal of supernatant liquor. When this inner area is filled with solids, a dragline is used to dredge gypsum onto the side of the starter dike. Thus, a cast gypsum perimeter dike is created, and the height of the structure is raised. As more slurry is added, the process is repeated.

Operation of a gypsum stack is generally much easier and simpler than

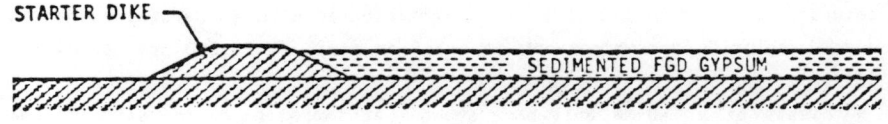

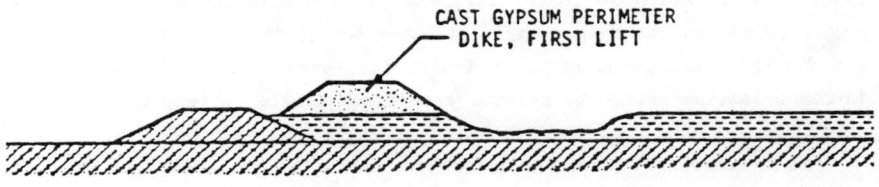

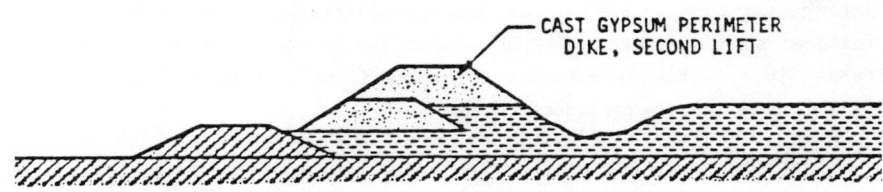

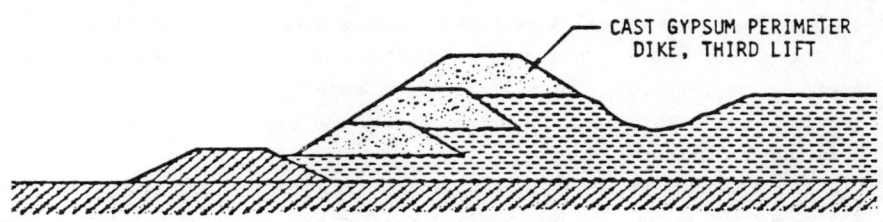

Figure D-6. Construction of an FGD gypsum stack (Golden 1980).

operation of a landfill for conventional unoxidized FGD wastes. Because gypsum can be pumped to the disposal area in a slurry, the problems of daily handling and transportation of wastes to a landfill are avoided. Further, the stacking method dewaters gypsum by gravity and thus eliminates the need for mechanical dewatering. Within the stacking area, gypsum quickly dewaters and forms a stable material; thus, additional compaction is unnecessary.

Compared with conventional unoxidized FGD wastes, FGD gypsum offers several advantages. For example, it can be stacked and stored in a smaller area than that required for landfilling of calcium sulfite sludge. If pure enough, FGD gypsum can also be used in agriculture and for the manufacture of wallboard and portland cement. Research-Cottrell and others are designing forced oxidation systems that will produce this pure gypsum for sale.

Contamination of nearby surface and ground waters by FGD wastes (both unoxidized sludges and gypsum stacks) is possible. Process water can contain concentrations of sulfate, calcium, chloride, and magnesium several orders of magnitude greater than those usually found in natural surface and ground waters and those allowed by drinking water standards. Also, such trace elements as arsenic, chromium, and selenium can be present in process water at levels greater than those permitted by drinking water standards. Thus, seepage from FGD gypsum stacks must be controlled or prevented, and surface and ground waters near stacks must be monitored.

Although gypsum from the phosphate fertilizer industry has been stacked for many years, the geotechnical and environmental feasibility of stacking FGD gypsum has been investigated only recently. A prototype FGD gypsum stack was constructed and operated at the Scholz Station of Gulf Power Company from October 1978 to June 1979. Gypsum was produced by the prototype Chiyoda Thoroughbred 121 scrubber evaluated at the same time. Analysis indicates that FGD gypsum has settling, dewatering, and structural characteristics similar to, and in some ways more favorable than, those of phosphate gypsum (Radian Corporation 1980). Limited monitoring of ground water showed no increase in concentrations of trace elements, but did reveal increases in concentrations of calcium, sulfate, and total dissolved solids (Radian Corporation 1980).

Adipic Acid Addition

Tests at the 0.1-MW IERL-RTP pilot plant and at the 10-MW prototype units of the Shawnee test facility have indicated that addition of adipic acid to a limestone wet scrubbing system significantly enhances SO_2 removal efficiency (Head et al. 1979). Adipic acid is a dicarboxylic organic acid [$HOOC(CH_2)_4COOH$] in powder form; it is available commercially and is used as a raw material in the nylon manufacturing and food processing industries. When adipic acid concentrations in limestone slurry ranged between 700 and 1500 ppm, the IERL-RTP and Shawnee tests consistently showed SO_2 removal efficiencies greater than 90 percent. Head et al. (1979) report that adipic acid effectively enhanced SO_2 removal efficiency even with chloride concentrations as great as 10,000 ppm at Shawnee and 17,000 ppm at the pilot plant. Addition of adipic acid improved SO_2 removal efficiencies equally in systems with and without forced oxidation, caused only minor differences in the dewatering and handling properties of oxidized sludge, and did not produce scaling.

One problem has been the decomposition of adipic acid at ordinary scrubber operating conditions, especially in systems with forced oxidation. In earlier Shawnee tests, 8 to 9 pounds of adipic acid were consumed per ton of limestone to maintain adipic acid concentration in the slurry at 1500 ppm (Head et al. 1979). Recent Shawnee tests suggest that maintenance of inlet slurry pH below 5.0 minimizes adipic acid decomposition (Burbank et al. 1980).

In the spring of 1980, the EPA contracted with Radian Corporation to evaluate adipic acid enhancement of limestone scrubbing by a full-scale FGD system. This program is being conducted at the Southwest Station of City Utilities near Springfield, Missouri. In the late summer of 1980, tests began at Southwest Unit 1, a 194-MW unit firing high-sulfur bituminous coal (Hicks, Hargrove, and Colley 1980).

Initial results at Southwest Unit 1 have been encouraging. Before adipic acid addition, SO_2 removal efficiency averaged roughly 65 percent at the normal operating pH of 5.5. When 800 to 1000 ppm of adipic acid was added to the scrubbing liquor, SO_2 removal efficiency increased to more than 90 percent; and at full load, SO_2 removal efficiency reached 95 percent (Hicks, Hargrove, and Colley 1980). Limestone utilization also improved. A test at an operating pH of 5.0 and an adipic acid concentration of 1500 ppm yielded an SO_2 removal efficiency of more than 90 percent and limestone utilization of 99 percent (Hicks, Hargrove, and Colley 1980).

Also, the EPA has sponsored tests at Rickenbacker Air National Guard Base near Columbus, Ohio, to show that adipic acid enhances the SO_2 removal efficiency of an industrial-size limestone scrubbing system. Adipic acid addition increased SO_2 removal efficiency from a marginally effective level to a high removal efficiency.

Magnesium Addition

The addition of modest amounts of magnesium can improve the operation of a limestone scrubbing system, especially the SO_2 removal efficiency

(Josephs 1980). Depending on pH and ionic effects, magnesium is 100 to 1000 times as soluble as calcium in the scrubbing liquor. Magnesium addition thereby improves liquid phase alkalinity and SO_2 absorption and often reduces overall slurry pumping rates. Enrichment with magnesium also provides more soluble alkali for chloride to combine with and prevents pH drop, although chloride interference requires greater magnesium addition. If enough magnesium is added to absorb most SO_2 in the flue gas, limestone dissolution is limited, and scaling potential is greatly reduced.

In 1976, the effects of magnesium oxide addition were studied at the Shawnee test facility. Limestone slurry was used in the packed mobile-bed scrubber to scrub flue gas containing fly ash. Under typical operating conditions, SO_2 removal efficiencies were 77, 84, and 94 percent at effective magnesium ion concentrations of 0, 5000, and 9000 ppm, respectively (Head 1977). Head found that magnesium oxide addition at the levels tested did not always produce gypsum-subsaturated operation, but that lower saturation levels tended to increase liquor sulfite concentration and SO_2 removal efficiency.

Pullman Kellogg (now a division of Wheelabrator-Frye) offers a commercially proven process that uses magnesium addition to enhance the SO_2 removal efficiency of limestone scrubbers. A Pullman Kellogg system will soon be operational at Associated Electric Cooperative's Thomas Hill Station in Moberly, Missouri. The reader is referred to Appendix F for further details on other Pullman Kellogg magnesium-enhanced systems.

ECONOMIC EVALUATION

New types of scrubbers and new process modifications offer economic advantages over conventional limestone scrubbing. Often, innovations can be combined to increase potential savings. Under EPA sponsorship, TVA is evaluating the economics of current and future limestone scrubbing systems. Complete results of the evaluation will be published in 1981.

A preliminary economic comparison has been made of conventional, improved, and advanced limestone scrubbing systems (McGlamery et al. 1980). The conventional system is defined as a packed mobile-bed scrubber with onsite ponding of calcium sulfate sludge; the improved system is defined as a spray tower with forced oxidation and landfilling of gypsum; and the advanced system is defined as identical to the improved system, but with adipic acid addition. Table D-1 presents design conditions for these systems, Table D-2 shows capital investments, and Table D-3 lists annual revenue requirements (McGlamery et al. 1980). Although capital investments for the improved and advanced systems are slightly greater in most areas

TABLE D-1. PROCESS DESIGN CONDITIONS FOR LIMESTONE SYSTEMS[a]

	Conventional system	Improved system	Advanced system
Type of scrubber	Mobile-bed	Spray tower	Spray tower
Forced oxidation	No	Yes	Yes
Adipic acid addition	No	No	Yes (1000 to 2000 ppm)
Waste disposal	Pond	Thickener, filter, landfill	Thickener, filter, landfill
Scrubber gas velocity, ft/s	12.5	10	10
L/G ratio, gal/10^3ft^3	58	90	80
Limestone stoichiometry	1.3	1.3	1.2
Air stoichiometry	0	2.5	2.5
Sulfite oxidation, %	30	95	95
Type of fan	Induced draft	Induced draft	Induced draft
Spare scrubber	Yes	Yes	Yes
Filter cake solids, wt. %	40	80	80
Pond settled solids, wt. %			
Spare ball mill	Yes	Yes	Yes
Reheat	In-line steam	In-line steam	In-line steam
Bypass available	50% emergency	50% emergency	50% emergency

[a] Source: McGlamery et al. 1980.

TABLE D-2. CAPITAL INVESTMENTS FOR LIMESTONE SYSTEMS ON 500-MW PLANTS[a]
($10³ except as indicated)

	Conventional system[b]	Improved system[c]	Advanced system[d]
Direct investment			
Material handling	3,498	3,497	3,503
Feed preparation	3,485	3,484	3,490
Gas handling	9,600	11,129	10,821
SO_2 absorption	19,830	22,988	22,351
Reheat	2,851	3,304	3,213
Solids disposal	2,063	2,868	2,850
Total	41,327	47,270	46,228
Services, utilities, and miscellaneous	2,480	2,836	2,774
Total	43,807	50,106	49,002
Landfill or pond construction	13,960	2,076	1,983
Landfill equipment		500	495
Total	57,767	52,682	51,480
Indirect investment			
Engineering design and supervision	3,346	3,663	3,579
Architect and engineering contractor	1,016	1,028	1,005
Construction expense	8,126	8,378	8,187
Contractor fees	2,888	2,608	2,549
Contingency	7,315	7,158	6,990
Total fixed investment	80,458	75,517	73,790
Other capital investment			
Allowance for startup and modifications	5,012	5,732	5,606
Interest during construction	12,551	11,781	11,511
Land	1,905	641	611
Working capital	3,104	3,161	3,090
Total capital investment	103,030	96,832	94,608
Total capital investment, $/kW	206	194	189

[a] Source: McGlamery et al. 1980.
Basis: Plant is in upper midwest. Project begins mid-1980 and ends mid-1983. Average cost basis is mid-1982. Spare pumps, one spare scrubbing train, and one spare ball mill are included. Disposal pond and landfill are located 1 mile from plant. Investment includes FGD feed plenum, but excludes stack plenum and stack.

[b] The conventional system is a mobile-bed scrubber with onsite ponding of calcium sulfite sludge.

[c] The improved system is a spray tower with forced oxidation and landfilling of gypsum.

[d] The advanced system is identical to the improved system, but with adipic acid addition.

TABLE D-3. ANNUAL REVENUE REQUIREMENTS FOR LIMESTONE SYSTEMS*
($10^3 except as indicated)

	Conventional System**	Improved System**	Advanced System†
First-year direct costs			
Raw materials			
Limestone	1,128	1,128	1,041
Adipic acid	–	–	216
Total raw materials cost	1,128	1,128	1,257
Conversion costs			
Operating labor and supervision			
FGD	460	658	658
Solids disposal	–	529	517
Utilities			
Process water	35	26	26
Electricity	1,732	2,018	1,874
Steam	1,273	1,365	1,367
Fuel	–	199	189
Maintenance			
Labor and material	3,923	4,025	3,937
Analyses	104	104	104
Total conversion costs	7,527	8,924	8,672
Total direct costs	8,655	10,052	9,929
First-year indirect costs			
Overheads			
Plant and administrative (60% of conversion costs less utilities)	2,692	3,057	2,998
Total first-year operating and maintenance (O&M) costs	11,347	13,109	12,927
Levelized capital charges (14.7% of total capital investment)	15,145	14,234	13,907
Total first-year annual revenue requirements	26,492	27,343	26,834
Levelized first-year O&M costs (1.886 x first-year O&M costs)	21,401	24,724	24,381
Levelized capital charges (14.7% of total capital investment)	15,145	14,234	13,907
Levelized annual revenue requirements	36,545	38,958	38,288
Total first-year annual revenue requirements, mills/kWh	9.63	9.94	9.76
Levelized annual revenue requirements, mills/kWh	13.29	14.17	13.92

*Source: McGlamery et al, 1980.
Basis: Upper midwest plant location, 1984 revenue requirements
New plant life—30 yr
Power unit time on stream—5,500 h/yr
Coal burned—1,116,500 tons/year
Boiler heat rate—9,500 Btu/kWh
Total capital investment:
  Conventional $103,030,000
  Improved $ 96,832,000
  Advanced $ 94,608,000

**The conventional system is a mobile-bed scrubber with onsite ponding of calcium sulfate sludge.
***The improved system is a spray tower with forced oxidation and landfilling of gypsum.
†The advanced system is identical to the improved system, but with adipic acid addition.

than those for the conventional system, the conventional system has disposal site (pond) construction costs much greater than the disposal site (landfill) construction costs of the other systems. Thus, overall capital investments for the improved and advanced processes are smaller. Annual revenue requirements for the improved and advanced systems, however, are slightly greater than those for the conventional system, primarily because of the lower costs of labor, supervision, and electricity for the conventional system.

REFERENCES FOR APPENDIX D

Borgwardt, R. H. 1977. Effect of Forced Oxidation on Limestone/SO_x Scrubber Performance. In: Proceedings: Symposium on Flue Gas Desulfurization, Hollywood, Florida, November 8-11, 1977. Vol. I. EPA-600/7-78-058a. NTIS No. PB-282 090.

Borgwardt, R. H., et al. 1979. Limestone Type-and-Grind Tests at EPA/IERL-RTP. Presented at the 5th Shawnee Industry Briefing Conference, Raleigh, North Carolina, December 5, 1979.

Burbank, D. A., et al. 1980. Test Results on Adipic Acid-Enhanced Limestone Scrubbing at the EPA Shawnee Test Facility--Third Report. Presented at the U.S. EPA Sixth Symposium on Flue Gas Desulfurization, Houston, Texas, October 28-31, 1980.

Burbank, D. A., and S. C. Wang. 1980. EPA Alkali Scrubbing Test Facility: Advanced Program - Final Report (October 1974 - June 1978). EPA-600/7-80-115. NTIS No. PB 80-204 241.

Crowe, J. L., G. A. Hollinden, and T. Morasky. 1978. Status Report of Shawnee Cocurrent and Dowa Scrubber Projects and Widows Creek Forced Oxidation. In: Proceedings of the Industry Briefing on EPA Lime/Limestone Wet Scrubbing Test Programs, August 29, 1978. EPA-600/7-79-092. NTIS No. PB-296 517.

Dauerman, L., and K. Rao. 1979. Double Alkali Process for Flue Gas Desulfurization Optimizing for the Regeneration of Sodium Sulfite; Part I: Lime as Regenerant, and Part II: Limestone as Regenerant. Presented at the 72d Annual Meeting of the Air Pollution Control Association, Cincinnati, Ohio, June 24-29, 1979.

Golden, D. M. 1980. EPRI FGD Sludge Disposal Demonstration and Site Monitoring Projects. Presented at the U.S. EPA Sixth Symposium on Flue Gas Desulfurization, Houston, Texas, October 28-31, 1980.

Head, H. N. 1977. EPA Alkali Scrubbing Test Facility: Advanced Program, Third Progress Report. EPA-600/7-77-105.

Head, H. N., et al. 1979. Recent Results From EPA's Lime/Limestone Scrubbing Programs--Adipic Acid as a Scrubber Additive. In: Proceedings: Symposium on Flue Gas Desulfurization, Las Vegas, Nevada, March 1979. Vol. 1. EPA-600/7-79-167a. NTIS No. PB80-133168.

Head, H. N., S. C. Wang, and R. T. Keen. 1977. Results of Lime and Limestone Testing With Forced Oxidation at the EPA Alkali Scrubbing Test Facility. In: Proceedings: Symposium on Flue Gas Desulfurization, Hollywood, Florida, November 8-11, 1977. Vol. I. EPA-600/7-78-058a. NTIS No. PB-282 090.

Hicks, N. D., O. W. Hargrove, and J. D. Colley. 1980. FGD Experiences, Southwest Unit 1. Presented at the U.S. EPA Sixth Symposium on Flue Gas Desulfurization, Houston, Texas, October 28-31, 1980.

Jackson, S. B. 1980. Cocurrent Scrubber Tests: Shawnee Test Facility. Presented at the U.S. EPA Sixth Symposium on Flue Gas Desulfurization, Houston, Texas, October 28-31, 1980.

Jackson, S. B., C. E. Dene, and D. B. Smith. 1980. Dowa Process Tests, Shawnee Test Facility. Presented at the U.S. EPA Sixth Symposium on Flue Gas Desulfurization, Houston, Texas, October 28-31, 1980.

Josephs, D. X. 1980. Magnesium Enrichment Improves Flue Gas Scrubbing. Power Engineering, 84(9):71-72.

LaMantia, C. R., et al. 1977. Dual Alkali Test and Evaluation Program. 3 vols. EPA-600/7-77-050a-c. NTIS No. PB-269 904, PB-272 770, PB-272 109.

Laseke, B. A., Jr., et al. 1979. Electric Utility Steam Generating Units--Flue Gas Desulfurization Capabilities as of October 1978. EPA-450/3-79-001. NTIS No. PB-298 509.

Martin, J. R., K. W. Malki, and N. Graves. 1979. The Results of a Two-Stage Scrubber/Charged Particulate Separator Pilot Program. In: Second Symposium on the Transfer and Utilization of Particulate Technology. Vol. I. Control of Emissions from Coal Fired Boilers. EPA-600/9-80-039a.

Massey, C. L., et al. 1980. Forced Oxidation of Limestone Scrubber Sludge at TVA's Widows Creek Unit 8 Steam Plant. Presented at the U.S. EPA Sixth Symposium on Flue Gas Desulfurization, Houston, Texas, October 28-31, 1980.

McGlamery, G. G., et al. 1980. FGD Economics in 1980. Presented at the U.S. EPA Sixth Symposium on Flue Gas Desulfurization, Houston, Texas, October 28-31, 1980.

Morasky, T. M., et al. 1980. Evaluation of Gypsum Waste Disposal by Stacking. Prepared for, but not presented at, the U.S. EPA Sixth Symposium on Flue Gas Desulfurization, Houston, Texas, October 28-31, 1980.

Oberholtzer, J. E., et al. 1977. Laboratory Study of Limestone Regeneration in Dual Alkali Systems. EPA-600/7-77-074. NTIS No. PB-272 111.

Radian Corporation. 1980. Evaluation of Chiyoda Thoroughbred 121 FGD Process and Gypsum Stacking. Volume 1: Chiyoda Evaluation. EPRI CS-1579.

Appendix E

Material and Energy Balances

This appendix exemplifies the procedures for performing a material balance and estimating the energy requirements for a limestone FGD system. An understanding of the material and energy balances will enable the Project Manager to verify limestone usage, makeup water requirements, and energy demands of the system.

The process selected for this illustration and the process assumptions are described in the following section. The process chemistry is simplified for the purpose of illustration. A set of conditions is specified, and the material balance is developed in a stepwise fashion. Sample calculations for the material balance are followed by an estimation of energy consumption. Two sets of calculations are performed: the first for flue gases resulting from combustion of high-sulfur eastern coal (3.7% S), and the second, for low-sulfur western coal (0.7% S).

For both coals, the SO_2 removal for the flue gas actually treated in the scrubbers is 90 percent. For the low-sulfur coal, however, the overall system requirement is 70 percent SO_2 removal. Therefore, 22 percent by volume of the flue gas can be allowed to bypass the FGD system untreated and be used for reheat. This untreated gas provides an energy savings of about 50 percent for a 500-MW plant, as compared to the high-sulfur coal case.

PROCESS DESCRIPTION

A flow diagram for the limestone scrubbing FGD process is shown in Figure E-1, along with stream characteristics. The process depicted and the values shown in this figure are those of the high-sulfur coal model, for which the material balance calculations are performed later in this section. In this example, 210 ton/h of coal is fired to generate approximately 500 MW (gross) of electricity. Although an FGD system of this size usually consists of several modules, it is assumed in these stream calculations that all the modules are combined.

The major components of coal are carbon, oxygen, nitrogen, hydrogen, sulfur, free moisture, and ash. Chloride, a minor component, is important

342   Handbook for FGD Scrubbing with Limestone

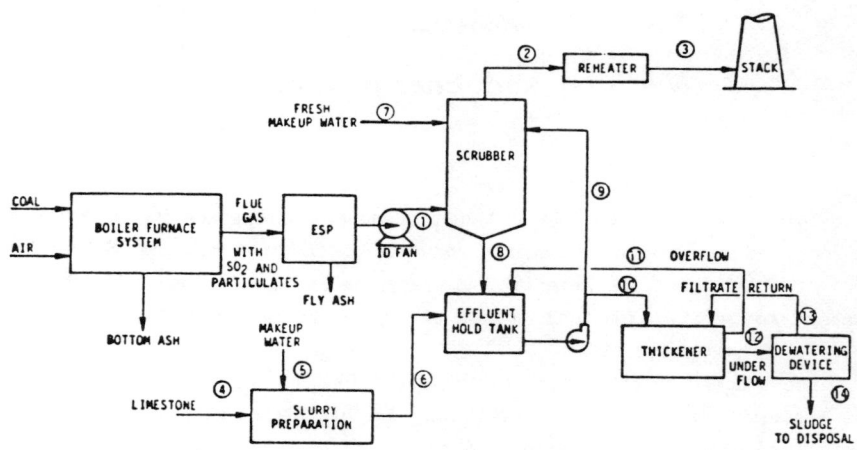

Stream characteristics[a]

Gas stream No.	1	2	3
Flow, 1000 acfm	1615.6	1376.7	1470.5
Flow, 1000 lb/h	5328.	5534.4	5534.4
Temp., °F	290	128	168
SO_2, lb/h	31,080	3,108	3,108
HCl, lb/h	432	0	0
CO_2, 1000 lb/h	942.5	961.7	961.7
H_2O, 1000 lb/h	293.5	499.5	499.5
Particulates, lb/h	138	138	138

Liquid/solid stream No.	4	5	6	7	8	9	10	11	12	13	14
Flow, gal/min	-	197	343	300	124,700	124,846	838	701	357	213	106
Flow, 1000 lb/h	50.9	98.5	149.4	150.0	56,700	56,748	456.7	350.5	213.1	106.6	106
Temp., °F	70	70	70	70	128	128	128	100	100	100	100
$CaCO_3$, lb/h	47,825	0	47,825	0			4,430	0	4,430	0	4,430
H_2O (free), 1000 lb/h	0	98.5	98.5	150.0	48,747	48,803	392.8	350.5	149.2	106.6	42
$CaSO_3 \cdot 1/2 H_2O$, lb/h	-	0	0	0			35,837	0	35,837	0	35,837
$CaSO_4 \cdot H_2O$, lb/h	-	0	0	0			2,986	0	2,986	0	2,986
$CaSO_4 \cdot CaSO_3 \cdot 1/2 H_2O$, lb/h	-	0	0	0			18,403	0	18,403	0	18,403
Inerts, lb/h	2,290	0	2,290	0			2,290	0	2,290	0	2,290

[a] A blank indicates an unknown value; a dash shows that an item does not apply.

Figure E-1. Model example limestone FGD system or 500-MW plant, 3.7 percent sulfur coal.

because of its corrosion effect and its impact on process chemistry. Chloride in the coal forms hydrochloric acid during combustion. At steady state, this acid is assumed to be completely absorbed by the scrubbing solution and removed with interstitial water in the waste sludge. In some coals, alkalinity in the ash reacts with sulfur dioxide or sulfur trioxide; this effect is neglected here. A typical analysis of high-sulfur coal is presented in Table E-1. The heating value (HV) of this coal is 11,150 Btu/lb.

The fly ash in the flue gas (Stream 1) is removed in a cold-side ESP ahead of the scrubber. The maximum particulate emission rate must be in compliance with the NSPS promulgated by EPA in September 1979, which is 0.03 lb/million Btu heat input. After passing through the ESP the flue gas (Stream 2) enters the scrubber at 290°F, and the SO_2 is removed by limestone scrubbing. The current NSPS requirement for SO_2 removal is 90 percent, which translates to 0.66 lb SO_2/million Btu heat input at the outlet of the scrubber.

The temperature of the saturated flue gas from the scrubber (Stream 3) is increased by 40°F in a reheater. It is assumed that there is no carryover of mist droplets in the reheated gas. (For a well-designed mist eliminator, the accepted carryover rate is 0.1 gr/scf.) The cleaned and reheated flue gas (Stream 4) is discharged to the atmosphere through a stack. In calculation of gas flow rates the ideal gas law is assumed.

Typical pressure drop data are shown in Table E-2.

A 60 percent solids limestone slurry is prepared in a ball mill from water and limestone. A typical limestone analysis is presented in Table E-3. In this example, it is assumed that the $MgCO_3$ available for reaction with SO_2 is 1.5 percent and that the unavailable portion is confined to inerts. The actual percentage of $MgCO_3$ in the limestone supplied could be higher, but only a portion is available for reaction.

This 60 percent solids slurry (specific gravity = 1.50) is diluted to 20 percent solids (specific gravity = 1.12) with makeup water in a slurry feed tank (not shown), then pumped to the effluent hold tank (EHT), which maintains a 14 percent solids concentration (specific gravity 1.09).

Slurry from the EHT is recirculated through the scrubber for removal of SO_2, and a portion of the slurry (to be determined by the material balance) is bled off to the thickener. The thickener underflow, containing 30 percent solids (Stream 13), is dewatered to 60 percent solids in a dewatering device such as vacuum filter or centrifuge. The dewatered sludge (Stream 15) is sent to the disposal area, and the filtrate is returned to the thickener. A portion of thickener overflow is sent to the mist eliminator.

The liquid-to-gas (L/G) ratio in the scrubber normally ranges from 40

TABLE E-1. DESIGN PREMISES: HIGH-SULFUR COAL CASE

Plant capacity (gross)		500 MW
Boiler:	Type	Pulverized-coal-fired
Coal:	Type	Bituminous
	Source	Pennsylvania
	Consumption	210 tons/h
	Heating value	11,150 Btu/lb
	Sulfur content	3.7%
	Oxygen content	7.3%
	Hydrogen content	4.3%
	Nitrogen content	1.2%
	Carbon content	61.2%
	Chloride content	0.1%
	Moisture content	8.5%
	Ash content	13.7%
Limestone:	Stoichiometric ratio (SR)[a]	1.10
	Utilization ($\frac{1}{SR}$)	90.9%
Scrubber:	Liquid-to-gas ratio	75 gal/1000 acf (saturated)
	Inlet gas temperature	290°F
Sulfite-to-sulfate oxidation:		20%
Solids:	Limestone slurry feed tank	~35%
	Effluent hold tank	14%
	Thickener underflow	30%
	Dewatered sludge	60%
SO_2 inlet loading:		6.6 lb/10^6 Btu input
Max. emission: (NSPS)	SO_2	0.66 lb/10^6 Btu input
	Particulates	0.03 lb/10^6 Btu input
Reheat:	Indirect in-line	40°F
Flue gas treated:	No bypass for reheat	100%
SO_2 removal:	Treated gas	90%

Flue gas components (inlet to scrubber)	Flow rate, lb/h	Composition, wt. %	mol %
Particulates	138		
Carbon dioxide	942,500	17.66	11.76
Hydrogen chloride	432	0.01	<0.01
Nitrogen	3,745,500	70.17	73.45
Oxygen	324,500	6.08	5.57
Sulfur dioxide	31,080	0.58	0.27
Moisture	293,500	5.50	8.94
Total	5,337,650	100.00	100.00

[a] Defined as moles of $CaCO_3$ and $MgCO_3$ (if available) fed per mol of SO_2 and HCl (if present) absorbed.

TABLE E-2. TYPICAL PRESSURE DROP DATA

	Pressure drop, in. H_2O
SO_2 scrubber	
(mobile bed)	6-8
(spray type)	2-3
Mist eliminator	0.3-1[a]
Reheater	1
Duct work	2

[a] Depends on generic type and whether it includes a wash tray. Pressure drop through the Shawnee mist eliminator is typically 0.3 in. H_2O.

TABLE E-3. LIMESTONE ANALYSIS
(percent by weight)

$CaCO_3$	94.0
$MgCO_3$	1.5
Inerts	4.5

to 100 gallons of slurry per 1000 acf of gas, depending on the sulfur content of the coal, type of scrubber, SO_2 removal efficiency, and water availability. In these examples, the L/G ratios are 55 gal/1000 acf (saturated) for the low-sulfur coal and 75 gal/1000 acf (saturated) for the high-sulfur coal. The spent slurry from the scrubber downcomer is collected in the EHT along with spent mist eliminator wash water.

The chemistry of limestone scrubbing is discussed in detail in Appendix A. For the sake of simplicity, only overall reactions are presented here.

In the scrubber, the SO_2 is absorbed by the reaction with $CaCO_3$. The overall reaction is:

$$CaCO_3 + SO_2 + H_2O \rightarrow CaSO_3 \cdot \tfrac{1}{2}H_2O + \tfrac{1}{2}H_2O + CO_2 \tag{1}$$

Available $MgCO_3$ also reacts with SO_2 in a similar fashion and gives magnesium sulfite (hexahydrate or trihydrate).

Some of the calcium sulfite formed in reaction (1) is oxidized to sulfate with the oxygen in the flue gas. The degree of oxidation in this example is 20 percent. The reaction is expressed by:

$$CaSO_3 \cdot \tfrac{1}{2}H_2O + \tfrac{1}{2}O_2 + 1\tfrac{1}{2}H_2O \rightarrow CaSO_4 \cdot 2H_2O^* \qquad (2)$$

All the hydrogen chloride (HCl) is absorbed and subsequently neutralized by the alkaline species to give magnesium chloride. The solids in the waste stream 15 are mainly $CaSO_3 \cdot \tfrac{1}{2}H_2O$, unreacted $CaCO_3$, inerts, coprecipitates ($CaSO_4 \cdot CaSO_3 \cdot \tfrac{1}{2}H_2O$), and $CaSO_4 \cdot 2H_2O$.

The fresh makeup water, which includes pump seal water, mist eliminator wash water, and slurry feed preparation water, is supplied through Stream 6 and Stream 8.

The following summarizes the assumptions on which calculations are based:

1. Flue gas flow and composition remain constant for a particular case.

2. The composition of the limestone supplied is 94 percent $CaCO_3$, 1.5 percent available $MgCO_3$, and 4.5 percent inerts.

3. Conversion of sulfur in the coal to SO_2 in flue gas is assumed to be 100 percent; however, 95 percent is a typical value.

4. Any contribution due to SO_3 in the flue gas is neglected.

5. Actual alkalinity supplied in the form of limestone for absorption of SO_2 and HCl is 1.10 times the stoichiometric amount.

6. Any HCl generated by chlorine in the coal is completely captured in the FGD system.

7. Degrees of sulfite-to-sulfate oxidation are 20 and 50 percent for the high-sulfur coal and low-sulfur coal cases, respectively.

8. About 16 mol percent of total SO_2 removed to the sludge forms coprecipitates of calcium sulfate with calcium sulfite hemihydrate when rate of oxidation (sulfite-to-sulfate) is 16 percent or more.

9. Although a large FGD system usually consists of several scrubbing modules, it is assumed in stream calculations that all modules are combined.

10. The fly ash in the fuel gas is removed primarily by a cold-side ESP and has no effect on SO_2 absorption. There is no secondary particulate removal in the scrubbing module(s).

11. The L/G ratio applied in the scrubbing module(s) are 55 gal/1000 acf (saturated) for the low-sulfur coal and 75 gal/1000 acf (saturated) for the high-sulfur coal.

12. The pressure drop across the entire FGD system is 11 in. H_2O.

13. There is no mist carryover in the gas leaving the FGD system.

---

* At 20% oxidation, 16 mol % of calcium sulfate will be in the form of a solid solution $CaSO_4 \cdot CaSO_3 \cdot \tfrac{1}{2}H_2O$.

Appendix E: Material and Energy Balances

MATERIAL BALANCE CALCULATIONS: HIGH-SULFUR COAL CASE

The basis for material balance calculations is depicted schematically in Figure E-2. Inputs to the FGD system include flue gas from the ESP, limestone slurry, and makeup water, as shown in Figure E-2a. The outputs are cleaned flue gas and dewatered sludge. Figure E-2b depicts the SO_2 and fly ash particulate balance with the ESP and the FGD system. The ESP reduces the fly ash particulates* from the flue gas to below the maximum allowable particulate emission (0.03 lb/million Btu heat input). The FGD system removes the required amount of SO_2. Figure E-2c shows the overall inputs and outputs to the boiler-furnace system, the ESP, and the FGD system.

Beginning with the known amount and composition of the flue gas entering the FGD system and the emission regulation, material balance calculations are performed in five steps, as follows:

- SO_2 removal requirement
- Limestone requirement/slurry preparation
- Humidification of flue gas
- Recirculation loop and sludge production
- Makeup water requirement

Once the makeup water requirement is known, the overall water utilization is established on the basis of the mist eliminator (ME) wash procedure. The important interplay of ME wash requirements and balance of water in the limestone scrubbing system is discussed later.

SO_2 Removal Requirement (Step 1)

Under current NSPS regulation, the FGD system must remove 90 percent of the inlet SO_2 for this high-sulfur coal case. The allowable SO_2 emission from the plant is:

31,080 lb/h SO_2 x (10%) SO_2 emission allowed = 3,108 lb/h SO_2

SO_2 removed by scrubber = SO_2 input - maximum allowable SO_2 emission

$$= 31,080 - 3,108$$
$$= 27,972 \text{ lb/h } SO_2$$

or $\times \dfrac{1 \text{ lb-mol } SO_2}{64.06 \text{ lb } SO_2}$

$$= 436.65 \text{ lb-mol/h } SO_2$$

---

* Based on 46,000 lb/h particulates entering the ESP, the particulate removal efficiency is 99.7%.

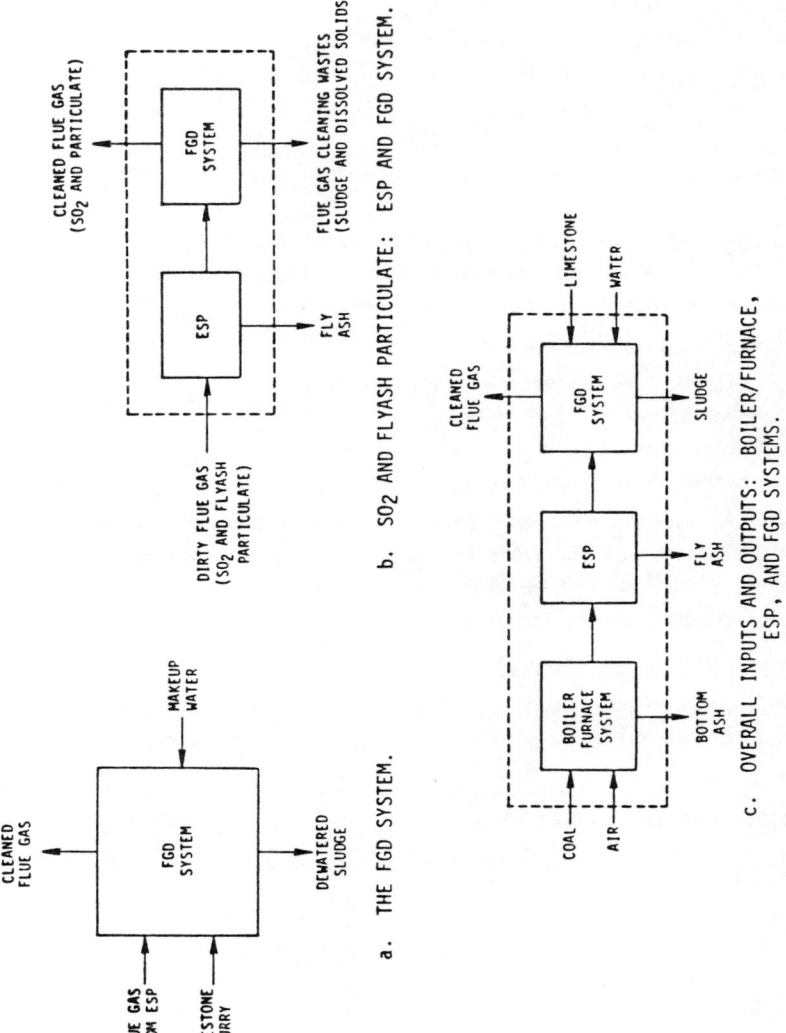

Figure E-2. Schematics of basis for material balance calculations.

Limestone Requirement/Slurry Preparation (Step 2)

The theoretical limestone requirement depends on the amounts of SO_2 and HCl to be removed from the FGD system. All of the HCl from the flue gas is removed and leaves the system with the interstitial water of the dewatered sludge:

$$\text{lb-mol HCl removed/h} = 432 \frac{\text{lb}}{\text{h}} \times \frac{\text{lb-mol}}{36.5 \text{ lb}}$$

$$= 11.84 \frac{\text{lb-mol}}{\text{h}} \text{ HCl}$$

Since 1 mol SO_2 requires 1 mol alkalinity (as $CaCO_3$ and $MgCO_3$), and 1 mol HCl requires ½ mol alkalinity, the theoretical alkalinity requirement is 436.65 + ½(11.84) = 442.57 lb-mol/h. The actual alkalinity supplied is 110 percent of theoretical value. Thus:

$$\text{Actual alkalinity} = 442.57 \frac{\text{lb-mol}}{\text{h}} \times \frac{110}{100}$$

$$= 486.83 \frac{\text{lb-mol}}{\text{h}}$$

It is supplied from limestone containing 94 percent $CaCO_3$, 1.5 percent $MgCO_3$, and 4.5 percent inerts.

The composition of the available limestone for SO_2 absorption is summarized in Table E-4.

TABLE E-4. COMPOSITION OF AVAILABLE LIMESTONE FOR SO_2 ABSORPTION
(Based on 100 lb limestone supplied)

	Molecular weight	Weight, lb	lb-mol	Mol percent
$CaCO_3$	100.09	94.0	0.9392	98.15
$MgCO_3$	84.33	1.5	0.0178	1.85
Total		95.5	0.9570	100.00

The amount of $CaCO_3$ in the limestone is the amount that satisfies the following equation:

$$\text{Actual alkalinity} \times \frac{\text{mol } CaCO_3}{\text{mol alkalinity}}$$

$$= 486.83 \frac{\text{lb-mol}}{\text{h}} \times \frac{98.15 \text{ lb-mol}}{100 \text{ lb-mol available alkalinity}}$$

$$= 477.82 \text{ lb-mol/h } CaCO_3 \text{ and } \times \frac{100.09 \text{ lb } CaCO_3}{\text{lb-mol } CaCO_3} \text{ to convert to pounds}$$

$$= 47,825 \text{ lb/h } CaCO_3$$

Similarly, the amount of $MgCO_3$ available for SO_2 absorption in the limestone supplied is:

$$\text{Available } MgCO_3 = \text{Actual alkalinity} \times \frac{\text{mol available } MgCO_3}{100 \text{ mol available alkalinity}}$$

$$= 486.83 \frac{\text{lb-mol}}{\text{h}} \times \frac{1.85 \text{ lb-mol } MgCO_3}{100 \text{ lb-mol alkalinity}}$$

$$= 9.01 \frac{\text{lb-mol}}{\text{h}} MgCO_3 \times \frac{84.33 \text{ lb } MgCO_3}{\text{lb-mol } MgCO_3}$$

$$= 760 \frac{\text{lb}}{\text{h}} MgCO_3$$

Thus:

Total alkalinity = $CaCO_3 + MgCO_3$

= 47,825 lb/h + 760 lb/h

= 48,585 lb/h

which is 95.5 percent of total limestone supplied. Therefore:

$$\text{Total limestone supplied} = \text{total alkalinity} \times \frac{\text{total limestone supplied}}{\text{available alkalinity}}$$

$$= 48,585 \text{ lb alkalinity} \times \frac{100 \text{ lb limestone}}{95.5 \text{ lb alkalinity}}$$

= 50,875 lb/h limestone, or 25.45 tons/h

which contains 2290 lb/h inerts.

Limestone requirement summary (Stream 5, lb/h):

$CaCO_3$	47,825
$MgCO_3$	760
Inerts	2,290
Total	50,875

The amount of excess makeup water used to prepare limestone solids slurry is established as the difference between that required by the mist eliminator wash procedure and the total makeup water required by the system.

Water remaining for slurry preparation = 197 gpm

or 197 gpm × 500 $\frac{\text{lb/h}}{\text{gpm}}$

= 98,500 lb/h of water for limestone slurry

And the flow rate of slurry (Stream 7) = 149,400 lb/h

Humidification of Flue Gas (Step 3)

Use of a psychrometric chart permits rapid estimation of the humidity and temperature of the wet flue gas leaving a scrubbing system. These values are then used to determine the amount of water required to saturate the flue gas.

When unsaturated, hot flue gas is introduced into a scrubbing system, water evaporates into the flue gas under adiabatic conditions at constant pressure and cools the gas. The wet-bulb temperature remains constant throughout the period of vaporization.

If evaporation continues until the flue gas is saturated with water vapor, the final temperature of the gas will be the same as its initial wet-bulb temperature (dew point). For air-water vapor mixtures, the wet-bulb temperature and adiabatic cooling lines are practically the same.

The humidity of the inlet flue gas is 0.0582 lb water per lb dry gas at 290°F. As vaporization takes place, the humidity of the gas is increased and the dry-bulb temperature must correspondingly decrease along the wet-bulb temperature line (adiabatic cooling line). The psychrometric chart in Figure E-3 is prepared for an "air-water" system.

The humidity of flue gas is defined as follows:

$$\text{humidity, } W_x = \frac{(18.02 \text{ lb})/(\text{lb-mol } H_2O)}{(\text{molecular wt of dry gas})/(\text{lb-mol dry gas})}$$

The mols of SO_2 removed from the flue gas are replaced with equal mols of CO_2 per equation (1) on page E-7. Based on the design premises, the molecular weights of dry flue gas at the inlet and outlet of the FGD system are calculated as 30.51 and 30.38 respectively.

For use with the psychrometric chart, the humidity of inlet flue gas must be corrected for air:

$$\text{humidity of air} = 0.0582 \, \frac{\text{lb } H_2O}{\text{lb dry gas}} \times \frac{\text{molecular wt. of dry gas (inlet)}}{\text{molecular wt. of dry air}}$$

$$= 0.0582 \times \frac{30.51}{28.97}$$

$$= 0.0613 \text{ lb } H_2O/\text{lb dry air}$$

In Figure E-3, point A corresponds to humidity 0.0613 lb H_2O/lb dry air at 290°F. Point B is the intersection of the adiabatic cooling line with the 100 percent saturation line, which is 0.104 lb H_2O/lb dry air at 128°F. The corrected value for saturation humidity of flue gas is

$$= 0.104 \, \frac{\text{lb } H_2O}{\text{lb dry air}} \times \frac{\text{molecular wt. of dry air}}{\text{molecular wt. of dry gas (outlet)}}$$

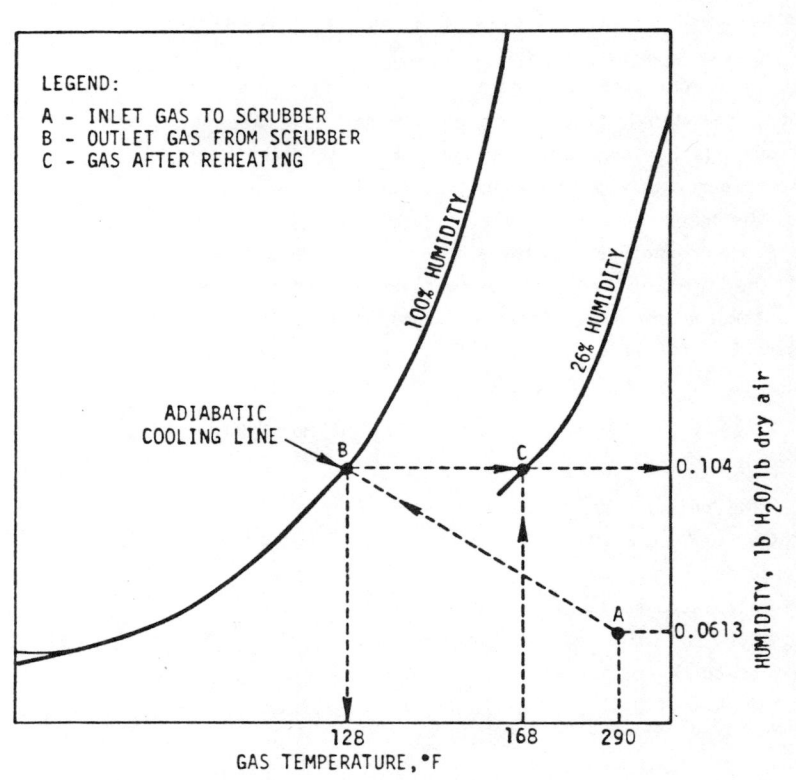

Figure E-3. Psychrometric chart (not to scale).

$$= 0.104 \times \frac{28.97}{30.38}$$

$$= 0.0992 \text{ lb } H_2O/\text{lb dry gas @ } 128°F$$

Note that the molecular weight of outlet dry flue gas is lower because 90 percent of the SO_2 removed is replaced by CO_2.

The amount of water vapor in the outlet gas is 0.0992 lb water/lb dry gas x 5,034,980 lb/h dry gas = 499,500 lb/h water or 999 gpm.

The total mass flow rate of the gas at the outlet is

5,034,980 lb/h dry gas + 499,500 lb/h water

= 5,534,400 lb/h gas

The amount of water required for humidification is

Saturation - Inlet

499,500 lb/h - 293,500 lb/h

= 206,000 lb/h water or 412 gpm

On a molar flow rate basis, the outlet gas contains 165,760 lb-mols of dry gas per hour and 27,710 lb-mols H_2O per hour.

The volumetric flow rate of outlet gas from the scrubber at 128°F and 2 in. H_2O is

$$(165,760 + 27,710) \frac{\text{lb-mol}}{\text{h}} \times \frac{(460 + 128)°F}{(460 + 32)°F} \times \frac{\text{h}}{60 \text{ min}} \times \frac{359 \text{ ft}^3}{\text{lb-mol}}$$

$$\times \frac{407 \text{ in. } H_2O^*}{(407 + 2) \text{ in. } H_2O}$$

$$= 1,376,700 \text{ acfm at } 128°F$$

* Atmospheric pressure.

The composition of the cleaned flue gas leaving the scrubber is summarized in Table E-5.

Flow rate of gas leaving the reheater (Stream 4) at 168°F

$$= 1,376,700 \text{ acfm} \times \frac{(460 + 168)°F}{(460 + 128)°F}$$

$$= 1,470,350 \text{ acfm}$$

When the gas is heated, the humidity of the gas remains constant. Therefore, the gas property moves along the dotted line from B to C in Figure E-3. Point C represents the gas leaving the reheater. The relative humidity (H_R) at this point, as read from the figure, is 26 percent.

TABLE E-5. COMPOSITION OF CLEANED FLUE GAS (STREAM 3)

	Mass flow rate, lb/h	Composition, wt. %
Particulate	138	0.0
CO_2	961,654	17.38
N_2	3,746,000	67.68
O_2	324,000	5.85
SO_2	3,108	0.06
H_2O	499,500	9.03
Total	5,534,400	100.00

Before reheat

1,376,700 acfm
at 128°F
H_R = 100%

After reheat

1,470,350 acfm
at 168°F
H_R = 26%

Recirculation Loop and Sludge Production (Step 4)

The overall effect of the scrubbing system is that 90 percent of the incoming SO_2 is removed from the flue gas and transferred to the effluent sludge. The mols of SO_2 removed from the flue gas equal the mols of sulfur in the sludge. The hydrogen chloride is assumed to be removed by $MgCO_3$ as $MgCl_2$. The remainder of the $MgCO_3$ reacts with SO_2 to form $MgSO_3 \cdot 3H_2O$. Some of the $MgSO_3$ may oxidize to $MgSO_4$, but for the sake of simplicity the formation of $MgSO_4$ is neglected here.

$$\text{Excess } CaCO_3 \text{ supplied} = \text{supply - use by system}$$
$$= [477.82 - (442.57 - 9.01)] \frac{\text{lb-mol}}{\text{h}}$$
$$= 44.26 \text{ lb-mol/h } CaCO_3 \text{ or } 4,430 \text{ lb/h}$$

This excess $CaCO_3$ leaves the system with the waste sludge.

In determination of sludge composition, the sulfite formation is determined as follows:

$CaSO_3$ formed by the precipitation reaction = 433.56 lb-mol/h $CaSO_3$

Likewise, the $MgSO_3$ formed = 3.09 lb-mol/h $MgSO_3$.

At 20 percent oxidation of sulfite to sulfate:

$CaSO_4$ formed by precipitation = 433.56 $\frac{\text{lb-mol}}{\text{h}}$ x 0.20 = 86.71 lb-mol/h $CaSO_4 \cdot 2H_2O$

16 mol percent $CaSO_4$ is in the form of $CaSO_4 \cdot CaSO_3 \cdot \frac{1}{2}H_2O$ (solid solution)

That is, $CaSO_4$ coprecipitates with $CaSO_3 \cdot \frac{1}{2}H_2O$ (16 mol percent of total sulfur in the sludge).

$CaSO_4 \cdot 2H_2O$ crystals formed = $86.71 \; \frac{lb\text{-}mol}{h} \times \frac{20-16}{20}$

$= 17.34 \; \frac{lb\text{-}mol}{h} = 2986 \; \frac{lb}{h} \; CaSO_4 \cdot 2H_2O$

$CaSO_3 \cdot \frac{1}{2}H_2O$ left in the product = $(433.56 - 86.71)$ lb-mol/h

= 346.85 lb-mol/h $CaSO_3 \cdot \frac{1}{2}H_2O$ = 44,744

$MgSO_3 \cdot 3H_2O$ crystals formed = 3.09 lb-mol/h $MgSO_3 \cdot 3H_2O$ = 490 lb/h

$CaSO_4 \cdot CaSO_3 \cdot \frac{1}{2}H_2O$ formed = $69.37 \; \frac{lb\text{-}mol}{h} \; CaSO_4 \cdot CaSO_3 \cdot \frac{1}{2}H_2O$
= 18,403 lb/h

$CaSO_3 \cdot \frac{1}{2}H_2O$ crystals formed = (346.85 - 69.37) = 277.48 lb-mol/h
= 35,837 lb/h $CaSO_3 \cdot \frac{1}{2}H_2O$

Formation of CO_2 is as follows per equation (1):

By the precipitation reaction = 433.56 lb-mol/h = 19,081 lb/h
By $MgCO_3$ = 3.09 lb-mol/h = 136 lb/h
Total CO_2 formed = 19,217 lb/h

The composition of sludge solids is summarized in Table E-6.

TABLE E-6. WASTE SLUDGE SOLIDS FOR HIGH-SULFUR COAL

	Mass flow rate, lb/h	Composition, wt. % (dry basis)
$CaCO_3$	4,430	6.93
$CaSO_3 \cdot \frac{1}{2}H_2O$	35,837	56.04
$CaSO_4 \cdot CaSO_3 \frac{1}{2}H_2O$	18,403	28.78
$CaSO_4 \cdot 2H_2O$	2,986	4.67
Inerts	2,290	3.58
Total	63,946	100.00

The amount of water in the 60 percent solids sludge (Stream 15) is:

63,946 lb/h solids $\times \frac{40 \; lb \; water}{60 \; lb \; solids}$

= 42,630 lb/h or 85 gpm.

Since the solids content in waste stream 13 from the thickener is 30 percent, the total slurry flow rate is:

63,946 lb/h solids $\times \frac{100 \; lb \; slurry}{30 \; lb \; solids}$

= 213,153 lb/h or 358 gpm
at 1.19 sp. gravity

and the water content is 149,207 lb/h by difference or 298 gpm.

The amount of filtrate returned from dewatering device to the thickener is:

Inlet - Outlet

(149,207 - 42,630) lb/h

= 106,577 lb/h water or 213 gpm.

Effluent slurry to thickener (Stream 11) from the recycle loop contains 14 percent solids and the total slurry flow rate is

63,946 lb/h solids $\times \dfrac{100 \text{ lb slurry}}{14 \text{ lb solids}}$

= 456,757 lb/h or 838 gpm
at 1.09 sp. gravity

and water content is 392,811 lb/h by balance or 786 gpm.

The amount of return of thickener overflow (Stream 12) to the recycle tank is

(392,811 + 106,577 - 149,207) lb/h

≅ 350,500 lb/h or 701 gpm.

Figure E-4 depicts the material balance around the thickener and dewatering device.

Slurry flow to the scrubber (Stream 10) is shown in Figure E-5. The L/G ratio for the scrubber in this process is given as 75 gallons of slurry (14 percent solids) per 1000 ft^3 of gas. Since the gas flow rate from the scrubber is 1,376,700 acfm, the slurry flow rate at 1.09 sp. gravity or 9.16 lb/gal is:

$$\dfrac{75 \text{ gal}}{1000 \text{ ft}^3} \times \dfrac{1,376,700 \text{ ft}^3 \text{ gas}}{\text{min}} \times \dfrac{60 \text{ min}}{\text{h}} \times \dfrac{9.16 \text{ lb}}{\text{gal}}$$

= 56,747,625 lb/h (104,124 gal/min) of slurry

which contains 14 percent solids (7,944,665 lb/h) and 86 percent water (48,802,960 lb/h) or 97,606 gpm of water.

The amount of SO$_2$ transferred to the liquid phase from the flue gas is 27,972 lb/h. The amount of CO$_2$ transferred from the liquid phase to the flue gas is 19,235 lb/h. This is based on a transfer of 1 mol of CO$_2$ for every mol of SO$_2$ absorbed in the scrubbing slurry. The amount of water lost to saturate the flue gas is 205,820 lb/h or 412 gpm.

Amount of solids in = 7,944,665 lb/h (Stream 10)

+ (27,972 - 19,360) lb/h

= 7,953,277 lb/h

Appendix E: Material and Energy Balances 357

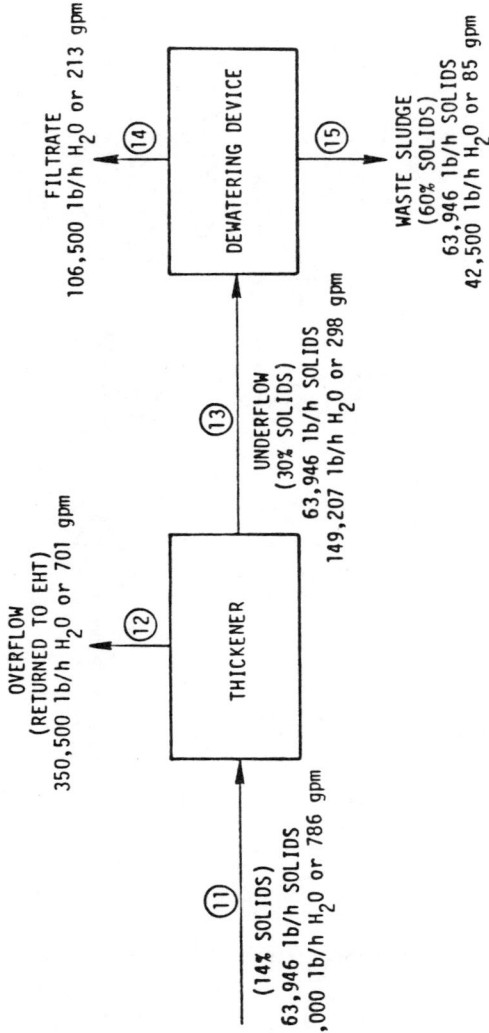

Figure E-4. Material balance around thickener and solids dewatering system.

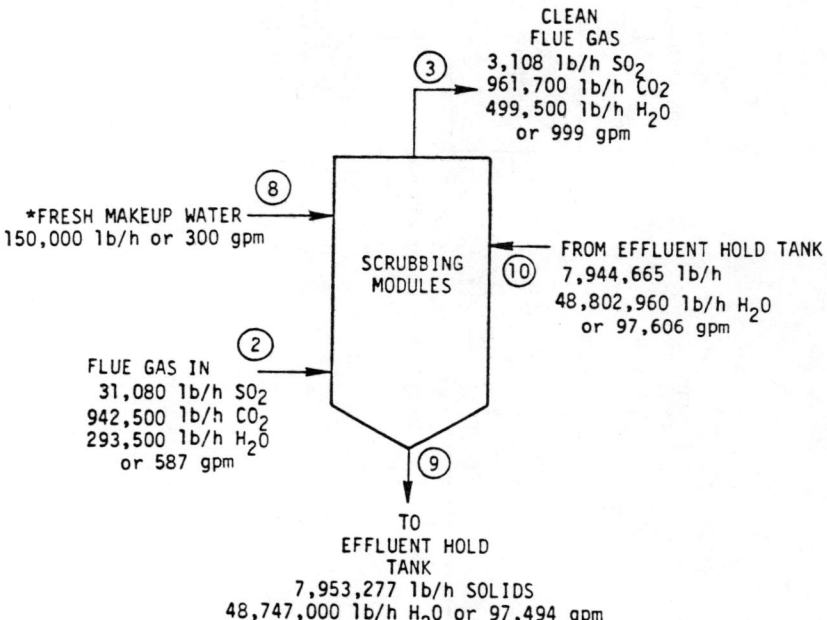

Figure E-5. Material balance around scrubbers.

Amount of water in = 293,500 lb/h (Stream 2) or 587 gpm

+ 105,000 lb/h (Stream 8) or 300 gpm

+ 48,802,960 lb/h (Stream 10) or 97,606 gpm

$\cong$ 49,246,500 lb/h or 98,493 gpm of water

Amount of water out = (49,246,500 - 499,320)(Stream 3) lb/h
(Stream 9)   = 48,747,180 lb/h or 97,494 gpm

Amount of solids out = 7,953,277 lb/h.
(Stream 9)

Total Makeup Water Required (Step 5)

The amount of water leaving with the saturated flue gas and with the waste sludge is the amount that must be made up. The makeup water consists of the amount of water required for humidification plus the water leaving with the sludge, i.e., 412 + 85 = 497 gpm.

Mist Eliminator Wash

After calculation of the total makeup water requirement of 497 gpm, the most important item is calculation of the fresh water requirements for mist eliminator wash. On the basis of work at Shawnee, the EPA has developed recommended guidelines (Burbank and Wang 1980) for a mist eliminator wash procedure that satisfies the needs of keeping a closed-loop water balance and maximizing limestone utilization while maintaining scale-free operation.

At limestone utilization above 85 percent, the guidelines recommend intermittent fresh water wash on both the top and bottom of the mist eliminator. The top should be washed sequentially with fresh water at 0.54 gpm/ft^2 for 3 min/h at 13 psig using six nozzles per 50 ft^2 of area. The bottom wash should be intermittent and full face with 1.5 gpm/ft^2 for 4 min/h at 45 psig using ten nozzles per 50 ft^2.

At limestone utilizations below 85 percent, a full-face continuous bottom wash should be used with clarified thickener overflow at a 33 to 50 percent blend with fresh water. The wash rate should be 0.4 gpm/ft^2 at 12 psig using four nozzles per 50 ft^2.

Scrubber operation at limestone utilization below 70 percent is not recommended.

For this high-sulfur coal case it follows that:

$$\frac{\text{Volumetric flow rate of scrubber outlet gas}}{\text{Gas velocity through eliminator}} = \text{Cross-sectional area of eliminator}$$

$$1,376,700 \, \frac{ft^3}{min} \left(\frac{MW}{60 \, sec} \right) \div \frac{10 \, ft}{sec} = 2295 \, ft^2$$

then fresh makeup water is on a continuous hourly basis: $\frac{\text{continuous}}{\text{hourly basis}}$

Top wash required

$$2295 \text{ ft}^2 \times 0.54 \frac{\text{gpm}}{\text{ft}^2} \times \frac{3 \text{ min}}{\text{h}} \times \frac{\text{h}}{60 \text{ min}} = 62 \text{ gpm}$$

Bottom wash required

$$2295 \text{ ft}^2 \times 1.50 \frac{\text{gpm}}{\text{ft}^2} \times \frac{4 \text{ min}}{\text{h}} \times \frac{\text{h}}{60 \text{ min}} = 230 \text{ gpm}$$

Total fresh water for ME: 292 gpm

This fresh water is used intermittently, but the calculation gives 292 gpm on a continuous hourly basis. Therefore, an even 300 gpm is adequate for scale-free operation at greater than 85 percent utilization.

Since the total system makeup requirement is 497 gpm, supplying 300 gpm for ME wash leaves an excess of 197 gpm. Normal practice is to use this water to slurry the limestone. The amount of water remaining will establish the solids content of the limestone slurry feed in weight percent. Slurry solids can range from as low as 20 weight percent to as high as 50 weight percent, as practiced at Shawnee. In this case: 50,900 lb/h limestone required is mixed with 197 gpm (98,500 lb/h) water to give 50,900 + 98,500 or 149,400 lb/h total solution and $\frac{50,900}{149,400}$ = 34 weight percent solids.

The overall water balance that provides for a good mist eliminator wash procedure is shown in Figure E-6.

ESTIMATION OF ENERGY CONSUMPTION: HIGH-SULFUR COAL CASE

Energy is consumed by the scrubber fans that pass the flue gas through the scrubbing system, by slurry recirculation pumps and other pumps, and by the reheater. Comparatively small amounts of energy are also consumed by the thickener, dewatering device, agitators, conveyors, bucket elevators, and other components. The energy demand of flue gas fans, slurry recirculation pumps, and reheater are calculated as a basis for estimates of total energy consumption. Energy consumption in the other areas of the system is assumed to be 20 percent of that used by fans and recirculation pumps. Table E-7 gives equations for use in determining the energy requirement of fans, slurry recirculation pumps, and reheater.

Flue Gas Fans

The gas entering the scrubbing system undergoes pressure drops across the scrubber, mist eliminator, reheater, and ductwork. In this example, assume a pressure drop of 7 in. H_2O across the scrubber, which could vary depending on the scrubber type (Table E-2). For instance, spray-type scrubbers require a high slurry recirculation rate and the nozzles add to the pump head and power requirement, but gas pressure drop is lower than for the other types. For spray-type scrubbers there is a variable trade off

Appendix E: Material and Energy Balances 361

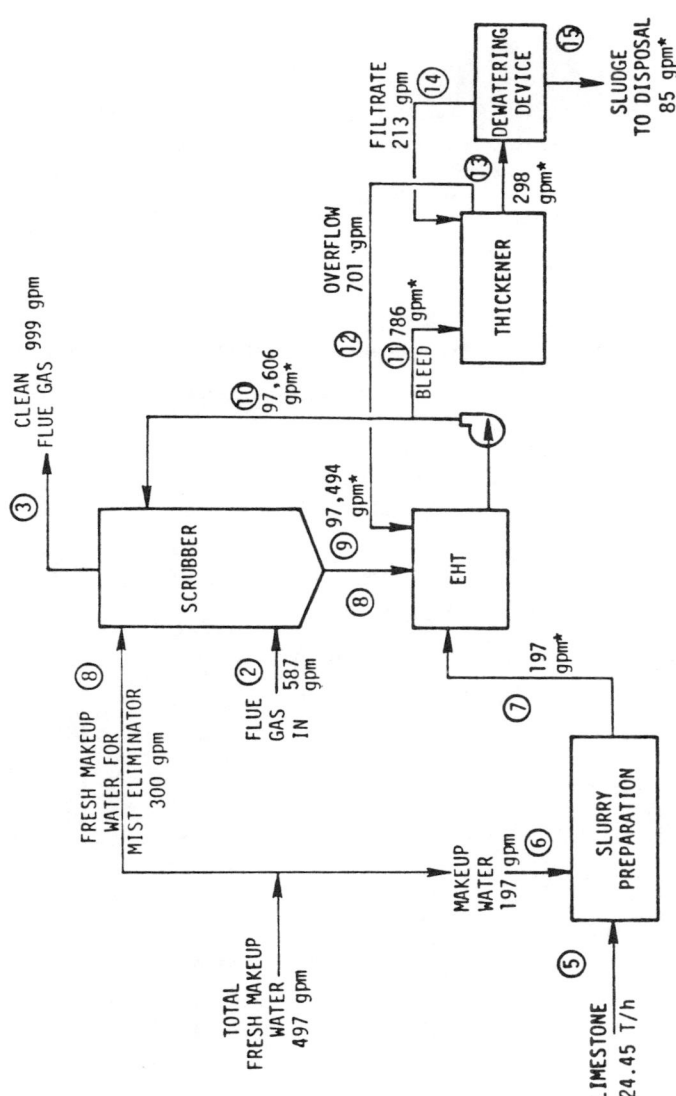

Figure E-6. Overall water balance.

TABLE E-7. ENERGY REQUIREMENT CALCULATIONS

C_p = Specific heat, Btu/(lb)(°F)

P = Energy required, Btu/h

E = Head energy, Btu/h

H_s = Head, ft.

L/G = Ratio of liquor flow to flue gas rate, gpm/1000 acf at the outlet

$\overset{\circ}{m}$ = Air flow rate at the inlet of reheat section, lb/h

ΔP = Pressure drop through FGD system, in. H_2O

Q = Gas flow rate at the outlet of scrubber, acfm

ΔT = Degree of reheat, °F

Flue gas fans (70% fan efficiency assumed)

$\quad$ P = 0.573 × ΔP (in. H_2O) × Q (acfm)

Slurry recirculation pumps (70% pump efficiency assumed)

$\quad$ P = 0.918 × H_s (ft) × (L/G) (gal/1000 acf)

$\qquad$ × (sp.gr.) × $\dfrac{Q\ (acfm)}{1000}$

or $\quad H_s$ = (L/G) (sp.gr) Q × (9.18 × 10^{-4})

$\quad$ P = H_s (ft) × L (gpm) × (sp.gr.) × (0.918)

Reheat of scrubber flue gas

$\quad$ E = $\overset{\circ}{m}$ ($\dfrac{lb}{h}$) × C_p ($\dfrac{Btu}{lb°F}$) × ΔT (°F) between lower fan power (low pressure drop) and higher pump power (more L/G). No clear advantage is apparent for the spray-type scrubber in regard to power requirement.

The total pressure drop accross the FGD system is 11 in. H_2O.

Power required by fans, $P_1 = 0.573 \times \Delta P$ (in. H_2O) $\times Q$ (acfm)
(70% efficiency) $P_1 = 10.18 \times 10^6$ Btu/h

Rapid estimates of fan energy requirement can be made for various plant capacities, as shown in Figure E-7. As the figure shows, the power required for high pressure drop (ΔP) at a 500-MW plant is high.

Slurry Recirculation Pumps

The pumping of slurry from the EHT to the top of scrubber requires considerable energy. This energy consumption depends on L/G ratio, saturated gas flow rate, slurry density, and pump delivery head. At an assumed delivery head of 90 ft, the power required by slurry recirculation pumps at 70 percent efficiency is

$P_2 = H_s$ (ft) $\times L$ (gpm) $\times$ (sp.gr.) $\times 0.918$
$P_2 = 90 \times 104,962 \times 1.09 \times 0.918$
$P_2 = 9.45 \times 10^6$ Btu/h.

Similarly, the power required by other pumps can be determined from the above equation. The graph in Figure E-8 is useful for a quick estimate of the recirculation pump energy requirement. Comparison with Figure E-7 shows that it is less energy-intensive to increase L/G than to increase the system gas pressure drop.

The total power consumed by fans and recirculation pumps is

$(P_1 + P_2)$
$= 10.18 \times 10^6$ Btu/h $+ 9.45 \times 10^6$ Btu/h
$= 19.63 \times 10^6$ Btu/h

The power consumed in other areas is assumed to be

$= 0.20 \times (P_1 + P_2)$
$= 0.20 \times 19.63 \times 10^6$
$= 3.93 \times 10^6$ Btu/h

Thus the total power consumption except reheat as electrical power

$= 23.56 \times 10^6$ Btu/h

Then 23.56×10^6 Btu/h divided by 35 percent is equivalent to the amount of energy that the coal must supply to the plant to generate this electrical power.

$$\frac{23.56 \times 10^6 \text{ Btu/h}}{0.35} = 67.31 \times 10^6 \text{ Btu/h}$$

Therefore, 67.31×10^6 Btu/h must be supplied as coal to the power plant.

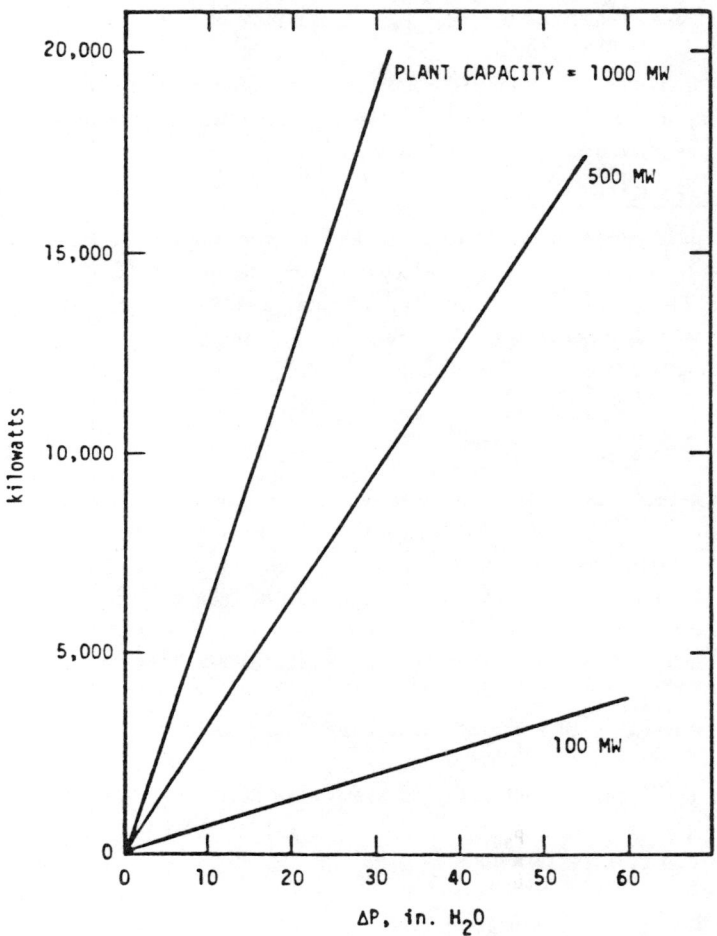

Figure E-7. Fan energy requirement (to obtain energy consumed as coal fired to produce this electricity, divide by 35 percent efficiency).

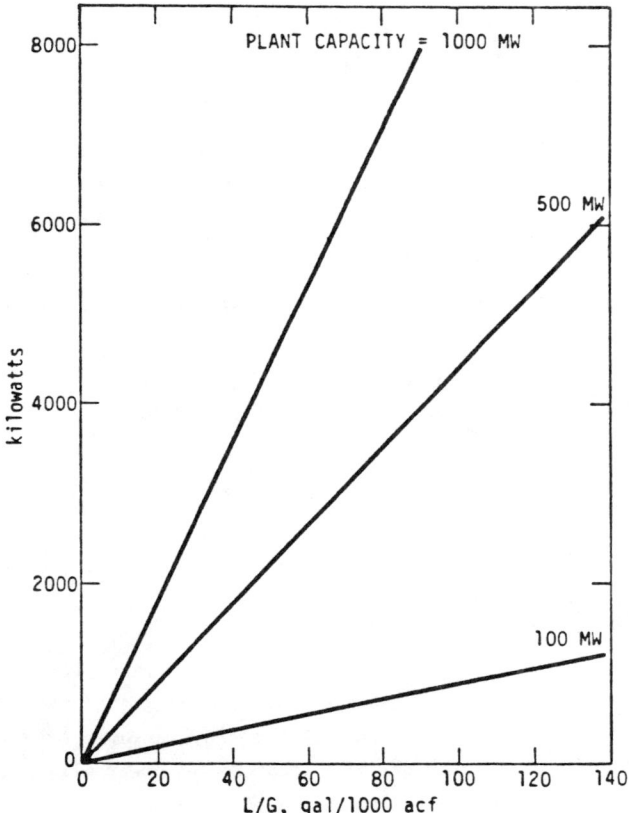

Figure E-8. Recirculation pump energy requirement (to obtain actual energy consumed as coal fired to produce this electricity, divided by 35 percent efficiency).

Reheat of Scrubber Flue Gas

The heat input needed to reheat the saturated flue gas from 128°F (Stream 3) to 168°F (Stream 4) is given by

$$E = \overset{\circ}{m}\left(\frac{lb}{h}\right) \times C_p \frac{Btu}{lb °F} \times \Delta T \text{ °F} \quad \text{(Table 7)}$$

In this case

$$E = 5534.3 \times 10^3 \text{ lb/h} \times 0.25 \frac{Btu}{lb °F} \times 40°F$$

$$E = 53.13 \times 10^6 \text{ Btu/h}$$

Then 55.35×10^6 Btu/h is used in a 90 percent thermally efficient unit ($55.35 \times \frac{1}{0.9} = 61.5 \times 10^6$ Btu/h). The total reheat energy required is 61.5 $\times 10^6$ Btu/h, which must be supplied as coal input to the powerplant.

Total Energy for FGD System as Percent of Plant Input

The percentage of energy input to the power plant on a coal-fired basis that is consumed by the FGD system is calculated as follows:

```
Total power consumption    =  67.31 x 10^6 Btu/h
plus total reheat energy   =  61.50 x 10^6 Btu/h
        Total FGD energy   = 128.81 x 10^6 Btu/h
```

The total energy input to the plant is given by
210 T/h × 23 × 10⁶ Btu/T = 4830 × 10⁶ Btu/h
(from Table E-1)

and percentage consumed by the FGD system is

$$\frac{\text{Total FGD energy (128.81)} \times 10^6 \text{ Btu/h}}{\text{Total plant consumption (4830)} \times 10^6 \text{ Btu/h}} \times 100 = 2.70\%$$

For this 500-MW plant firing a 3.7 percent sulfur Pennsylvania bituminous coal, the energy demand of the FGD system is 2.7 percent of the total energy input (as coal) to the power plant. The energy consumed by the reheater alone is 1.3 percent of the total heat input to the power plant, or approximately half of the total consumption by the FGD system.

The energy input to the power plant to produce this amount of electric power (assuming 35 percent overall efficiency) is

$$\frac{23.33 \times 10^6 \text{ Btu/h}}{0.35} = 66.66 \times 10^6 \text{ Btu/h}$$

MATERIAL BALANCE CALCULATIONS: LOW-SULFUR COAL CASE

Design premises for an FGD system at a 500-MW plant firing 0.7 percent sulfur coal are presented in Table E-8. According to the current NSPS for utility power plants, the SO_2 removal requirement for the boiler firing 0.7 percent sulfur coal is 70 percent (outlet emissions to be less than 0.6 lb

TABLE E-8. DESIGN PREMISES: LOW-SULFUR COAL CASE

Plant capacity (gross)		500 MW
Boiler:	Type	Pulverized-coal-fired
Coal:	Type	Sub-bituminous
	Source	Wyoming
	Consumption	267.6 tons/h
	Heating value	8,750 Btu/lb
	Sulfur content	0.7%
	Oxygen content	10.75%
	Hydrogen content	4.5%
	Nitrogen content	1.2%
	Carbon content	62.6%
	Chloride content	0.05%
	Moisture content	13.0%
	Ash content	7.2%
Limestone:	Stoichiometric ratio (SR)[a]	1.10
	Utilization ($\frac{1}{SR}$)	90.9%
Scrubber:	Liquid-to-gas ratio	55 gal/1000 acf (saturated)
	Inlet gas temperature	285°F
Sulfite-to-sulfate oxidation		50%
Solids:	Limestone slurry feed tank	33%
	Effluent hold tank	14%
	Thickener underflow	30%
	Dewatered sludge	60%
SO_2 inlet loading:		1.6 lb/10^6 Btu input
Max. emission: (NSPS)	SO_2	0.48 lb/10^6 Btu input
	Particulates	0.03 lb/10^6 Btu input
Reheat:	Flue gas bypassed	34°F
Flue gas treated:	Flue gas not bypassed	78%
SO_2 removal:	Required for treated flue gas	90%

Flue gas components (inlet to scrubber)	Flow rate, lb/h	Composition, wt. %	Composition, mol %
Particulates	138		
Carbon dioxide	1,228,440	18.00	11.93
Hydrogen chloride	280	0.00	0.00
Nitrogen	4,760,000	69.70	72.65
Oxygen	423,000	6.19	5.65
Sulfur dioxide	7,495	0.11	0.05
Moisture	409,700	6.00	9.72
Total	6,829,053	100.00	100.00

[a] Defined as moles of $CaCO_3$ and $MgCO_3$ (if available) fed per mol of SO_2 and HCl (if present) absorbed.

$SO_2/10^6$ Btu heat input)*. In such a case, it is possible to reheat with bypass gas instead of using an indirect in-line reheater. A portion of the flue gas is bypassed around the scrubbing module(s) and mixed with the cleaned gas. The degree of reheat is limited by the inlet SO_2 concentration of the gas and the SO_2 removal requirement.

Figure E-9 shows the basic flow diagram for the model limestone FGD system on a 500-MW plant burning low-sulfur coal. The flow is similar to that in Figure E-1 except that the flue gas leaving the FGD system is reheated with untreated flue gas. Basic process chemistry remains the same. The following discussion describes a procedure for determining the portion of inlet flue gas that bypasses the system and summarizes the steps involved in calculations of stream characteristics. All the process assumptions listed earlier are applicable here.

As depicted in Figure E-9, the ESP collects particulates[†] from the flue gas to a level below the maximum allowable particulate emission (0.03 lb/million Btu heat input) under the current NSPS regulation.

Determination of SO_2 Removal Requirement and Bypass Gas Fraction (Step 1)

The over-all SO_2 removal required is 70 percent. Because 22 percent by volume of the inlet flue gas bypasses the system, higher SO_2 removal (90 percent) from the remainder of the gas is needed to achieve an overall SO_2 removal efficiency of 70 percent.

$$\text{Allowable } SO_2 \text{ emission} = 7495 \text{ lb/h } SO_2 \times 0.30$$
$$= 2248.5 \text{ lb/h } SO_2$$

By difference the total SO_2 removed = 5246.5 lb/h SO_2 or 81.9 lb mol/h SO_2

Let x be the fraction of inlet gas treated. Equating the amount of SO_2 removed

$$(x)(0.90) = (1)(0.70)$$
$$x = \frac{0.70}{0.90} = 0.778 \text{ or } 77.8 \text{ percent}$$

Therefore, 100 - 77.8 = 22.2 percent of the incoming flue gas bypasses the system and is used for reheat purposes.

The amount of flue gas treated (dry basis) is

$$6,419,350 \text{ lb/h} \times 0.778 = 4,994,300 \text{ lb/h dry flue gas}$$

---

* Federal Register, June 11, 1979.

[†] Based on 30,800 lb/h particulates entering the ESP, the particulate removal efficiency is 99.55 percent.

Appendix E: Material and Energy Balances   369

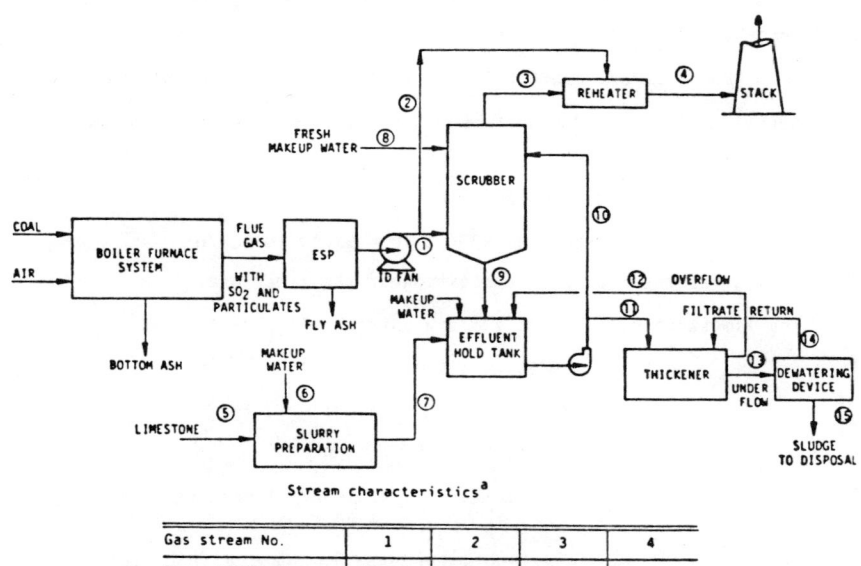

Stream characteristics[a]

Gas stream No.	1	2	3	4
Flow, 1000 acfm	1653.8	465.5	1372.65	1852.2
Flow, 1000 lb/h	5313.0	1516.0	5507.0	7023.0
Temp., °F	295	285	128	162
SO_2, lb/h	5830	1665	583	2248
HCl, lb/h	220	60	0	60
CO_2, 1000 lb/h	955.76	272.68	959.42	1232.10
H_2O, 1000 lb/h	318.75	90.95	514.40	608.35
Particulates, lb/h	107	31	107	138

Liquid/solid stream No.	5	6	7	8	9	10	11	12	13	14	15
Flow, gal/min	-	39	66	370	91,182	91,285	171	100	74	44	29
Flow, 1000 lb/h	9.76	19.5	29.3	185.0	41,484	41,493	94.0	50.1	43.8	21.9	21.9
Temp., °F	70	70	70	70	128	128	128	100	100	100	100
$CaCO_3$, lb/h	9,175.5	0	9175.5	0			850	0	850	0	850
H_2O (free), 1000 lb/h	0	19.5	19.5	185.0	35,673	35,684	80.8	50.1	30.7	21.9	8.8
$CaSO_3 \cdot 1/2 \, H_2O$, lb/h	-	0	0	0			3,597	0	3,597	0	3,359
$CaSO_4 \cdot H_2O$, lb/h	-	0	0	0			4,795	0	4,795	0	4,795
$CaSO_4 \cdot CaSO_3 \cdot 1/2 \, H_2O$, lb/h	-	0	0	0			3,475	0	3,475	0	3,475
Inerts, lb/h	439	0	439	0			439	0	439	0	439

[a] A blank indicates an unknown value; a dash shows that an item does not apply.

Figure E-9. Model limestone FGD system on 500-MW plant, 0.7 percent sulfur coal.

Limestone Requirement/Slurry Preparation (Step 2)

The theoretical limestone requirement depends on the amounts of SO_2 and HCl to be removed from the flue gas. The amount of HCl in flue gas Stream 1 (100 percent removal assumed) is

$$(280 \times 0.78) \text{ lb/h}/(36.46 \text{ lb per lb mol}) = 6.0 \text{ lb-mol HCl/h}.$$

The theoretical alkalinity requirement is based on the amount of SO_2 and HCl present in the gas stream:

$$81.90 + \tfrac{1}{2}(6.0) = 84.90 \text{ lb-mol/h}.$$

Actual alkalinity required = $84.92 \times 1.10 = 93.40$ lb-mol/h

Table E-4 gives the required composition of the limestone.

The limestone requirement summary (Stream 5) is as follows:

$CaCO_3$	9175.5 lb/h
$MgCO_3$	146.5
Inerts	439.0
Total	9761 lb/h limestone or 4.88 tons/h

The amount of excess makeup water used to prepare the limestone solids slurry is established by the mist eliminator wash procedure and the total makeup water required:

$$= 39 \text{ gpm}$$
$$= 39 \text{ gpm} \times 500 \frac{\text{lb/h}}{\text{gpm}}$$
$$= 19{,}500 \text{ lb/h of water}$$

and the flow rate of slurry (Stream 7) = 29,261 lb/h

Humidification of Flue Gas (Step 3)

The procedure followed is similar to that in the high-sulfur coal case. The flow rate of water in the flue gas entering the scrubbing modules (after bypass) at 285°F is 318,750 lb/h or ~ 638 gpm. The dry flue gas flow rate (into scrubber) is 4,994,300 lb/h.

The molecular weights of dry flue gas at the inlet and outlet of the scrubbing modules are 30.40 and 30.38, respectively.

$$\text{Humidity (inlet)} = \frac{318{,}750 \text{ lb/h } H_2O}{4{,}994{,}300 \text{ lb/h dry gas}} \times \frac{30.40 \text{ (molecular wt of dry gas)}}{28.97 \text{ (molecular wt of air)}}$$

$$= 0.067 \text{ lb } H_2O/\text{lb dry air}$$

This conversion facilitates the use of a psychrometric chart based on air-water. The saturation temperature of the gas is 128°F. The humidity of saturated gas is calculated as follows.

Saturation humidity = 0.108 lb H$_2$O/lb dry air at 128°F

To correct for flue gas multiply by ratio of molecular weights

$$0.108 \ \frac{\text{lb H}_2\text{O}}{\text{lb dry air}} \times \frac{28.97}{30.38} = 0.103 \ \text{lb H}_2\text{O/lb dry gas}$$

The amount of water vapor in the outlet gas is

0.103 lb H$_2$O/lb dry gas × 4,994,300 lb/h dry gas

= 514,400 lb/h of water or 1029 gpm.

The amount of water required for 100 percent saturation

Saturation - Inlet
(514,400 - 318,750) lb/h
= 195,650 lb/h or 391 gpm.

On a molar flow rate basis, the outlet gas contains 164,340 lb-mol/h dry gas and 28,560 lb-mol/h H$_2$O.

The volumetric flow rate of outlet gas (before mixing with untreated gas) at 128°F and 2 in. H$_2$O is

$$(164{,}340 + 28{,}560) \ \frac{\text{lb-mol}}{\text{h}} \times \frac{(460+128)°F}{(460+32)°F} \times \frac{\text{h}}{60 \ \text{min}} \times \frac{359 \ \text{ft}^3}{\text{lb-mol}}$$

$$\times \frac{407 \ \text{in. H}_2\text{O}^*}{409 \ \text{in. H}_2\text{O}}$$

= 1,372,650 acfm at 128°F

* Atmospheric pressure.

The amount of cleaned flue gas entering the mixing chamber (Stream 3) is 5,507,050 lb/h, which accounts for the SO$_2$ replaced by the evolved CO$_2$ in the gas. The flow rate of untreated gas (Stream 2) is 1,516,000 lb/h. Let T(°F) be the temperature of the mixed gas. The energy balance gives

$$5{,}507{,}050 \ \frac{\text{lb}}{\text{h}} \times 0.25 \ \frac{\text{Btu}}{\text{lb}°F} \times (T - 128)°F = 1{,}516{,}000 \ \text{lb/h} \times 0.25 \ \frac{\text{Btu}}{\text{lb}°F} \times (285 - T)°F$$

T is computed from this relationship to be 162°F. Hence the degree of reheat is 162° - 128° = 34°F. This amount of reheat is assumed to be adequate.

The flow rate of mixed gas leaving the system (Stream 4) is

$$244{,}860 \ \frac{\text{lb-mol}}{\text{h}} \times \frac{622°F}{492°F} \times \frac{\text{h}}{60 \ \text{min}} \times \frac{359 \ \text{ft}^3}{\text{lb-mol}}$$

= 1,852,200 acfm at 162°F

Recirculation Loop and Sludge Production (Step 4)

The mols of SO_2 removed from the flue gas equal the mols of sulfur in the sludge. The hydrogen chloride is assumed to be removed by $MgCO_3$ and $CaCO_3$ as $MgCl_2$ and $CaCl_2$.

The excess $CaCO_3$ supplied = $[91.67 - (84.92 - 1.74)]\ \frac{\text{lb-mol}}{\text{h}}$

$$= 8.49\ \text{lb-mol/h or 850 lb/h }CaCO_3$$

This excess $CaCO_3$ leaves the system with the waste sludge.

In determination of sludge composition, the $CaSO_3$ precipitation = 81.9 lb-mol/h $CaSO_3$. At 50 percent oxidation of sulfite to sulfate the $CaSO_4$ formed by precipitation is

$$81.9\ \frac{\text{lb-mol}}{\text{h}} \times 0.5 = 40.95\ \text{lb-mol/h }CaSO_4.$$

Sixteen mol percent $CaSO_4$ is in the form of $CaSO_4 \cdot CaSO_3 \cdot \tfrac{1}{2}H_2O$ (solid solution) and the remainder is $CaSO_4 \cdot 2H_2O$

$$CaSO_4 \cdot 2H_2O\ \text{crystals formed} = 40.95\ \frac{\text{lb-mol}}{\text{h}}\left(\frac{50-16}{50}\right)$$

$$= 27.85\ \frac{\text{lb-mol}}{\text{h}}$$

$$= 4795\ \text{lb/h }CaSO_4 \cdot 2H_2O$$

$CaSO_3$ left in the product = 40.95 lb-mol/h $CaSO_3$

$CaSO_4 \cdot CaSO_3 \cdot \tfrac{1}{2}H_2O$ formed = 13.1 lb-mol/h

$$= 3475\ \text{lb/h }CaSO_4 \cdot CaSO_3 \cdot \tfrac{1}{2}H_2O$$

$CaSO_3 \cdot \tfrac{1}{2}H_2O$ crystals formed = (40.95 - 13.1) lb-mol/h

$$= 3597\ \text{lb/h }CaSO_3 \cdot \tfrac{1}{2}H_2O$$

83.2 lb-mol/h of CO_2 is formed in the scrubbing modules or 3661 lb/h.

Composition of the sludge solids is summarized in Table E-9.

TABLE E-9. WASTE SLUDGE SOLIDS FOR LOW-SULFUR COAL

	Mass flow rate, lb/h	Composition, wt. % (dry basis)
$CaCO_3$	850	6.46
$CaSO_3 \cdot \tfrac{1}{2}H_2O$	3597	27.34
$CaSO_4 \cdot CaSO_3 \cdot \tfrac{1}{2}H_2O$	3475	26.41
$CaSO_4 \cdot 2H_2O$	4795	36.45
Inerts	439	3.34
Total	13,156	100.00

The amount of water in the 60 percent solids sludge (Stream 15) is

$$13{,}156 \text{ lb/h solids} \times \frac{40 \text{ lb water}}{60 \text{ lb solids}}$$

$$= 8771 \text{ lb/h of water or } \sim 18 \text{ gpm}$$

Figure E-10 depicts the material balance around the thickener and solids dewatering device.

Slurry flow to the scrubber (Stream 10) is shown in Figure E-11.

The L/G ratio for the scrubbing modules in this process is given as 55 gallons of slurry (14 percent solids) per 1000 ft^3 of saturated gas. Since the saturated gas flow rate is 1,372,650 acfm at 128°F, the slurry flow rate at 1.09 specific gravity entering the scrubbing modules is

$$\frac{55 \text{ gal}}{1000 \text{ ft}^3} \times \frac{1{,}372{,}650 \text{ ft}^3 \text{ gas}}{\text{min}} \times \frac{60 \text{ min}}{\text{h}} \times \frac{9.16 \text{ lb}}{\text{gal}}$$

$$= 41{,}492{,}455 \text{ lb/h (75,440 gpm) of slurry}$$

which contains 14 percent solids (5,808,944 lb/h) and 86 percent water (35,683,511) or 71,367 gpm of water.

The amount of SO_2 transferred to the liquid phase is 5,247 lb/h. The amount of CO_2 transferred from the liquid phase to the flue gas is 3.661 lb/h. The amount of water lost to saturate the flue gas is 195,650 lb/h or 319 gpm.

As Figure E-11 illustrates:

    input streams: 2, 8, 10
    output streams: 3, 9

Amount of solids in = 5,808,944 lb/h (Stream 10)
                        + (5247 - 3661) lb/h
                        = 5,810,530 lb/h

Amount of water in = 318,750 lb/h (Stream 2)
                        + 185,000 lb/h (Stream 8)
                        + 35,683,511 lb/h (Stream 10)
                        = 36,187,301 lb/h

Amount of water out
  (Stream 9)   = [36,187,301 - 514,400 (Stream 3)] lb/h
                        = 35,672,901 or 71,346 gpm

Amount of solids out
  (Stream 9)   = 5,810,530 lb/h

Total Makeup Water Required (Step 5)

The amount of water leaving with the saturated flue gas and with the waste sludge is the amount that must be made up. The makeup water consists of the amount of water required for humidification plus the water leaving with the sludge, i.e., 391 + 18 = 409 gpm.

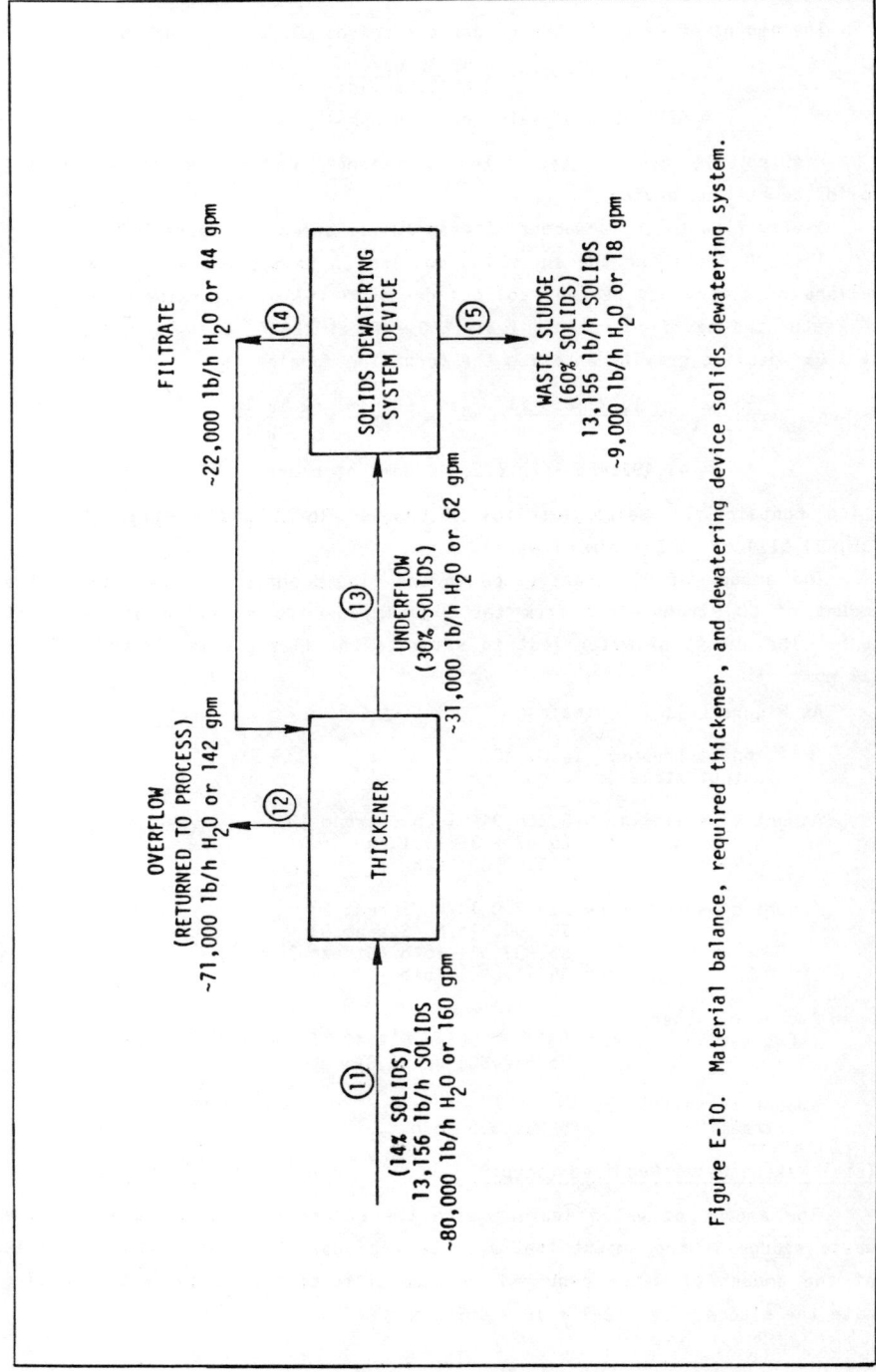

Figure E-10. Material balance, required thickener, and dewatering device solids dewatering system.

Appendix E: Material and Energy Balances 375

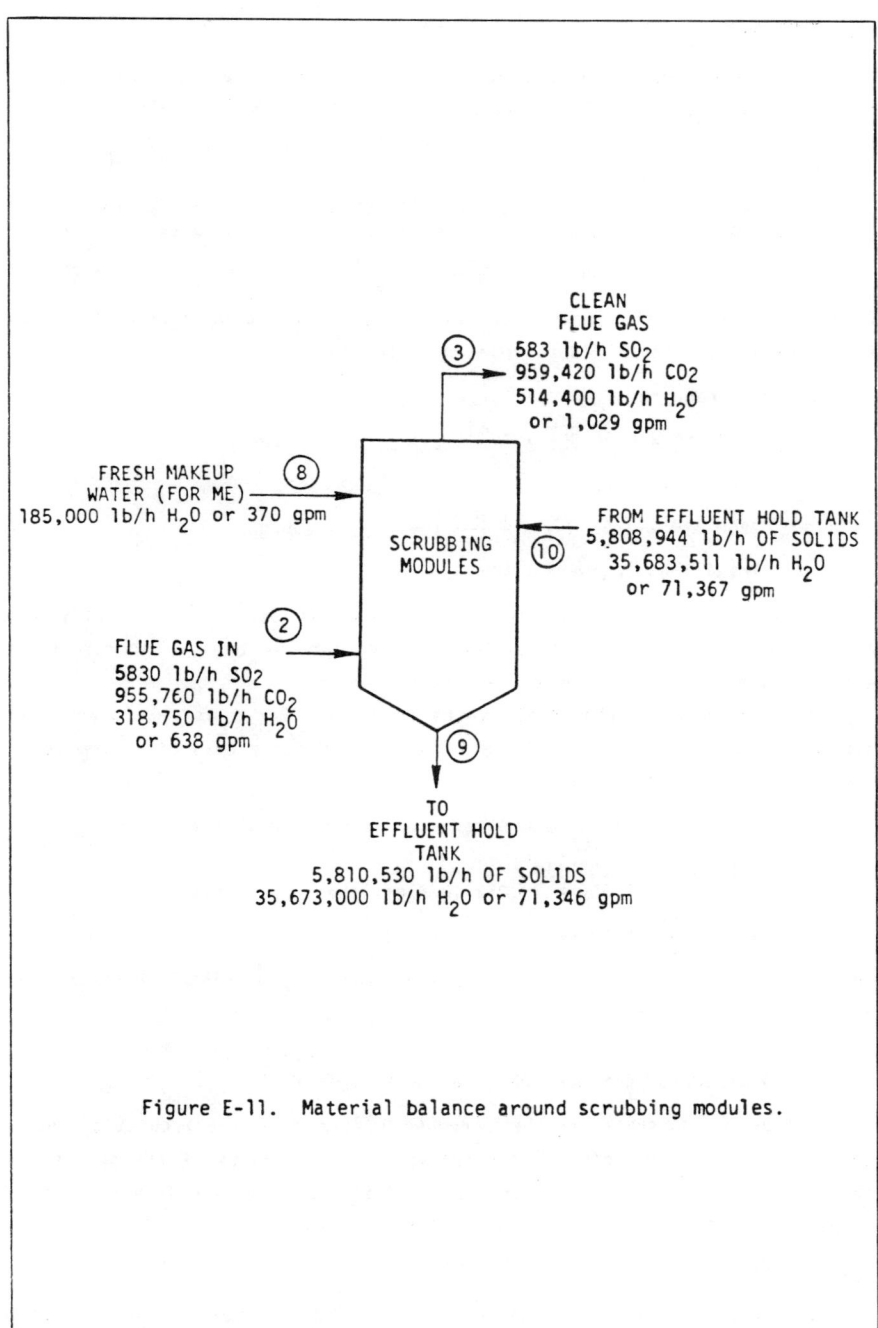

Figure E-11. Material balance around scrubbing modules.

Mist Eliminator Wash

As in the high-sulfur coal case, the 409 gpm of makeup water required by the system must be split between that needed for the mist eliminator wash and the excess that is used to prepare the limestone slurry feed.

For this low-sulfur coal case:

$$\frac{\text{Volumetric flow rate}}{\text{of scrubber outlet gas}} \div \frac{\text{Gas velocity through}}{\text{mist eliminator}} = \text{Cross-sectional area}$$

$$1,372,650 \; \frac{ft^3}{min} \; (\frac{min}{60 \; sec}) \div \frac{8 \; ft}{sec} = 2860 \; ft^2.$$

Then, according to the Shawnee guidelines (Burbank and Wang 1980) the fresh makeup water requirement is calculated as follows:

Top wash required:

$$2860 \; ft^2 \times 0.54 \; \frac{gpm}{ft^2} \times \frac{3 \; min}{h} \; (\frac{h}{60 \; min}) = 77 \; gpm$$

Bottom wash required:

$$2860 \; ft^2 \times 1.50 \; \frac{gpm}{ft^2} \times \frac{4 \; min}{h} \; (\frac{h}{60 \; min}) = 286 \; gpm$$

Total fresh water for ME: 363 gpm

As before, it is noted that this fresh water is used intermittently at a higher flow rate but is calculated on a continuous basis. Therefore, an even 370 gpm is adequate for scale-free operation.

Since this low-sulfur coal system requires 409 gpm as makeup water and 370 gpm is directed to the mist eliminator wash, an excess of 39 gpm can be used to slurry the limestone:

9,761 lb/h of limestone required is mixed with the 39 gpm

= 9,761 + 39 (500) =
= 9,761 + 19,500 = 29,261 lb/h of limestone slurry

and $\frac{9,761}{29,261} \cong 33$ weight percent solids.

The overall water balance that provides for a good mist eliminator wash procedure is shown in Figure E-12.

ESTIMATION OF ENERGY CONSUMPTION: LOW-SULFUR COAL CASE

Energy is consumed by the scrubber system fans, recirculation pumps, and other pumps. As before, comparatively small amounts of energy are also consumed by other components; the total of this consumption is assumed to be 20 percent of that by fans and recirculation pumps.

Flue Gas Fans

In this example the assumed pressure drops across the scrubbing modules, mist eliminator, ductwork, and mixing chamber are 7, 1, 2, and 1 in. H_2O, respectively.

Appendix E: Material and Energy Balances 377

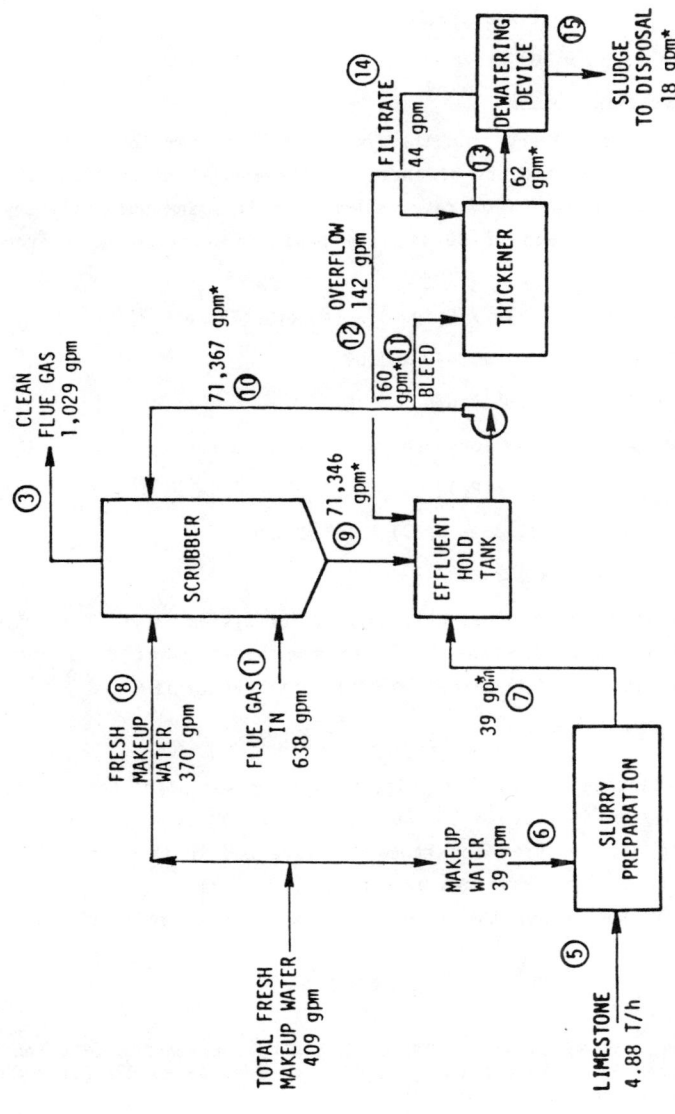

Figure E-12. Overall water balance.

*gpm for water content only.

Total pressure drop across the FGD system is 11 in. H_2O. Assuming 70 percent fan efficiency

P_1 = 0.573 x ΔP (in. H_2O) x Q (acfm)

= 0.573 x 11 x 2,119,300

P_1 = 13.36 x 10^6 Btu/h

Slurry Recirculation Pumps

The pumping of slurry from the recycle tank to the top of the scrubber requires considerable amounts of energy. This energy consumption depends on L/G ratio, saturated gas flow rate, slurry density, and pump delivery head. Assuming a delivery head of 90 ft, the power consumption by recirculation pumps is

P_2 = 0.918 H_s (ft) x L (gpm) x (sp.gr) (Table E-7)

= 0.918 x 90 x 75,440 x 1.09

P_2 = 6.79 x 10^6 Btu/h

The assumed power consumed in the other areas

= 0.20 (P_1 + P_2)

= 0.20 (13.36 + 6.79) x 10^6 Btu/h

= 4.03 x 10^6 Btu/h

Therefore the total power consumption by the FGD system is 24.18 x 10^6 Btu/h.

The energy input as coal fired to the power plant to produce this electric power (assuming 35 percent overall efficiency) is

$$\frac{24.18 \times 10^6 \text{ Btu/h}}{0.35} = 69.09 \times 10^6 \text{ Btu/h}.$$

Hence the percentage of the energy input to the power plant that is consumed by the FGD system is 69.09 x 10^6/4683 x 10^{6*} x 100 or 1.50 percent. Note that no energy is consumed for reheat of flue gas in this low-sulfur coal case because of the heat provided by untreated flue gas.

Table E-10 summarizes the energy demands for both low-sulfur and high-sulfur coals.

REFERENCE

Burbank, D. A. and S. C. Wang. 1980. EPA Alkali Scrubbing Test Facility: Advanced Program Final Report (October 1974 to June 1978) EPA-600/7-80-115. NTIS No. PB 80-204 241.

---

* 267.6 T/h coal-fired x 8750 $\frac{Btu}{lb}$ x $\frac{2 \times 10^3 \text{ lb}}{T}$ = 4683 x 10^6 Btu/h as the total heat input of powerplant.

TABLE E-10. ENERGY DEMAND FOR FGD ON A 500-MW PLANT

	Electrical energy, 10^3 kWh	Electrical power, 10^6 Btu/h	Power input, 10^6 Btu/h	Reheat input, 10^6 Btu/h	Total energy demand, percent of coal fired
Low-sulfur coal	7.1	24.2	69.1	Nil	1.5
High-sulfur coal	6.9	23.6	67.4	61.5	2.7

The table shows that use of the untreated flue gas for reheat in the low-sulfur coal case yields about a 50 percent energy savings relative to the high-sulfur coal case.

Appendix F

Limestone Utility FGD Systems in the United States

Table F-1 presents current information about wet limestone utility FGD systems in the United States. The main source of this information was the Flue Gas Desulfurization Information System (FGDIS), a computerized data base sponsored by the U.S. EPA and developed by PEDCo Environmental (see Appendix C).* Additional information came from Black & Veatch Consulting Engineers and from some scrubber suppliers. When discrepancies were found, FGDIS data were generally assumed to be accurate. The other sources, however, were contacted and were sometimes able to correct or supplement FGDIS data.

The high SO_2 removal efficiencies in Table F-1 for the flue gas treated indicate that limestone scrubbers can meet the stringent requirements of the 1979 New Source Performance Standards. Also, the large number of entries suggests that limestone scrubbing is a very reliable and economical means of SO_2 control. Limestone scrubbers are thus expected to remain the main type of SO_2 control system in the United States.

---

* Information about access to the FGDIS can be obtained from Mr. Walter L. Finch, Product Manager, National Technical Information Service, 5285 Port Royal Road, Springfield, Virginia 22161 [telephone (703) 487-4807].

Appendix F: Limestone Utility FGD Systems in the U.S. 381

TABLE F-1. WET LIMESTONE UTILITY FGD SYSTEMS IN THE UNITED STATES

Utility and Plant	Effective Scrubbed Capacity[a] (MW)	New or Retrofit	Startup Date	Sulfur Content of Coal (%)	Scrubber Supplier	Scrubber System Architect/ Engineer or Designer	Type of SO$_2$ Scrubber	Number of Modules	SO$_2$ Removal Efficiency (%)	Type of Mist Eliminator	Type of Reheat	Type of Particulate Removal Device	Thick-ener	Type of Dewater-ing[b]	Forced Oxida-tion	Sludge Stabiliza-tion	Sludge Fixa-tion	Means of Waste Trans-portation	Final Dis-posal
Alabama Electric Coop Tombigbee 2	179	New	9/78	1.15	Peabody	Burns & McDonnell	Spray tower	2	85	Chevron	Bypass	ESP	—	—	—	—	—	Pipeline	Pond
Alabama Electric Coop Tombigbee 3	179	New	6/79	1.15	Peabody	Burns & McDonnell	Spray tower	2	85	Chevron	Bypass	ESP	—	—	—	—	—	Pipeline	Pond
Arizona Electric Power Coop Apache 2	98	New	8/78	0.55	Research-Cottrell	Burns & McDonnell	Combination (spray/packed)	2	92[c]	Chevron	None	ESP	—	—	—	—	—	Pipeline	Pond
Arizona Electric Power Coop Apache 3	98	New	6/79	0.55	Research-Cottrell	Burns & McDonnell	Combination (spray/packed)	2	85	Chevron	None	ESP	—	—	—	—	—	Pipeline	Pond
Arizona Public Service Cholla 1	119	Retrofit	10/73	0.50	Research-Cottrell	Ebasco	Combination (spray/packed)	2	92[c]	Chevron	In-line	Venturi	×	SP	—	—	—	Pipeline	Pond
Arizona Public Service Cholla 2	264	New	4/78	0.50	Research-Cottrell	Ebasco	Combination (spray/packed)	4	97[c]	Chevron	In-line	ESP	×	SP	—	—	—	Pipeline	Pond
Arizona Public Service Cholla 3	126	New	6/81	0.50	Research-Cottrell	—	Combination (spray/packed)	—	—	—	—	ESP	—	—	—	—	—	Pipeline	Pond
Associated Electric Coop Thomas Hill 3	670	New	1/82	4.80	Pullman Kellogg	Burns & McDonnell	Spray tower (Weir)	4	91.5	Chevron	Bypass	ESP	×	C	×	×	—	Truck	Land-fill
Basin Electric Power Coop Laramie River 1	570	New	7/80	0.81	Research-Cottrell	Burns & McDonnell	Combination (spray/packed)	5	90	Chevron	None	ESP	×	C	×	×	—	Con-veyor	Land-fill
Basin Electric Power Coop Laramie River 2	570	New	6/81	0.81	Research-Cottrell	Burns & McDonnell	Combination (spray/packed)	5	90	Chevron	None	ESP	×	C	×	×	—	Con-veyor	Land-fill
Big Rivers Electric D.B. Wilson 1	440	New	7/84	—	Pullman Kellogg	—	Spray tower (Weir)	—	—	Chevron	—	ESP	×	—	—	—	—	—	—
Big Rivers Electric D.B. Wilson 2	440	New	1/86	—	Pullman Kellogg	—	Spray tower (Weir)	—	—	Chevron	—	ESP	×	—	—	—	—	—	—
Central Illinois Light Duck Creek 1	416	New	7/76	3.66	Riley Stoker	Gilbert/Common-wealth Assn.	Packed tower (rod deck)	4	85	Chevron	None	ESP	—	—	—	—	—	Pipeline	Pond

(continued)

382 Handbook for FGD Scrubbing with Limestone

TABLE F-1. (continued)

Utility and Plant	Effective Scrubbed Capacity[a] (MW)	New or Retrofit	Startup Date	Sulfur Content of Coal (%)	Scrubber Supplier	Scrubber System Architect/ Engineer or Designer	Type of SO$_2$ Scrubber	Number of Modules	SO$_2$ Removal Efficiency (%)	Type of Mist Eliminator	Type of Reheat	Type of Particulate Removal Device	Thickener	Type of Dewatering[b]	Forced Oxidation	Sludge Stabilization	Sludge Fixation	Means of Waste Transportation	Final Disposal
Central Illinois Light Duck Creek 2	450	New	1/86	3.30	Not selected	—	—	—	—	—	—	ESP	—	—	—	—	—	—	—
Colorado Ute Electric Assn. Craig 1	447	New	10/80	0.45	Peabody	Stearns-Roger	Spray tower	4	85	Chevron	In-line	ESP	X	C	—	—	—	Truck	Landfill[d]
Colorado Ute Electric Assn. Craig 2	400	New	12/79	0.45	Peabody	Stearns-Roger	Spray tower	4	85	Chevron	In-line	ESP	X	C	—	—	—	Truck	Landfill[d]
Commonwealth Edison Powerton 51	450	Retrofit	4/80	3.53	UOP	Sargent & Lundy	Packed tower (mobile bed)	3	74	Chevron	Indirect hot air	ESP	X	VF	—	—	IUCS	Truck	Landfill[e]
Deseret Generation & Transmission Coop Moon Lake 1	410	New	9/84	0.50	Combustion Engineering	Burns & McDonnell	Spray tower	3	95	Chevron	None	Fabric filter	X	VF	—	X	—	—	Landfill
Deseret Generation & Transmission Coop Moon Lake 2	410	New	0/88	0.50	Not selected	—	—	—	95	—	—	Fabric filter	—	—	—	—	—	—	Landfill
Hoosier Energy Division Merom 1	441	New	5/82	3.50	Mitsubishi	—	Packed tower	5	90	Chevron	Bypass	ESP	X	VF	—	—	IUCS	Truck	Landfill
Hoosier Energy Division Merom 2	441	New	9/81	3.50	Mitsubishi	—	Packed tower	5	—	—	Bypass	ESP	X	VF	—	—	IUCS	Truck	Landfill
Houston Lighting & Power Limestone 1	750	New	12/84	1.08	Combustion Engineering	Ebasco	Spray tower	5	90	Chevron	Bypass/ ambient air	ESP	X	VF	—	X	—	—	Pond
Houston Lighting & Power Limestone 2	750	New	12/85	1.08	Combustion Engineering	Ebasco	Spray tower	5	90	Chevron	Bypass/ ambient air	ESP	X	VF	—	X	—	—	Pond
Houston Lighting & Power W.A. Parrish B	492	New	11/82	0.60	Chemico	—	—	—	82	—	—	Bypass	—	—	—	X	—	—	Landfill
Indianapolis Power & Light Patriot 1	650	New	0/87	3.50	Not selected	—	—	—	—	—	—	—	—	—	—	—	—	—	—

(continued)

Appendix F: Limestone Utility FGD Systems in the U.S. 383

TABLE F-1. (continued)

Utility and Plant	Effective Scrubbed Capacity[a] (MW)	New or Retrofit	Startup Date	Sulfur Content of Coal (%)	Scrubber Supplier	Scrubber System Architect/ Engineer or Designer	Type of SO_2 Scrubber	Number of Modules	SO_2 Removal Efficiency (%)	Type of Mist Eliminator	Type of Reheat	Type of Particulate Removal Device	Thick-ener	Type of Dewater-ing[b]	Forced Oxida-tion	Sludge Stabiliza-tion	Sludge Fixa-tion	Means of Waste Trans-portation	Final Dis-posal
Indianapolis Power & Light Patriot 2	650	New	—	3.50	Not selected	—	—	—	—	—	—	—	—	—	—	—	—	—	—
Indianapolis Power & Light Patriot 3	650	New	—	3.50	Not selected	—	—	—	—	—	—	—	—	—	—	—	—	—	—
Indianapolis Power & Light Petersburg 3	532	New	12/77	3.25	UOP	Gibbs & Hill	Packed tower	4	85	Chevron	Indirect hot air	ESP	X	DVF	—	—	IUCS	Pipeline	Pond
Indianapolis Power & Light Petersburg 4	530	New	10/84	3.50	Research-Cottrell	Stone & Webster	Combination (spray/packed)	3	90	—	Bypass	—	X	DVF	—	—	IUCS	Pipeline	Pond
Iowa Electric Light & Power Guthrie County 1	720	New	11/84	0.40	Combustion Engineer-ing	Black & Veatch	Spray tower	4	85	Chevron	Bypass	ESP	X	—	X	—	—	—	Land-fill
Jacksonville Electric Authority New Project 1	600	New	12/85	3.00	Not selected	—	—	—	—	—	—	ESP	—	—	—	—	—	—	—
Jacksonville Electric Authority New Project 2	600	New	6/87	3.00	Not selected	—	—	—	—	—	—	ESP	—	—	—	—	—	—	—
Kansas City Power & Light La Cyne 1	820	New	2/73	5.39	Babcock & Wilcox	Black & Veatch	Tray tower (sieve)	8	80	Sieve tray and chevron	Indirect hot air	Venturi	X[f]	—	—	—	—	Pipeline	Pond
Kansas Power & Light Jeffrey 1	540	New	8/78	0.32	Combustion Engineer-ing	Black & Veatch	Spray tower	6	85	Chevron	Bypass	ESP	X	—	—	—	—	Pipeline	Pond
Kansas Power & Light Jeffrey 2	490	New	4/80	0.30	Combustion Engineer-ing	Black & Veatch	Spray tower	6	85	Chevron	Bypass	ESP	X	—	—	—	—	Pipeline	Pond
Kansas Power & Light Jeffrey 3	490	New	0/83	0.5	Combustion Engineer-ing	Black & Veatch	Spray tower	4	85	Chevron	Bypass	ESP	X	—	—	—	—	Pipeline	Pond
Kansas Power & Light Lawrence 4	125	Retrofit	1/76	0.55	Combustion Engineer-ing	Black & Veatch	Spray tower	2	73	Chevron	In-line	Venturi (rod)	X	—	—	—	—	Pipeline	Pond
Kansas Power & Light Lawrence 5	420	Retrofit	11/71	0.55	Combustion Engineer-ing	Black & Veatch	Spray tower	2	73	Chevron	In-line	Venturi (rod)	—	—	—	—	—	Pipeline	Pond
Lakeland Utilities McIntosh 3	364	New	8/81	2.56	Babcock & Wilcox	C.T. Main	Spray tower	2	85	Chevron	Bypass	ESP	X	VF	—	—	IUCS	Truck/con-veyor	Con-struc-tion base

(continued)

384 Handbook for FGD Scrubbing with Limestone

TABLE F-1. (continued)

Utility and Plant	Effective Scrubbed Capacity[a] (MW)	New or Retrofit	Startup Date	Sulfur Content of Coal (%)	Scrubber Supplier	Scrubber System Architect/ Engineer or Designer	Type of SO$_2$ Scrubber	Number of Modules	SO$_2$ Removal Efficiency (%)	Type of Mist Eliminator	Type of Reheat	Type of Particulate Removal Device	Thick-ener	Type of Dewater-ing[b]	Forced Oxida-tion	Sludge Stabiliza-tion	Sludge Fixa-tion	Means of Waste Trans-portation	Final Dis-posal
Louisville Gas & Electric Mill Creek 1	358	Retrofit	1/81	3.75	Combustion Engineer-ing	Fluor Pioneer	Spray tower	2	85	Chevron	In-line	ESP	X	VF	–	–	IUCS	Truck	Land-fill
Louisville Gas & Electric Mill Creek 2	350	Retrofit	12/81	3.75	Combustion Engineer-ing	Fluor Pioneer	Spray tower	2	85	Chevron	In-line	ESP	X	VF	–	–	IUCS	Truck	Land-fill
Michigan South Central Power Agency Project 1	55	New	6/82	2.25	Babcock & Wilcox	–	–	–	–	–	In-line	–	–	–	–	–	–	–	–
Middle South Utilities Arkansas Lignite 5	890	New	0/90	0.50	Combustion Engineer-ing	Sargent & Lundy	Spray tower	6	92	Chevron	Bypass	ESP	–	–	–	–	–	Pipeline	Pond
Middle South Utilities Arkansas Lignite 6	890.	New	0/92	0.50	Combustion Engineer-ing	Sargent & Lundy	Spray tower	6	92	Chevron	Bypass	ESP	–	–	–	–	–	Pipeline	Pond
Middle South Utilities Unassigned 1	890	New	0/89	0.50	Combustion Engineer-ing	Sargent & Lundy	Spray tower	6	92	Chevron	Bypass	ESP	–	–	–	–	–	Pipeline	Pond
Middle South Utilities Unassigned 2	890	New	0/93	0.50	Combustion Engineer-ing	Sargent & Lundy	Spray tower	6	92	Chevron	Bypass	ESP	–	–	–	–	–	Pipeline	Pond
Middle South Utilities Wilton 1	890	New	0/88	0.50	Combustion Engineer-ing	Sargent & Lundy	Spray tower	6	92	Chevron	Bypass	ESP	–	–	–	–	–	Pipeline	Pond
Middle South Utilities Wilton 2	890	New	0/91	0.50	Combustion Engineer-ing	–	Spray tower	6	92	Chevron	Bypass	ESP	–	–	X	–	–	Pipeline	Pond
Muscatine Power & Water Muscatine 9	166	New	9/82	3.00	Research-Cottrell	–	Combination (spray/packed)	–	94	Chevron	In-line	ESP	X	–	–	X	–	–	Land-fill
New York State Electric & Gas Somerset 1	625	New	6/84	2.20	Peabody	Ebasco	Spray tower	–	90	Chevron	Indirect hot air	ESP	–	DVF	–	–	–	Truck	Land-fill
Northern States Power Sherburne 1	740	New	3/76	0.80	Combustion Engineer-ing	Black & Veatch	Packed tower (mobile-bed)	12	50	Chevron	In-line	Venturi (rod)	X	–	X	X	–	Pipeline	Pond
Northern States Power Sherburne 2	740	New	3/77	0.80	Combustion Engineer-ing	Black & Veatch	Packed tower (mobile-bed)	12	50	Chevron	In-line	Venturi (rod)	X	–	X	–	–	Pipeline	Pond

(continued)

Appendix F: Limestone Utility FGD Systems in the U.S. 385

TABLE F-1. (continued)

Utility and Plant	Effective Scrubbed Capacity[a] (MW)	New or Retrofit	Startup Date	Sulfur Content of Coal (%)	Scrubber Supplier	Scrubber System Architect/ Engineer or Designer	Type of SO$_2$ Scrubber	Number of Modules	SO$_2$ Removal Efficiency (%)	Type of Mist Eliminator	Type of Reheat	Type of Particulate Removal Device	Thick-ener	Type of Dewater-ing[b]	Forced Oxida-tion	Sludge Stabiliza-tion	Sludge Fixa-tion	Means of Waste Trans-portation	Final Dis-posal
Northern State Power Sherburne 3	860	New	5/84	—	Combustion Engineer-ing	—	—	—	—	—	—	—	—	—	—	—	—	—	—
Pacific Gas & Electric Montezuma 1	800	New	6/89	0.80	Not selected	—	—	—	—	—	—	Fabric filter	—	—	—	—	—	—	—
Pacific Gas & Electric Montezuma 2	800	New	6/90	0.80	Not selected	—	—	—	—	—	—	Fabric filter	—	—	—	—	—	—	—
Plains Electric G&T Coop Escalante 1	233	New	12/83	0.80	Combustion Engineer-ing	Burns & McDonnell	Spray tower	3	95	Chevron	None	ESP	×	C	—	—	—	—	Land-fill
Public Service of Indiana Gibson 5	650	New	0/82	3.30	Pullman Kellogg	—	Spray tower	4	—	—	—	ESP	×	VF	—	—	IUCS	—	Land-fill
Salt River Project Coronado 1	280	New	11/79	1.00	Pullman Kellogg	Bechtel	Spray tower	2	82.5	Chevron	Bypass	ESP	×	—	—	—	—	Pipeline	Pond
Salt River Project Coronado 2	280	New	7/80	1.00	Pullman Kellogg	Bechtel	Spray tower	2	82.5	Chevron	Bypass	ESP	×	—	—	—	—	Pipeline	Pond
Salt River Project Coronado 3	280	New	0/89	0.60	Not selected	—	—	—	—	—	Bypass	—	—	—	—	—	—	—	—
San Miguel Electric Coop San Miguel 1	400	New	11/80	1.70	Babcock & Wilcox	Tippet & Gee	Packed tower	4	86	Chevron	In-line	ESP	×	DVF	—	×	—	Truck	Land-fill
Seminole Electric Seminole 1	620	New	3/83	2.75	Peabody	Burns & Roe	Spray tower	5	90	Chevron	None	ESP	×	BVF	—	—	IUCS	Truck	Land-fill
Seminole Electric Seminole 1	620	New	3/85	2.75	Peabody	Burns & Roe	Spray tower	5	90	Chevron	None	ESP	×	BVF	—	—	IUCS	Truck	Land-fill
Sikeston Board of Municipal Utilities Sikeston 1	235	New	3/81	2.80	Babcock & Wilcox	—	Tray tower (sieve)	3	—	Chevron	—	ESP/venturi	—	—	—	—	—	—	—
South Carolina Public Service Cross 1	500	New	3/85	1.80	Peabody	Burns & Roe	Spray tower	3	90	Chevron	Indirect air and bypass	—	×	DVF	—	—	×	IUCS	Off-site land-fill
South Carolina Public Service Cross 2	500	New	11/83	1.80	Peabody	Burns & Roe	Spray tower	3	90	Chevron	Indirect air and bypass	—	×	DVF	—	—	×	IUCS	Off-site land-fill

(continued)

TABLE F-1. (continued)

Utility and Plant	Effective Scrubbed Capacity[a] (MW)	New or Retrofit	Startup Date	Sulfur Content of Coal (%)	Scrubber Supplier	Scrubber System Architect/ Engineer or Designer	Type of SO$_2$ Scrubber	Number of Modules	SO$_2$ Removal Efficiency (%)	Type of Mist Eliminator	Type of Reheat	Type of Particulate Removal Device	Thick-ener	Type of Dewater-ing[b]	Forced Oxida-tion	Sludge Stabiliza-tion	Sludge Fixa-tion	Means of Waste Trans-portation	Final Dis-posal
South Carolina Public Service Winyah 2	140	New	7/77	1.70	Babcock & Wilcox	Burns & Roe	Tray tower (sieve)	1	90	Chevron	Bypass	ESP/venturi	x	—	—	—	—	Pipeline	Pond
South Carolina Public Service Winyah 3	280	New	5/80	1.70	Babcock & Wilcox	Burns & Roe	Tray tower (sieve)	2	90	Chevron	Bypass	ESP/venturi	x	—	—	—	—	Pipeline	Pond
South Carolina Public Service Winyah 4	280	New	7/81	1.70	American Air Filter	Burns & Roe	Spray tower	2	90	Chevron	Indirect hot air	ESP	x	—	—	—	—	Pipeline	Pond
Southern Illinois Power Coop Marion 4	173	New	5/79	3.75	Babcock & Wilcox	Burns & Roe	Spray tower	2	89.4	Chevron	None	ESP	x	—	—	x	—	Con-veyor	Land-fill
Southern Mississippi Electric Power R.D. Morrow 1	124	New	8/78	1.30	Riley Stoker	Burns & Roe	Packed tower	1	85	Chevron	Bypass	ESP	x	BVF	—	x	—	Truck	Land-fill
Southern Mississippi Electric Power R.D. Morrow 2	124	New	6/79	1.30	Riley Stoker	Burns & Roe	Packed tower	1	85	Chevron	Bypass	ESP	x	BVF	—	x	—	Truck	Land-fill
Southwestern Electric Power Dolet Hills 1	720	New	0/86	0.70	UOP	—	—	—	—	—	—	ESP	—	—	—	—	—	—	—
Southwestern Electric Power Dolet Hills 2	720	New	0/88	0.70	Not selected	—	—	—	—	—	—	—	—	—	—	—	—	—	—
Southwestern Electric Power Henry W. Pirkey 1	720	New	12/84	0.80	UOP	—	Spray tower	4	99+	Chevron	None	ESP	x	VF	—	—	IUCS	—	—
Springfield City Utilities Southwest 1	194	New	4/77	3.50	UOP	Burns & McDonnell	Packed tower (mobile-bed)	2	80	Chevron	None	ESP	x	DVF	—	x	—	Truck	Land-fill
Springfield Water, Light & Power Dallman 3	205	New	10/80	3.30	Research-Cottrell	Burns & McDonnell	Combination (spray/packed)	2	95	Chevron	None	ESP	—	CY	x	—	—	Truck	Land-fill
Tampa Electric Big Bend 4	475	New	12/84	2.35	Research-Cottrell	Stone & Webster	Combination (spray/packed)	4	90	Chevron	Indirect hot air	ESP	x	DVF	x	—	—	Truck	Wall-board manu-facture
Tennessee Valley Authority Paradise 1	704	Retrofit	3/82	4.20	Chemico	—	Spray tower	6	84.2	—	In-line	ESP/venturi	x	DVF	x	—	—	Truck	Land-fill

(continued)

Appendix F: Limestone Utility FGD Systems in the U.S. 387

TABLE F-1. (continued)

Utility and Plant	Effective Scrubbed Capacity[a] (MW)	New or Retrofit	Startup Date	Sulfur Content of Coal (%)	Scrubber Supplier	Scrubber System Architect/ Engineer or Designer	Type of SO_2 Scrubber	Number of Modules	SO_2 Removal Efficiency (%)	Type of Mist Eliminator	Type of Reheat	Type of Particulate Removal Device	Thick- ener	Type of Dewater- ing[b]	Forced Oxida- tion	Sludge Stabiliza- tion	Sludge Fixa- tion	Means of Waste Trans- portation	Final Dis- posal
Tennessee Valley Authority Paradise 2	704	Retrofit	6/82	4.20	Chemico	—	Spray tower	6	84.2	—	In-line	ESP/ venturi	X	DVF	X	—	—	Truck	Land- fill
Tennessee Valley Authority Shawnee 10A	10	Retrofit	4/72	2.90	UOP	Bechtel	Packed tower (mobile-bed)	1	85+	Chevron	Direct combus- tion	—	X	C	—	—	—	Pipeline	Pond
Tennessee Valley Authority Shawnee 10B	10	Retrofit	4/72	2.90	Chemico	Bechtel	Spray tower	1	85+	Chevron	Direct combus- tion	Venturi	—	DVF	X	—	—	Pipeline	Pond
Tennessee Valley Authority Widows Creek 7	575	Retrofit	9/81	3.70	Combustion Engineer- ing	TVA	Spray tower	4	84	Chevron	In-line	ESP/ venturi (rod)	X	VF	X	—	—	Pipeline	Pond
Tennessee Valley Authority Widows Creek 8	550	Retrofit	5/77	3.70	TVA	TVA	Packed (grid)	4	80	Chevron	Indirect hot air	ESP/ venturi	—	VF[i]	X[i]	—	—	Pipeline	Pond
Texas Municipal Power Agency Gibbons Creek 1	400	New	1/82	1.06	Combustion Engineer- ing	Tippet & Gee	Spray tower	3	90	Chevron	In-line	ESP	X	VF	—	—	IUCS	Truck	Land- fill
Texas Power & Light Sandow 4	382	New	11/80	1.60	Combustion Engineer- ing	Brown & Root	Spray tower	3	92	Chevron	Bypass	ESP	—	—	X	—	—	Pipeline	Pond
Texas Power & Light Twin Oaks 1	750	New	8/84	0.70	Chemico	United Engineers	Spray tower	3	92	Chevron	Bypass	ESP	—	—	X	—	—	Pipeline	Pond
Texas Power & Light Twin Oaks 2	750	New	8/85	0.70	Chemico	United Engineers	Spray tower	3	92	Chevron	Bypass	ESP	—	—	X	—	—	Pipeline	Pond
Texas Utilities[g] Martin Lake 1	595	New	4/77	0.90	Research- Cottrell	C.T. Main	Combination (spray/ packed)	6	98[c]	Chevron	Bypass	ESP	X	C	—	X	—	Rail	Land- fill
Texas Utilities[g] Martin Lake 2	595	New	5/78	0.90	Research- Cottrell	C.T. Main	Combination (spray/ packed)	6	98[c]	Chevron	Bypass	ESP	X	C	—	X	—	Rail	Land- fill

(continued)

TABLE F-1. (continued)

Utility and Plant	Effective Scrubbed Capacity[a] (MW)	New or Retrofit	Startup Date	Sulfur Content of Coal (%)	Scrubber Supplier	Scrubber System Architect/ Engineer or Designer	Type of SO$_2$ Scrubber	Number of Modules	SO$_2$ Removal Efficiency (%)	Type of Mist Eliminator	Type of Reheat	Type of Particulate Removal Device	Thickener	Type of Dewatering[b]	Forced Oxidation	Sludge Stabilization	Sludge Fixation	Means of Waste Transportation	Final Disposal
Texas Utilities[g] Martin Lake 3	595	New	2/79	0.90	Research-Cottrell	C.T. Main	Combination (spray/packed)	6	94[h]	Chevron	Bypass	ESP	X	C	–	X	–	Rail	Landfill
Texas Utilities Martin Lake 4	750	New	0/85	0.90	Research-Cottrell	C.T. Main	Combination (spray/packed)	8	94[h]	Chevron	Bypass	ESP	X	C	–	X	–	Rail	Landfill
Texas Utilities Monticello 3	800	New	5/78	1.50	Chemico	C.T. Main	Spray tower	3	92	Chevron	Indirect hot air	ESP	–	–	–	X	–	Pipeline	Pond
Utah Power & Light Hunter 3	400	New	6/83	0.55	Chemico	Brown & Root	Spray tower	3	90	Chevron	Bypass	Fabric filter	X	–	–	X	–	–	Landfill
Utah Power & Light Hunter 4	400	New	6/85	0.55	Chemico	Brown & Root	Spray tower	3	90	Chevron	Bypass	Fabric filter	X	–	–	X	–	–	Landfill

[a] Effective scrubbed capacity is defined as the gross generating capacity times the average fraction of the flue gas that is scrubbed.
[b] Abbreviations are as follows:
  BVF = belt vacuum filter
  C = centrifuge
  CY = cyclone
  DVF = drum vacuum filter
  SP = settling pond
  VF = vacuum filter
[c] Supplier information about actual performance.
[d] Within a year, mine disposal will be used.
[e] Ponding is currently used because insufficient sludge is produced for landfilling.
[f] Dry lime is used as a flocculant.
[g] Two more scrubber modules per unit are being added at Martin Lake Units 1, 2, and 3. These modules will increase SO$_2$ removal efficiency to 95%. Also, an ambient air reheat system and forced oxidation are being added.
[h] Supplier information about guaranteed performance.
[i] Only 25% of sludge is treated by forced oxidation then vacuum filtered.

Appendix G

Materials of Construction

The descriptions of scrubber equipment in Section 3 of this manual refer to materials of construction within the context of equipment function and operation. A brief analysis of the major materials categories is also given at the end of that section. This appendix gives additional information regarding the major categories of materials, as follows:

1. Base metals - composition of materials (carbon steels, stainless steels, and high-grade alloys); details of major test programs.

2. Protective linings, plastics, and ceramics - properties and performance characteristics of organic and inorganic nonmetallic materials.

This appendix also includes experience histories of the metals and protective lining materials used in limestone FGD systems.

BASE METALS

In a manner consistent with the common or trade names given in Section 3, Table G-1 provides the nominal chemical composition of base metals that have been used, are planned for use, or have been tested in limestone FGD systems.

Test Programs

To date, four major test programs have been performed for evaluation of base metals with respect to performance and corrosion resistance in wet scrubbing systems. These programs, sponsored by International Nickel Company (Inco), Stellite, EPA, and Combustion Engineering, have focused on performance in wet lime/limestone FGD systems. Results of these test programs are summarized in the paragraphs that follow.

Inco Test Program (Inco 1980). Inco has undertaken an extensive program to test the performance of the iron-nickel-chromium-molybdenum family of alloys in lime/limestone FGD systems. Some important trends discovered in this work are illustrated in Figure G-1, which shows correlations between the degree of localized attack on 316L and 317L stainless steels and

TABLE G-1. NOMINAL CHEMICAL COMPOSITION OF BASE METALS
(wt % except as indicated)

	Ni	Fe	Cr	Mo	Cu	C	Si	Mn	Al	S	P	Other elements
Carbon steel												
A-285		Balance				<0.35		<0.8		<0.05	<0.05	
AISI 1010		Balance				0.1				<0.05	<0.04	
HSLA	0.44	Balance	0.92		0.41	0.08	0.46	0.31		0.033	0.11	N, 0.03; Ti, 0.35
Stainless steel (wrought)												
430		Balance	16.18			0.12	1.00	1.00		<0.03	<0.04	
E-Brite 26-1		Balance	26.00	1.00		0.02	0.25	0.35				
304	9.5	Balance	18.5			<0.03	<1.0	1.5				
304L	10.0	Balance	17.0			<0.08	<1.0	1.3				
316	13.0	Balance	17.0	2.25		<0.03	<1.0	1.7				
316L	13.0	Balance	17.0	2.25		<0.08	<1.0	1.8				
317	14.0	Balance	19.0	3.25		<0.03	<1.0	<2.0				
317L	14.0	Balance	19.0	3.25		<0.03	<1.0	<2.0				
Stainless steel (cast)												
CD-4 MCu	5.5	61.0	26.0	2.0	3.0	<0.04	<1.0	<1.0		0.13	0.022	
High-grade alloy (wrought)												
Carpenter 7	4.25	66.0	27.0	1.40	0.10	0.05	0.38	0.50				Combined Cb and Ta, 0.06
Carpenter 20	34.0	39.0	20.0	2.50	3.30	<0.07	0.60	0.30				
Uddeholm 904L	25.0	47.0	20.0	4.50	1.50	<0.02	0.35	1.70				
Climax 18-2	0.4	72.0	18.5	2.0	0.20	0.016	0.40	0.4				N, 0.013; Ti, 0.33
Incoloy 825	42.0	30.0	21.0	3.0	1.80	0.03	0.35	0.65				Ti, 0.90
Hastelloy G	45.0	20.0	22.0	6.5	2.00	0.03	<1.0	1.30				W, 0.5; combined Cb and Ta, 2.12
Haynes 6B	2.0	2.0	29.75	1.0		<1.0	0.36	1.4				W, 4.3; Co, balance
Haynes 20	26.0	42.0	22.0	5.0		<0.05	<1.0	<2.5				Ti, minimum is four times carbon content
Jessup JS 700	25.0	46.0	21.0	4.5		0.03	0.5	1.7				Cb, 0.03
Allegheny AL-6X	24.0	46.0	20.0	6.5		<0.025	<0.5	<1.5				
Allegheny 29-4	0.1	66.0	29.0	4.0		0.004	0.01	0.10				N, 0.01
Nitronic 50	12.0	54.0	22.0	2.5		0.03	1.00	5.0				V, 0.20; N, 0.30; Cb, 0.20
Hastelloy C	54.0	5.0	15.5	16.0		<0.08	<1.0	<1.0		0.013	0.013	Co, <2.5; W, 4.0; V, <0.4
Hastelloy C-276	54.0	5.0	15.5	14.8		<0.02	<0.05	<1.0				Co, <2.5; W, 4.0; V, <0.4
Hastelloy C-4	Balance	5.0	15.7			0.008	0.04					
Inconel 625	60.0		21.5	9.0		<0.1	<0.5	<0.5		<0.005	0.007	Combined Cb and Ta, 3.65
High-grade alloy (cast)												
IN-862	24.0	44.0	21.0	5.0		<0.04	0.8	0.5				
Other metals												
Titanium		0.13				0.022						O₂, 0.12; Ti, balance
Zirconium 702		0.05	0.05			0.015						N, 0.05; Hf, <0.1; combined Zr and Hf, >99.2

the pH and chloride levels of the scrubbing liquor. Localized attack includes both pitting and crevice corrosion. The data indicate that localized corrosion increases as the chloride level increases and scrubbing slurry pH decreases. Moreover, at a given slurry pH, the environment becomes more aggressive as chloride level increases, or at a given chloride level as pH decreases. Thus, low-pH/high-chloride environments are the most aggressive with respect to corrosion. Although the data do indicate definite correlations, the scatter minimizes their usefulness in specific design.

Inco also conducted a series of intensive corrosion studies on Module A of the Cholla 1 FGD system of Arizona Public Service. In these studies various alloys were tested in six exposures at six locations in the FGD system. The test results showed that susceptibility to pitting and crevice corrosion is the principal criterion for judging the suitability of a metal to resist the corrosive action of these environments. Pitting was observed in 316 stainless steel, Carpenter 20, and Incoloy 825. Deep pitting was observed in 304 stainless steel. All the remaining materials displayed complete resistance to corrosion.

Stellite Test Program (Leonard 1978). The Stellite Division of Cabot Corporation has performed a series of spool tests under conditions similar to those in the tests performed by Inco. On the basis of the test results, Stellite developed a set of guidelines for service application of the materials tested. These guidelines are summarized in Table G-2.

EPA Test Program (Crow 1978). Two 10-MW lime/limestone scrubbing systems in service at the Shawnee test facility have been used in the evaluation of the materials of construction. Specimens of alloys and nonmetallic materials have been exposed at test locations in the scrubber systems having different degrees of severity of exposure. In the most recent tests, specimens were exposed to modified slurries, and alloys of different molybdenum contents were evaluated for resistance to pitting.

TABLE G-2. SERVICE APPLICATION GUIDELINES BASED ON STELLITE TEST RESULTS

Service condition	316L stainless steel	Haynes 20	Hastelloy G	Haynes 625	Hastelloy C-276
Maximum temperature, °F	150	150	110-160	110-160	>160
pH	Over 5.5	Over 4.5	3.5-6.0	3.5-6.0	Under 3.5
Chloride content, ppm	Under 1500	Under 3000	100-5000	1000-5000	Over 5000

392 Handbook for FGD Scrubbing with Limestone

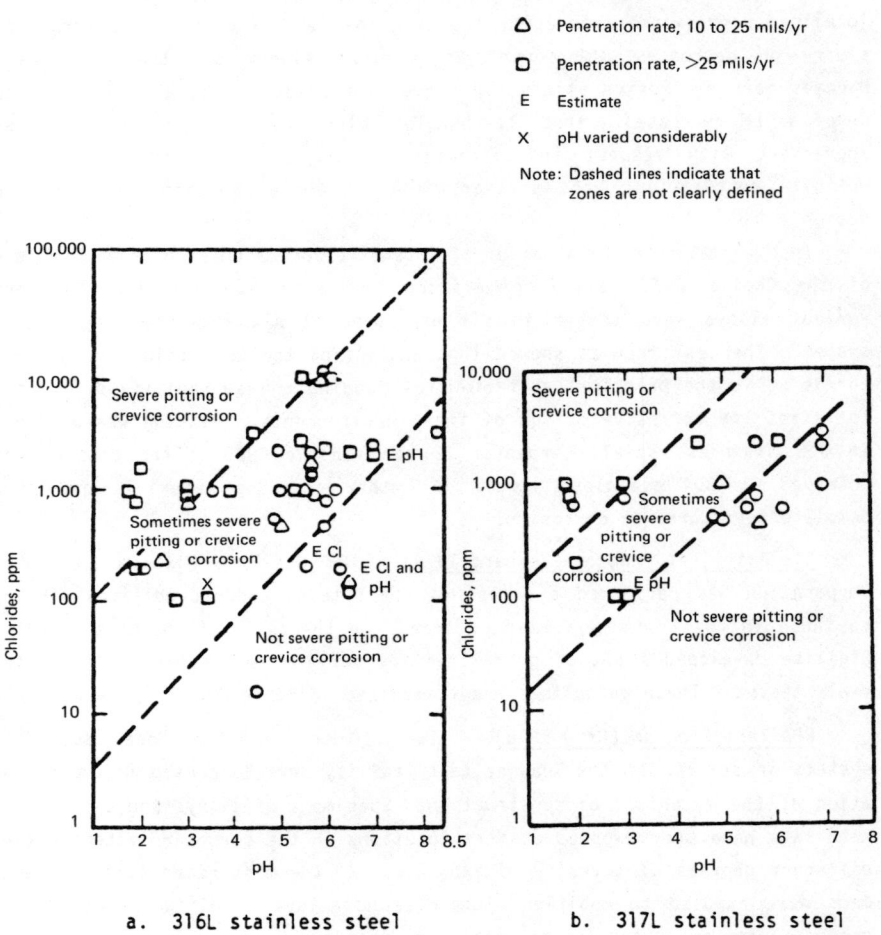

Figure G-1. Effects of pH and chlorides on stainless steel alloys.

Source: International Nickel Company, Inc.

The test conditions are given in Table G-3, and the test data are shown in Table G-4.

The alloy test specimens were least attacked by corrosion in the inlet flue gas duct, where the gas temperature was maintained above the dewpoint. The corrosion rates of carbon steel and Cor-Ten were only 3 mils per year. Cooled and humidified flue gas was much more corrosive. In an earlier test in which humidification sprays were used, the corrosion rate for carbon steel was more than 330 mils per year.

A section of 316L stainless steel inlet duct, which extends about 7 in. inside the chamber immediately above the venturi, has undergone chloride stress-corrosion cracking. The cracks originated on the outside surface of the duct that extends into the scrubber chamber, and some penetrated the wall. The residual stresses produced by cold-forming and welding and the cyclic stresses produced by vibrating equipment could have added to the severity of the cracking. The outside surface of the section that failed contained a tightly adhering scale approximately 51 mils thick. Also, pits to depths of 39 mils were found under the scale, which contained 0.85 weight percent of chloride as calcium chloride. Splashing of the process limestone slurry occurred at the adjustable plug, causing deposition of wet solids on the surface of the duct. The slurry contained 400 to 2440 ppm chlorides, and the high temperature of the duct wall (260° to 330°F) caused concentration of chlorides in the duct.

Below the venturi throat, the greatest attack on the specimen was by erosion-corrosion. The specimens of mild steel and Cor-Ten were completely destroyed, with penetration rates greater than 1850 mils per year. The most promising alloys in the order of decreasing resistance to erosion-corrosion were Zirconium 702, Haynes 6B, 317L stainless steel, 316L stainless steel, Allegheny 29-4, Allegheny AL-6X, and Climax 18-2.

In general, the attack on carbon steel and Cor-Ten has varied greatly in the scrubber tower during the test program. In earlier tests, the corrosion of specimens in the top of the tower was greater than that of specimens near the middle and bottom. During the fourth series, however, corrosion of the specimen near the mist eliminator diminished, very likely because the automatic spray system for mist eliminator washing also washed the spool of test specimens located immediately below it.

The scrubbed gas, which was reheated to about 250°F, corroded carbon steel and Cor-Ten at a rate of 4 mils per year. Minute pitting and crevice corrosion occurred on some of the stainless steels. The 316L stainless steel stacks were corroded at higher rates than specimens of the same material suspended inside the stack. This may be attributable to the operating conditions when the temperature of the specimens is a few degrees

TABLE G-3. OPERATING CONDITIONS DURING CORROSION TESTS AT THE SHAWNEE TEST FACILITY

Location	Physical parameters		
	Temperature, °F	Velocity, ft/s	Gas flow rate, 10^3 acfm at 330°F
Inlet duct	275-300	32-67	17-35
Venturi throat	80-170	40-100	15-30
Bottom of scrubber	125-130	4.5-9.4	15-30
Top of scrubber	125-130	4.5-9.4	15-30
Downstream of mist eliminator	125-130	4.5-9.4	15-30
Downstream of reheater	125-130	32-67	17-35

Chemical composition			Ionic composition	
Constituent	Content at inlet duct, vol %	Content downstream of reheater, vol %	Species	Scrubber effluent slurry,[a] ppm
SO_2	0.2-0.4	0.02-0.10	SO_3^{--}	40-3900
CO_2	10-18	11-19	CO_3^{--}	5-150
O_2	5-15	6-16	SO_4^{--}	400-1400
H_2O	8-15	9-16	Ca^{++}	540-3000
HCl	0.01		Mg^{++}	100-5500
N_2	74	69	$Na^+(K^+)$	60-110
Fly ash[b]	2-7	0.01-0.04	Cl^-	3300-5700

[a] Slurry pH ranged from 4.3 to 6.5, temperature ranged from 90° to 130°F, suspended solids content ranged from 8 to 18 wt %, and dissolved solids content ranged from 0.5 to 8.4 wt %.

[b] Measured in gr/scf.

TABLE G-4. DATA FROM CORROSION TESTS AT THE SHAWNEE TEST FACILITY

	No. of tests	No. of pitted specimens	Maximum depth of pits,[a] mil	No. of specimens with crevice corrosion	No. of specimens with pitting and/or crevice corrosion
Hastelloy C-276	12	0		0	0
Inconel 625	12	0		0	0
Hastelloy G	12	0		1	1
Haynes 6B	14	1	9	2	3
Multimet	10	1	4	0	1
Nitronic 50	9	1	b	3	4
AL 29-4	14	3	2	2	5
Jessop 700	12	3	7	2	5
317L	9	4	b	2	6
Nitronic 50M	18	6	4	5	11
Climax 18-2	12	8	14	5	13
316L (2.3% Mo)	18	10	8	8	18
Zirconium 702	15	0		0	0
AL 6X	13	3	17	3	6
316L (2.8% Mo)	14	9	2	8	17

[a] Values show the actual depth of penetration during the test period.
[b] Minute pit.

higher than the stack walls, even though the stack is insulated. Further, during any outage of the reheater, corrosive condensate drains down the stack walls without flowing over the specimens suspended in the center of the stack.

Combustion Engineering Test Program (Lewis 1978). The test program initiated by Combustion Engineering consisted of corrosion tests of numerous ferritic and austenitic iron- and nickel-base alloys in both the field and laboratory. In the field, these materials were exposed as coupons at several locations in commercial FGD systems. In addition, full-scale operational scrubber components made of carbon steel, 304 stainless steel, 316 stainless steel, and 316L stainless steel were inspected for corrosion. In the laboratory, loop and bench-scale tests were conducted to characterize the influence of high-chloride slurries on the same materials that were tested as spools in the field.

Various components of eight operating commercial-scale FGD systems and one demonstration system were systematically inspected for materials performance. The inspection included removal of selected component sections and subsequent metallographic analyses for susceptibility to pitting, stress-corrosion cracking, crevice corrosion, and general corrosion. The scrubber slurries typically contained 600 ppm chloride ions and 10 percent solids, with a pH range of 5 to 6.5 at 130°F. The inspections revealed that carbon steel had poor general corrosion resistance except in reheaters, where resistance was good. Some general corrosion and often severe pitting and stress-corrosion cracking were found in components fabricated of stainless steel. Components of 316 and 316L stainless steel were free of stress-corrosion cracking and had only minor pitting. The period of exposure of these materials in a commercial system was 2-1/2 years. The spool tests indicated that 309, 18-8-2, and 304L stainless steels were attacked more severely than 316L stainless steel, 317L stainless steel, 18 Cr-2Mo-Ti, Alloy 20 Cb3, and Alloy 825.

Performance, Economic, and Fabrication Considerations

In view of the test results just described, together with industrial experience, 316L stainless steel is finding widespread use as a material of construction; however, under certain conditions of scrubber liquor pH, temperature, and chloride content, this alloy can undergo localized attack. Under these more stringent conditions, nickel-based alloys with higher molybdenum and chromium content are superior to 316L stainless steel. It has been shown that resistance to stress-corrosion cracking is achieved with higher nickel contents because nickel accelerates repassivation of the metallic surface (Rhodin and Carson 1959). Although more expensive initially, these high-grade alloys may be economically justified for use in certain severe scrubber environments.

The beneficial effect of molybdenum content is shown in Figure G-2, which indicates that the resistance of alloys to pitting and crevice corrosion generally increases as the molybdenum content increases (Crow 1978). Moreover, molybdenum content alone does not ensure resistance to localized attack. Chromium content is also a consideration, as indicated in data from the Inco test program (Hoxie and Michaels 1978), shown in Figure G-3 and confirmed by the Shawnee test program results plotted in Figure G-4 (Crow 1978). These data seem to indicate that the resistance of various alloys to localized corrosion is a function of the combined molybdenum and chromium content.

The high nickel, molybdenum, and chromium contents of the high-grade alloys generally make them more expensive than 316L stainless steel. As shown in Table G-5, most of these alloys are stronger than 316L stainless steel and thus can be used in thinner plates with resultant cost reductions that reduce the price differential (Leonard 1978). Figure G-5 plots the weighted price ratio against the Shawnee test results, demonstrating the potential advantage of Nitronic 50 over 316L stainless steel (Crow 1978).

The use of high-grade alloys demands careful fabrication. Specifically, the welding recommendations of the alloy producer should be followed precisely. Failures of high-grade alloys generally occur because of faulty

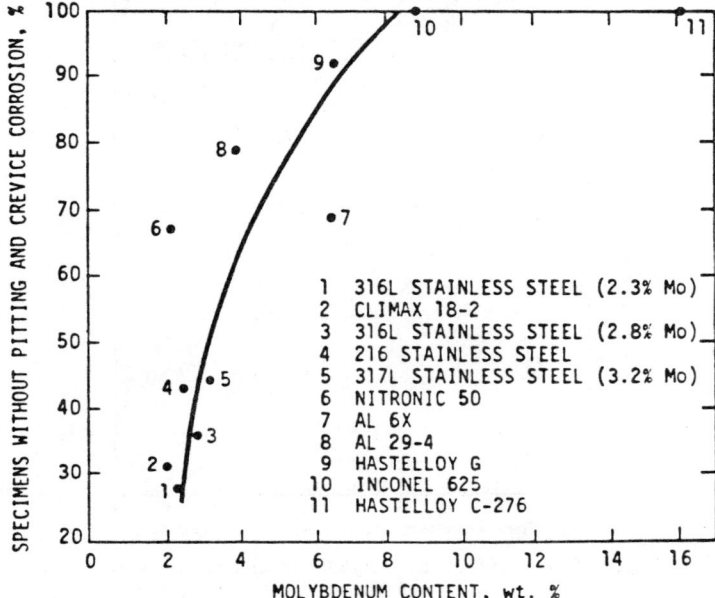

Figure G-2. Effect of molybdenum content on resistance to pitting and crevice corrosion.

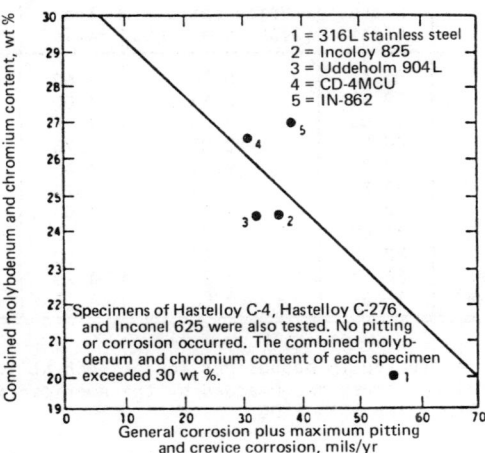

Figure G-3. Effect of molybdenum and chromium content on corrosion resistance (Inco tests).

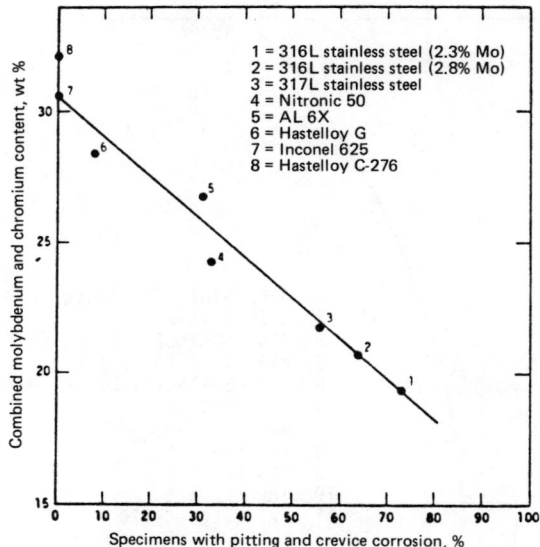

Figure G-4. Effect of molybdenum and chromium content on corrosion resistance (Shawnee tests).

TABLE G-5. PRICES OF HIGH-GRADE ALLOYS AND STAINLESS STEELS

	Price of 1/4-in. plate, $/ft^2	Price ratio (316L SS = 1.00)	Design stress ratio[a]	Weighted price ratio[b]	Weighted price,[c] $/ft^2
Hastelloy C-276	82.42	5.00	1.60	3.12	51.48
Haynes 625	59.95	3.63	1.60	2.27	37.45
Inconel 625	70.95	4.30	1.76	2.44	40.26
Incoloy 825	46.2	2.80	1.76	1.59	26.24
Hastelloy G	46.71	2.71	1.60	1.69	27.88
Haynes 20	30.36	1.84	1.20	1.53	25.24
Nitronic 50	22.28	1.35	1.60	0.84	13.86
JS-777	31.80	1.93	1.20	1.60	26.40
Uddeholm 904L	32.50	1.97	1.20	1.64	27.06
317L stainless steel	20.54	1.25	1.20	1.04	17.16
316L stainless steel	16.50	1.00	1.00	1.00	16.50

[a] Design stress is expressed in pounds per square inch at 100°F and represents the maximum operating stress recommended by the American Society for Testing and Materials. Type 316L SS = 1.00.

[b] Weighted price ratio represents the price ratio divided by the design stress ratio.

[c] Weighted price represents the price times the weighted price ratio.

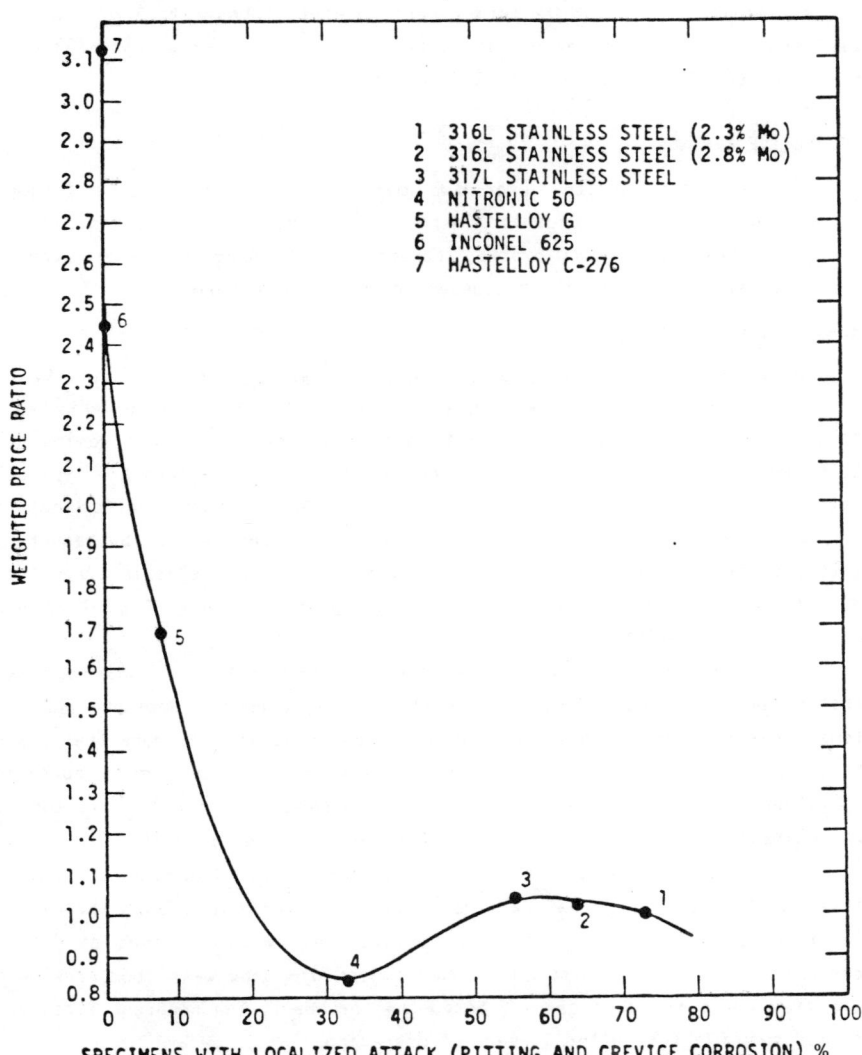

Figure G-5. Weighted price ratio and corrosion resistance.

welding rather than corrosive attack by the flue gas being treated. Failures in rotating parts can be traced to fatigue. Some alloys (Uddeholm 904L, Carpenter 20) may suffer pitting when welded. Furthermore, delivery and availability in suitable form (sheet, plates, tubes, etc.) on a commercial scale may present serious limitations with some other alloys--Allegheny 6X, Allegheny 29-4 (Uddeholm Steel Corporation 1977).

PROTECTIVE LININGS

The utility industry uses two major types of protective linings: organic and inorganic. Organic linings include resin, rubber, and plastic. Inorganic linings include bricks, ceramics, and concrete. These various lining materials are briefly discussed in the following paragraphs.

Resin and Rubber Linings

Polyester, vinyl ester, and epoxy are commonly used in utility FGD systems. Polyester resins are generally known for their excellent resistance to acid and good resistance to heat and abrasion. Vinyl resins have been improved to the point that properties of the vinyl esters are typically better than those of polyesters (Singleton 1978). Epoxy resin coatings generally have less resistance to acids than do other resins, but adhere to metals better and have higher tensile strength and good elastic properties. Bituminous and furan resins are less widely used. Table G-6 shows some physical characteristics of the resins.

Among several types of rubber liners, black natural rubber and synthetic neoprene rubber liners are most commonly used in limestone FGD systems. Natural rubber is softer and more resilient and has more tear resistance than neoprene rubber. Neoprene, however, provides more corrosion resistance and can withstand higher temperatures. Table G-7 shows some of the characteristics of both materials (Fontana and Greene 1967).

Tests of black natural rubber and neoprene rubber liners at the Shawnee test facility show that natural rubber is superior (Crow 1978). The natural liner withstood the design scrubber environment, with no signs of general corrosion or erosion. Neoprene rubber liners did show wear from erosion in the area where the flue gases entered the scrubber. The neoprene liner also formed some blisters after 3 years of operation.

Rubber liners do have disadvantages. They are susceptible to adhesion losses, mechanical damage, abrasion wear, and fire. Overheating can cause adhesion losses and exposure of the substrate to the corrosive environment. Rubber liners can be torn or cut by material in the flue gases or during operation, installation, or removal of other equipment. Natural rubber can withstand abrasion better than neoprene rubber, but neither material can

TABLE G-6. TYPICAL CHARACTERISTICS OF RESINS

	Resin alone	Resin with glass flakes	Resin with fabric mat
Furan			
Tensile strength, 10^3 psi	1.20	1.25	8.15
Coefficient of expansion, 10^5 in./in. per °F	2.0	1.40	1.50
Barcol hardness	Not reported	28	20
Temperature resistance, °F	350	125	125
Flexural strength, 10^3 psi	3.80	2.66	19.85
Abrasion resistance, Taber Wear Index	Not reported	83	57
Epoxy			
Tensile strength, 10^3 psi	1.80	3.35	3.40
Coefficient of expansion, 10^5 in./in. per °F	3.00	1.50	1.90
Barcol hardness	Not reported	40	45
Temperature resistance, °F	175	160	180
Flexural strength, 10^3 psi	3.80	6.74	9.50
Abrasion resistance, Taber Wear Index	Not reported	129	140
Vinyl ester			
Tensile strength, 10^3 psi	2.30	2.30	6.70
Coefficient of expansion, 10^5 in./in. per °F	1.60	1.50	1.50
Barcol hardness	Not reported	38	50
Temperature resistance, °F	180	160	160
Flexural strength, 10^3 psi	4.20	6.00	10.50
Abrasion resistance, Taber Wear Index	Not reported	167	185
Polyester			
Tensile strength, 10^3 psi	2.30	2.05	6.60
Coefficient of expansion, 10^5 in./in. per °F	1.90	1.50	1.50
Barcol hardness	Not reported	42	52
Temperature resistance, °F	225	160	160
Flexural strength, 10^3 psi	4.80	6.10	12.20
Abrasion resistance, Taber Wear Index	Not reported	177	187

TABLE G-7. CHARACTERISTICS OF BLACK NATURAL RUBBER AND NEOPRENE RUBBER

	Natural rubber	Neoprene rubber
Hardness range[a]	40-100	30-90
Tensile strength,[a] 10^3 psi	4.50	3.50
Maximum elongation, %	900	1000
Abrasion resistance	Excellent	Very good
Maximum ambient temperature allowable, °F	160	225
Resilience	Excellent	Very good
Aging resistance	Good	Excellent
Flame resistance	Poor	Good
Tear resistance	Excellent	Good

[a] Indicates values for soft rubber; values run higher for hard rubber.

withstand the abrasion in the venturi throat. Because rubber liners are not flame-resistant, extreme care must be exercised when welding near them.

The most important considerations in selection and use of organic coating materials include testing, surface preparation, coating application, and engineering design. As with the metallic materials, a series of tests should be conducted in both the laboratory and the field. The three variables to be considered in testing are type of coating, thickness of the coating, and degree of surface preparation. Surface preparation is probably the most crtical factor in performance of protective coatings. Two prime sources of information regarding the preparation of steel for application of a coating are the National Association of Corrosion Engineers (NACE, Houston, Texas) and the Steel Structures Painting Council (Pittsburgh, Pennsylvania). Detailed specifications for surface preparation are summarized in published literature. The advantages and disadvantages of various techniques for application of the coating after the surface is prepared are summarized in Table G-8. The key to the life and effectiveness of an organic lining surface under corrosive conditions is design. The following elements of good design can extend the performance of coating materials. Flat surfaces are preferred, and round or curved surfaces are preferred to sharp angles. Crevices should be avoided, and surfaces made smooth. For example, welds should be ground smooth, and splatter removed. Similarly, scaffolding brackets and other items should be removed, and the

TABLE G-8. COATING APPLICATION TECHNIQUES

	Positive features	Negative features
Brush	Maximum paint saving; no overspray or fogging; minimum masking; minimum capital investment	Questionable film control; requires human effort and skill; very costly on large areas (1000 square feet of coating per day can be applied)
Roller	Greater area of coverage than brush; minimum masking	Questionable film control; requires human effort and skill; costly (2000 square feet of coating per day can be applied)
Conventional	Safe; minimum film thickness available; equipment is strong; minimum solvent entrapment; pattern can be changed by adjustment of air flow; mist coats and feathering possible, parts less expensive than airless	Maximum material loss; maximum power loss; maximum heat loss and cleanup time; pressure head loss (lose 0.6 psi per foot vertical rise); control in hands of operator; compressor and pots are heavy
Airless	Maximum production; reasonable thickness control; cleaner air, less fogging; simple pattern control; simple rigging; generally applicable without thinning	Inherent problem called spitting; hand trigger control; pressure variations

surfaces from which they came should be ground smooth. Welding is preferred to riveting, and continuous welding is preferred to skip or spot welding. Lapped edges should be welded. Where rivets are used, they should be countersunk. Faying surfaces (surfaces that rub against one another) should be avoided. Flexing surfaces should be avoided or should be given special coatings that will not crack or spall. All surfaces should be well drained.

Although difficulties with organic lining materials are sometimes due to improper application and poor design, selection of inappropriate materials is a principal cause of failures. Experience in recent years has enabled designers to categorize the basic types of failures and to recommend selection of materials on the basis of specific properties. Table G-9 summarizes the types and causes of failure and the basis for appropriate selections.

Plastic Linings (Kensington 1978; Furman 1977)

Plastic linings can be made of tetrafluoroethylene (TFE, or Teflon), fiber-reinforced plastic (FRP), and Armalon. The most chemical-resistant plastic commercially available is TFE, which is unaffected by all alkalis and acids except fluorine and chlorine gas at elevated temperatures. It retains its properties up to 500°F. Fluorinated ethylene-propylene (FEP), a copolymer of TFE and hexafluoropropylene, has properties similar to those of TFE except that it is not recommended for continuous exposures above 400°F. An advantage of FEP is that it can be extruded on conventional equipment, whereas TFE parts must be made by complicated powder-metallurgy techniques. Chlorotrifluoroethylene (CFE), another derivative of TFE, also possesses excellent corrosion resistance to all acids and alkalis up to 690°F.

TABLE G-9. CAUSES OF ORGANIC LINING FAILURES AND
PROCEDURES RECOMMENDED TO PREVENT FAILURE

Type of Failure	Causes of Failure	Procedures Recommended to Prevent Failure
Chalking	Action of actinic rays from sun on basic resin; improper pigmentation	Select a coating composed of weather-resistant resins such as acrylics and high-hiding, noncatalytic pigments
Checking and cracking	Gradual coating shrinkage caused by weathering, heating, and cooling; continued polymerization or oxidation	Select coatings formulated with reinforcing pigments in addition to colored pigments
Alligatoring or mud cracking	Application of hard coating over a softer coating; continued polymerization and shrinkage of a coating from the surface toward the interior	Select a coating with high adhesion; never apply hard, tough coating such as an epoxy on a softer primer or undercoat such as a petroleum resin
Blistering	Moisture vapor transmission (MVT), osmosis, electroendosmosis	Select coatings such as epoxies and vinyls with high adhesion and low MVT
Peeling	Poor inherent adhesion, reaction with substrate, high MVT, dirty or contaminated substrate	Select a coating with strong adhesion and low MVT that is inert to substrate; assure clean surface for coating application; apply over primer with strong adhesion to substrate such as inorganic zinc primers
Undercutting	Poor inherent adhesion; high permeability to water, oxygen, and salts; smooth, nonporous substrate	Select a coating with low permeability to moisture, oxygen, and chemicals; assure maximum adhesion by clean, sandblasted, or otherwise abraded surface; apply over primer with strong adhesion such as inorganic zinc primer
Intercoat delamination	Very rapid and complete curing of coating, strong crosslinking to insolubility, contamination of surface	Select coating that is soluble in its own solvents such as vinyl lacquer; apply second coat before first coat is cured to insolubility, or before exterior contamination occurs
Abrasion or physical	Impact, continued rubbing, or abrasion from vibration; rubber-tired vehicles on floors; movement or handling of equipment	Select a coating with high abrasion resistance such as polyurethane, with very strong adhesion such as an epoxy, or with resilience and impact resistance such as a vinyl
Chemical exposure	Reaction or solution of coating caused by fumes from or contact with acids, alkalis, solvents, etc.	Select a coating for the specific chemical exposure; all ingredients should be resistant to the chemical involved; a coating test is suggested for positive results
Fungal or bacterial attack	Coating formulated with biodegradable oil, plasticizers, or resins (bacteria and fungus organisms feed on these ingredients)	Select coatings that are completely inert to biological degradation or coatings that contain permanent bactericidal or fungicidal additives
Discoloration	Sunlight, ultraviolet light, and weather; coating resins with reactive parts that break down or change in sunlight	Select coatings with light-resistant resins such as acrylics, or those containing opaque pigmentation or ultraviolet absorbers

Plastics are reinforced by addition of fiberglass, which is supplied as roving or reinforcing mat. Roving fiberglass is a rope-like bundle of continuous strands, and the reinforcing mat consists of chopped strands. Complete FRP construction is being used in FGD systems, especially for piping, tanks, and scrubber internals (liquid distributor, mist eliminators, spray headers, etc.). The most commonly used resins are epoxy, vinyl ester, furan, polyester, and chlorinated polyester. During pilot-plant tests at Shawnee (Crow 1978), 12 specimens of each of these five types were tested. All the specimens made with epoxy and vinyl esters were in good condition after the test. Of the specimens made with furan, 11 were in good condition and 1 in fair condition; of those made with polyester, 8 were in good condition and 4 in fair condition; and of those made with chlorinated polyester, 2 were in fair condition and 10 in poor condition.

The basic causes of FRP failure include lack of specifications and inspection, poor workmanship, incorrect resin selection, mishandling of equipment during shipment or installation, and use of equipment beyond rated conditions (Kennington 1978). It is recommended that Standard PS 15-69 of the National Bureau of Standards (a product of the joint effort of the resin suppliers, the fabricators, and the Society of Plastic Industry) be set as the minimum specification. This standard may be modified to satisfy particular requirements for obtaining quality FRP equipment. The bids must be evaluated on the basis of a thorough review of the supplied design. Plant inspection of equipment before shipment is advisable.

The purchase cost of FRP is less than that of metal. In addition, FRP weighs less than metal alloys and thus reduces freight and installation costs. Unlike steel structures, FRP constructions are not usually degraded by exterior corrosion. The maintenance cost for exterior protection is therefore reduced. In spite of these advantages, FRP has some limitations. The temperature restriction, based on currently available resins, is 250°F. Under current American Society of Mechanical Engineers (ASME) Codes, FRP tanks are not available for use at pressure above 1 atmosphere. Although FRP products have good chemical resistance, FRP is not recommended for use in the presence of nitric acid, hydrofluoric acid, sodium hydroxide (30% or more), and sulfuric acid (50% or more).

Despite outstanding physical properties that make them desirable corrosion barriers, TFE and FEP cannot be bonded to a metallic tank wall readily, cannot be used under vacuum or in conditions of agitation, and are very expensive. A recent composite material, designated Armalon, is being developed for production of equipment with an inner surface of TFE and an outer surface of FRP (Furman 1977). The essential features of Armalon are its integral TFE-fabric bond and its ability to form a tight bond with either FRP or metal. Armalon is claimed to allow service under vacuum or agitation and under extreme thermal cycling without linear delamination. Armalon/FRP is reported to have withstood combinations of hydrochloric acid, hydrobromic acid, and traces of hydrofluoric acid at 194°F for over 1 year with no signs of degradation. It also handled sodium hydroxide (20%) and sulfuric acid (10%) at 176°F for 6 months with no apparent deterioration. The material is claimed to be inert at temperatures up to 300°F, in wide variations of pH, and in the presence of chlorides. Also, the initial price of Armalon/FRP equipment is claimed to be very competitive with that of Hastelloy G equipment (Furman 1977).

Bricks (Brova 1977; Mellan 1976)

Inorganic linings can be made of bricks, ceramics, and concrete. The types of bricks most commonly used in FGD systems are red shale, fire clay, and silicon carbide. Each type of brick has limitations that restrict its use. Red shale should be used where minimum permeation of liquor through the brick is required and thermal shock is not a factor. Fire clay should be used where thermal shock is a factor and minimum absorption is not required. Silicon carbide brick should be used where high abrasion-resistance is required.

In the venturi throat, silicon carbide brick in conjunction with furan resin mortar has proved to be a suitable construction material. It can withstand abrasion caused by fly ash in the flue gases. Fire clay brick can be used above the mist eliminator and at the inlet to the scrubber. In the scrubber inlet, fire clay brick with a furan resin mortar is recommended because slurry from the sprays does not contact the gases and because mist in the flue gases is minimal above the mist eliminator. Red shale brick can be used in the main body of the scrubber. This section is normally in contact with the slurry, and the temperature of the flue gases is reduced. A furan resin should be used as the mortar lining.

Brick alone will not prevent the scrubber shell from corroding. An impervious membrane must be applied between the brick and the scrubber shell. The purpose of the brick is to protect the membrane from abrasion and excessive heat. The membranes are made from vinyl resins, natural and synthetic rubbers, or asphaltic materials.

Ceramics (Gleekman 1978)

Selection of ceramics for construction is governed by their physical properties, which result in objects of relatively thick cross section, heavy weight, and lack of resistance to impact. Because ceramic shapes are usually made by extrusion or casting, the shapes that are symmetrical (such as nozzles, cylindrical scrubbers, ductwork, and piping) are most suitable, particularly on a production basis. Because vitrification is required to achieve maximum resistance, the limitation on size is that of the furnace used to vitrify the object. It is virtually impossible to make adaptations, changes, or repairs in the field that will provide resistance equivalent to that of the as-fired material.

Because ceramic is composed of siliceous material, the shapes are susceptible to corrosion or deterioration in hot alkaline media and in certain inorganic acids, particularly concentrated hydrochloric and hydrofluoric acid. Fluorine compounds rapidly attack the glass bond in ceramics and break down the body structure. At high temperatures, concentrated sulfuric acid and steam attack ceramics. Though ceramics are fired at 2200° to 2400°F, the operating temperature to which they are subjected is of utmost importance to service life because of their low heat transfer rate and resultant poor resistance to thermal shock. The temperature differential across the thickness of a ceramic item should not be greater than 90°F.

Armorizing is used with ceramics as well as with many other materials, to improve the resistance of the material to impact. Armorizing generally consists of applying polyester or epoxy resin reinforced with glass cloth or mat to the exterior of the nonmetallic material. In tower construction, ceramic shapes are joined with bell and spigot joints as well as bolted flange joints. Because of the fragility of ceramic materials, care must be taken to ensure the proper use of torque wrenches with flanged joints and to observe the pressure limitations of bell and spigot joints.

Concrete (Gleekman 1978)

Concrete is made up of a fused and ground clay-limestone product with an aggregate. Concrete need not be fired as a formed shape to achieve strength and resistance, which result from hydration, usually as a function of time and temperature. Concrete is usually reinforced with steel and more recently with fibrous materials to meet strength requirements. In limestone FGD systems, concrete is used in presaturators, tanks, and piping.

Concrete has obvious disadvantages of porosity and susceptibility to attack by alkaline and certain acidic media. For this reason, much attention is given to protective coatings, including thin films of liquid such as silicones and penetrating oils. As with ceramics, weight, poor resistance to thermal shock, and joining problems must be considered.

Some concretes have excellent thermal resistance and can be used at relatively high temperatures without danger of spalling. An additional advantage is that concrete can be gun-applied and does not have to be cast. In the broad category of Gunite, selection of the cement and the aggregate can produce variations in properties of the material. Often, Gunite is used for temporary repairs of structures such as packed towers because operators cannot afford the long cure cycles needed with ordinary materials of construction.

OPERATING EXPERIENCE (PEDCo Environmental 1981; Rosenberg 1980)

The balance of this appendix reviews currently available information on operating experience with construction materials for limestone FGD systems. This review first discusses major equipment items in the gas circuit, including prescrubbers, scrubbers, mist eliminators, reheaters, fans, ducts, dampers, expansion joints, and stacks. It then examines items in the slurry and solids circuit, including pumps; storage silos; ball mills; spray nozzles; piping; spray headers; valves; tanks; thickeners, vacuum filters, and centrifuges; agitators and rakes; and pond linings.

Prescrubbers

Included in the category of prescrubbers are quenchers, presaturators, and venturi scrubbers. All these items of equipment wet and cool the gas. In addition, venturi scrubbers effect particulate removal (primary or secondary) and some SO_2 removal.

Because of hot and possibly erosive operating conditions in quenchers and presaturators, organic linings have been used with a concrete protective layer (Marion 4). Other materials used include hydraulically bonded, concrete-lined carbon steel (Petersburg 3, Winyah 2, and Southwest 1), Carpenter 20 (Tombigbee 2 and 3), and Uddeholm 904L (Dallman 3). Petersburg 3, Southwest 1, and Dallman 3, all of which use high-sulfur coal, have reported materials problems. Petersburg 3 has experienced erosion of the concrete layer after less than 1 year of operation. At Southwest 1, the concrete-lined carbon steel has been replaced with Uddeholm 904L, which performed adequately for only a limited period of time. After approximately 2 years, patching with Hastelloy G was necessary. Following these repairs, further deterioration of the Uddeholm 904L was observed. Additional corrective actions have not yet been determined. At Dallman 3, the presaturator is also constructed of Uddeholm 904L. Pitting caused by attack from chloride was observed shortly after startup. Chloride levels of 15,000 ppm were measured at the time the pitting was first noticed.

The hot, wet, erosive environments of venturi scrubbers require erosion- and corrosion-resistant materials. Mat-reinforced epoxy phenolics covered with hydraulically bonded concretes are used at Will County 1 and La Cygne 1. Organic linings covered with prefired ceramic bricks or blocks are used in the venturi throats at Cholla 2, Will County 1, La Cygne 1, and Widows Creek 8. No major problems have been reported with these materials, which have operated up to 8 years at Will County 1.

Organic linings used without protective coatings (Cholla 1 and 2, Sherburne 1 and 2, Widows Creek 8) have failed because of disbonding. Concrete linings without organic linings underneath have worked well at Winyah 2 (hydraulically bonded) and Widows Creek 8 (silicon carbide castable).

Unlined 316L stainless steel has been used at the modified scrubbers on Lawrence 4 and 5, Cholla 1, Sherburne 1 and 2 (throat), and Winyah 2 (wear plate). The chloride level (2000 ppm) in the Cholla 1 system has caused rapid attack of the 316L stainless steel and prompted a switch to mat-reinforced, epoxy-lined mild steel (vessel) and Hastelloy C (externals) for Cholla 2. The chloride attack problem at low pH levels is leading to the selection of alloys more resistant to corrosion (e.g., 317L stainless steel, Uddeholm 904L, and Incoloy 825) for scrubbers now in the design stage.

Scrubbers

Various designs and many materials have been used in scrubbers. Virtually every scrubber incorporates a combination of materials. Major types of material have included 316L stainless steel, rubber-lined carbon steel, organic-lined carbon steel, and ceramic-lined carbon steel.

The scrubbers at Cholla 1, Duck Creek 1, and La Cygne 1 and the modified scrubbers at Lawrence 4 and 5 are constructed primarily of 316L stainless steel. This material typically requires a higher initial investment than lined carbon steel, but may require less maintenance and downtime. Weld patching and replacement of new plates can typically be accomplished with more ease than lining repair or reapplication. Components made of 316L stainless steel (e.g., trays, plates, supports, and fasteners) have been used in many other scrubbers. High chloride concentration, however, can present a corrosion problem, and abrasion by scrubber slurries can sometimes cause wear failure. The operating experience thus far ranges from 1-1/2 years (Lawrence 5) to 6-1/2 years (La Cygne 1).

The only major problems reported to date occurred at Cholla 1, where corrosion and pitting of the 316L stainless steel occurred because of the relatively high chloride content (2000 ppm) of the scrubbing liquor. Also, some patches have been made in the stainless steel sidewalls of the new scrubbers at Lawrence 4 and 5.

Rubber-lined carbon steel costs less than stainless steel, but more than carbon steel protected with organic linings. Both natural and synthetic rubbers (neoprene and chlorobutyl) have been used. Natural rubber is typically superior to neoprene in both abrasion and chemical resistance, but synthetic rubber is less flammable. The rubber is applied in sheets, which are bonded to the steel with adhesive. Care must be taken in lapping the rubber because the laps are the areas most subject to failure by debonding. Thus, it is extremely important that the rubber be lapped so that liquids flowing over the surface do not get under the lap. Disadvantages in the use of rubber linings are the difficulty of repair and the possibility of fires caused by welders' torches. Besides having excellent resistance to chemicals in the gas and slurry, rubber is outstanding in abrasion resistance. Because rubber provides such a high degree of abrasion resistance, it might be used to advantage in localized, high-abrasion areas in scrubbers. This has been accomplished at Tombigbee 2 and 3, where a natural rubber lining is used in the spray zone.

The scrubbers at Petersburg 3, Southwest 1, and Widows Creek 8 are constructed primarily of neoprene-lined carbon steel; the scrubbers at R. D. Morrow 1 and 2 are constructed of chlorobutyl-lined carbon steel; and the scrubbers at Will County 1 and Winyah 2 are constructed primarily of natural-rubber-lined steel. After 7000 hours of operation at Widows Creek 8, the neoprene lining in the tapered hopper bottom portion of the scrubbers disbonded and was replaced with 316L stainless steel cladding. Also, sparks from a welder's torch caused the neoprene lining to catch fire at Widows Creek 8. The longest operating experience thus far has been roughly 8 years at Will County 1, where the original linings are still in service.

Organic-lined carbon steel has the lowest initial cost of construction. The performance of lined carbon steel has varied from satisfactory to poor. Reasons for the variation in performance include selection of marginal linings and improper application of linings, or inadequate surface preparation. Linings selected for some first-generation limestone FGD systems were poor. Little experience was available regarding linings in SO_2 scrubbers when these first systems were constructed. Wear resulting from abrasive slurries was a common cause of failure. In some cases, the problem was solved by relining with a more abrasion-resistant material such as a mat-reinforced epoxy lining in place of a glass-flake-filled polyester lining. Blistering and disbonding of the linings have also occurred in some cases.

Several scrubbers are presently in operation with their original organic linings, which include mica- and glass-flake-filled polyesters and mat-reinforced linings. The only problems reported were at Sherburne 1, where some patching of the mica-flake-filled lining was required.

Some organic linings in scrubbers have required replacement. The original scrubbers at Lawrence 4 and 5 were relined several times with various linings, none of which provided both the abrasion resistance and chemical resistance needed. The modified scrubbers were constructed of 316L stainless steel. A mica-flake-filled polyester lining at Sherburne 2 disbonded between layers, resulting in bubbles and blisters. The lining was replaced with an inert-flake-filled epoxy. At Hawthorn 3 and 4, the glass-flake-filled polyester lining failed because of temperature and pH excursions, and because of damage from welding. The scrubbers were relined with 316L stainless steel cladding. At R. D. Morrow 1 and 2, the original glass-flake-filled polyester lining failed because of an accumulation of spot failures and pinhole leaks. The scrubbers were relined with chlorobutyl rubber, which has held up well since application.

Although linings have been used in many scrubbers, their time in service in most instances has been too short to obtain an effective appraisal of their performance. From the information available at present, the following conclusions are possible regarding the use of organic linings:

High-quality lining materials should be selected. The minimum requirements are trowel-applied, glass-flake-filled polyester linings of 80-mil nominal thickness in areas subject to normal abrasion and heavy-duty, fiber-mat-reinforced materials of 1/8-inch nominal thickness with abrasion-resistant fillers in areas subject to high abrasion (wherever slurry is projected against linings).

Lining materials must be applied by skilled, experienced applicators who will stand by their work.

Applicators must understand metal surface preparation procedures and use them properly.

Careful quality control procedures must be used.

Although very few scrubbers have been constructed predominantly of ceramic-lined carbon steel, many have used some type of ceramic lining at specific high-abrasion areas such as venturi throats and sumps. The use of ceramics has been limited because of high cost, high weight, and occasional problems caused by brittleness. The bottom of the scrubber sumps at La Cygne 1 and Winyah 1 are lined with hydraulically bonded concrete. Acid-resistant brick is used to protect the bottom section of the scrubbers at Huntington 3 (interval effluent hold tanks). Although good performance has generally been reported for the ceramic linings, the experience to date is too limited in most instances to draw reliable conclusions regarding long-term performance and maintenance requirements.

Mist Eliminators

Practically all mist eliminators in use are of the chevron type with a variety of vane shapes. The vanes are most often constructed of some form of plastic with or without fiberglass reinforcement, but alloys have also been used. In general, mist eliminators have been satisfactory from the materials standpoint. Because of plugging problems, mist eliminators require frequent cleaning plus spray washing on a continuous or intermittent basis. Personnel often break FRP mist eliminators during cleaning by striking them with hammers or walking on them. This breakage can be prevented by care in cleaning and use of sufficiently thick FRP. Also, unreinforced plastic mist eliminators are subject to warping, sagging, and melting during temperature excursions.

Base metal mist eliminators are used at Duck Creek 1 (Hastelloy G) and Widows Creek 8 (316L stainless steel). The only problem reported is at Widows Creek 8, where mud deposits lower pH and thus cause chloride corrosion. An FRP mist eliminator, however, would not have survived the cleaning that has been necessary.

Reheaters

If flue gas is not reheated before leaving the scrubber, acid condensation can occur in the downstream ductwork, fan (if present), and stack. The main types of reheat are in-line, indirect, and bypass.

If an in-line reheater is used, it is installed either in the scrubber or in the discharge duct after the mist eliminator. The heat medium is either steam or hot water. The materials of construction for the tubes or coils range from carbon steel to Inconel 625. Acid corrosion from sulfuric acid condensation is a problem with carbon steel, and stress corrosion from chlorides is a problem with stainless steels and alloys. Therefore, the severity of corrosion problems with carbon steel is related to the concentration of SO_2 in the scrubbed flue gas, whereas the severity of corrosion problems with stainless steels and alloys depends on the chloride content of scrubbing liquor or wash water that may be carried over from the mist eliminator. Plugging of the tube banks is also a problem, particularly if finned tubes are used. Soot blowers are required to maintain performance with in-line reheaters.

Circumferential-finned carbon steel tubes containing hot water under pressure have been used for heat at Lawrence 4 and 5, Sherburne 1 and 2, and Hawthorn 3 and 4. Some tube failures occurred at Lawrence 4 and 5 after 6 years of operation, but there have been no serious problems since the station switched to firing low-sulfur coal, presumably because of less SO_2 in the scrubbed flue gas. At Sherburne 1 and 2, which also burn low-sulfur coal, four weld failures occurred early in the operation because of excessive stress. Only infrequent leaks have occurred during the past 3-1/2 years, since the stress was removed. The tubes at Hawthorn 3 and 4, which burn a blend of low- and high-sulfur coal, have not experienced corrosion problems, although plugging with deposits caused them to be replaced with smooth carbon steel tubes to facilitate cleaning.

At Will County 1, steam in smooth tubes is used for reheat. The top tube banks were originally Cor-Ten, and the bottom tube banks were originally 304L stainless steel. Leaks developed in 6 months, and the original tube banks were replaced with carbon steel banks at the top and 316L stainless steel banks at the bottom. Pinhole attack on the carbon steel requires tube replacement approximately every year, and stress-corrosion cracking of the stainless steel requires tube replacement approximately every 1-1/2 years. No remedial measures are planned because the flue gas cleaning system, which is now used primarily for particulate control, may be replaced by an electrostatic precipitator.

Smooth 316L stainless steel tubes containing steam are also used for reheat at Cholla 1 and La Cygne 1. At Cholla 1, the tubes needed replacement after 6 years of operation because of corrosion. At La Cygne 1, the tubes are a replacement for the original 304 stainless steel, which failed because of corrosion. The 316L tubes have a service life of at least 2 to 3 years at La Cygne 1 before replacement is required.

Inconel 625 tubes with Uddeholm 904L baffles are used at Cholla 2. No corrosion problems have occurred after 1-1/2 years of operating experience.

In the indirect method of reheat, air is heated separately from the scrubber system. Thus, corrosion and fouling problems are circumvented. Carbon steel steam coils have been used at Petersburg 3 and Widows Creek 8 (where tubes have copper fins) without any materials problems. The longest operating experience thus far is 2-1/2 years at Widows Creek 8, where operation is continuing.

When the entire flue gas output does not have to be scrubbed to meet emission regulations, partial bypass can be used as a reheat method. This approach has been used at Tombigbee 2 and 3, Jeffery 1 and 2, Coronado 1 and 2, Winyah 2, and R. D. Morrow 1 and 2. Materials problems associated with bypass reheat are discussed in the subsection on ducts.

Fans

The majority of the fans used in limestone FGD systems are located upsteam of the scrubber. These forced draft fans operate on hot flue gases and can be constructed of ordinary carbon steel. Cor-Ten is used at Cholla 1, Hawthorn 3 and 4, and Lawrence 4; and stainless steel blades are used at Cholla 2. In some cases these hot-side fans have been eroded by the fly ash in the flue gas stream, particularly if the electrostatic precipitators are not operating at maximum efficiency. This has occurred at Widows Creek 8, where erosion has required rotor replacement every 10 weeks even though there are chromium carbide wear plates.

Several limestone FGD systems have the fans installed downstream of the scrubber. Corrosion-resistant alloys are required if these induced draft fans are exposed to wet flue gas at the scrubber exit temperature. If the fans are located after the reheater, however, less-resistant materials are used, particularly if the exit SO_2 concentrations are low. Fans are located after reheaters at Will County 1, La Cygne 1, and Sherburne 1 and 2. The fans at Will County 1 are constructed of Cor-Ten, and the other fans are constructed of carbon steel. Some polymer coatings and Inconel 625 rotors have been used at La Cygne 1, where corrosion and erosion problems have been encountered since startup. These fans are also unique in that they are washed every 4 days to remove deposits caused by particulate carryover from the scrubbers.

Ducts

Inlet and bypass ducts are generally not major problem areas for scrubbers. The outlet duct, however, has been a major problem area, particularly for units with duct sections that handle both hot and wet gases. These sections are for the most part beyond the bypass junction on units without reheat. Acidic conditions become more severe in systems with bypass reheat as the temperature is raised and other corrosive species in the unscrubbed flue gas (chlorine and fluorine) are introduced.

Carbon steel or Cor-Ten is used as the material of construction at the inlet duct of all installations. Hawthorn 3 and 4 were originally designed for boiler injection of limestone with simultaneous removal of SO_2 and fly ash in a marble-bed absorber. The inlet ducts have a Gunite-applied refractory concrete lining to protect the Cor-Ten steel from potential erosion caused by the high fly ash loading in the flue gas.

Outlet duct materials include unlined carbon steel (in the reheated zone of Sherburne 1 and 2 and the breeching of the modified Lawrence 5), Hastelloy G (at Duck Creek 1), and Inconel 625 (at Marion 4). Material selection depends on such factors as the location and use of reheaters and the type of coal burned. At most stations, outlet ducts are exposed to different flue gas conditions; i.e., before and after the reheater and/or before and after the bypass junction.

Systems with wet outlet duct sections predominantly use organic linings to protect the carbon steel or Cor-Ten structure. An exception is the use of Hastelloy G in the outlet duct at Duck Creek 1, where pitting has been severe enough to cause penetration of the duct walls. Deluge sprays are used in the ductwork before the stack to protect the organic lining on the flue, and condensate accumulates in the outlet duct, which has a low pH. There have been some lining failures. At R. D. Morrow 1 and 2, one glass-flake-filled polyester lining was replaced with another glass-flake-filled polyester lining from a different manufacturer. The second lining also failed and was replaced with Hastelloy G cladding, which was applied to the outlet duct, the scrubbed-gas/bypass-gas mix zone (partial scrubbing and bypass reheat are used), and the breeching. Failure of the Hastelloy G also occurred in some spots in the gas mix zone, resulting in replacement with Hastelloy C-276 (applied over the top of the Hastelloy G). Subsequent operation with the Hastelloy C-276 indicates that it too is failing (because of pitting from chloride attack). Southern Mississippi Electric Power is presently considering the use of an organic or inorganic lining at R. D. Morrow 1 and 2. A glass-flake-filled polyester at Southwest 1 was replaced by inert-flake-filled vinyl ester. For the most part, however, organic linings are providing several years of service regardless of whether high- or low-sulfur coal is fired. The longest operating life of an organic lining has been 8 years (for a mat-reinforced epoxy phenolic at Will County 1). Fluoroelastomer linings are used at Apache 1 and 2 and Tombigbee 2 and 3.

Stations with outlet duct sections exposed only to reheated gas have had a good record of materials performance. Problems have been reported only at Module B of Cholla 1, where flue gas is not treated for SO_2 removal (limestone slurry is not circulated through Module B) and water has a high chloride content. Module B is operated for only particulate removal because there is no electrostatic precipitator and the SO_2 emission regulation can be met with one module on line. No problems have been reported with a similar mica-flake-filled polyester lining on the Module A where SO_2 is scrubbed. Because the service lives of unlined carbon steel and Cor-Ten are 6-1/2 to 8 years at units firing high-sulfur and low-sulfur coals (La Cygne 1 and Will County 1), organic or inorganic linings may not be required (except as a precautionary measure) in outlet duct sections exposed only to reheated gas.

Although outlet duct sections handling both reheated gas and hot gases have fewer problems than those handling both wet and hot gases, a few problems have occurred. Severe corrosion problems were experienced with the carbon steel outlet ducts of the original Lawrence 4 and 5 FGD systems. Cor-Ten, however, has been used successfully for more than 7 years at Will County 1. The failure of a glass-flake-filled polyester lining at Petersburg 3 may be related to the large size of the unit. Other organic linings that have been used with no problems in outlet duct sections handling both reheated and hot gases include fluoroelastomers (for 1-1/2 years at Tombigbee 2 and 3 and 1 year at Apache 1 and 2) and an inert-flake-filled vinyl ester (for 1-1/2 years at Cholla 2). Inorganic linings (hydraulically bonded concretes) have thus been used successfully for 7 years at Hawthorn 3 and 4 and for 1-1/2 years at Huntington 3.

Dampers

Mechanical problems with dampers, caused by deposition of solids from the flue gas, have outweighed any materials problems. Dampers are utilized in three locations: the inlet duct to the scrubber, the outlet duct from the scrubber, and the bypass duct from the air preheater or dry particulate collector to the stack. Some installations have no bypass duct, and all of the flue gas goes through the scrubber system.

The inlet dampers used in the limestone systems are about evenly divided between guillotine and louver types, with a few butterfly dampers in use. Because inlet dampers are subjected to the hot, relatively dry flue gases ahead of the scrubber, such dampers can be constructed of unlined carbon steel. This material is used at several sites, particularly those where SO_2 concentrations are relatively low. Damper seals, however, are usually made of base metals, such as 316L stainless steel, Hastelloy G, or Inconel 625. Corrosion of carbon steel inlet dampers has been reported only in the original systems at Lawrence 4 and 5. The corrosion was probably caused by acid condensation resulting from the high concentration of SO_2 (high-sulfur coal was burned at the time). The modified systems at Lawrence 4 and 5 use Cor-Ten inlet dampers (low-sulfur coal is presently burned).

The 316L stainless steel damper seals used in conjunction with carbon steel or Cor-Ten suffered corrosion damage at Widows Creek 8 as a result of high chloride concentrations, and the seals were successfully replaced with Inconel 625. In spite of hot, relatively dry conditions at the inlet, most systems have inlet dampers constructed of Cor-Ten or clad with stainless steel to avoid possible corrosion. Inlet dampers made of 316L stainless steel have operated without major problems at Cholla 1 for 6 years, and inlet dampers made of Hastelloy G have operated without major problems at Duck Creek 1 for 3-1/2 years.

Dampers on the scrubber discharge are subject to the wet flue gas and thus are often made of stainless steels, either as a cladding or for the entire construction. Duck Creek 1 has Hastelloy G outlet dampers, and R. D. Morrow 1 and 2 have Hastelloy G cladding on the outlet side of the scrubber discharge dampers. The original carbon steel outlet dampers at Lawrence 4 and 5 corroded and were replaced with Cor-Ten equipment. Carbon steel outlet dampers at La Cygne 1 also corroded and were replaced with 316L stainless steel equipment. As a result of high chloride concentrations, 316L stainless steel dampers at Southwest 1 corroded badly in 3 months and were replaced with an Uddeholm 904L frame, Inconel 625 seals, and Hastelloy G fasteners. To date, the replacement has been reported as satisfactory.

Where bypass dampers are used, they are generally constructed of the same materials as the outlet damper, because they can be exposed to the wet flue gas. An exception is at Apache 1 and 2, where Inconel 625 cladding is used on the wet side of the bypass damper, as compared with 317L stainless steel on the outlet damper. No serious corrosion problems have been reported with bypass dampers; the chief difficulties have been clogging of seals and problems of mechanical operation.

Expansion Joints

Expansion joints are generally U-shaped and constructed of an elastomer with fabric (fiberglass or asbestos) reinforcement. Some metal bellows expansion joints are also used. The major problem with wet-side expansion joints has been to choose the proper metal for attachment, rather than the fabric. The kind of expansion joint selected depends upon the highest temperature to be encountered. Suppliers suggest fabric-reinforced neoprene at 250°F or below, fabric-reinforced chlorobutyl rubber at 300°F or below, fabric-reinforced fluoroelastomer at 400°F or below, and layered asbestos above 400°F.

Metal expansion joints have been used successfully in some cases, especially under dry conditions. Even stainless steels, however, can pose problems if condensate contains high concentrations of acid or chloride. Several FGD installations have replaced metal expansion joints with elastomer joints because of corrosion problems.

Stacks

The performance of a stack lining depends on whether the scrubbed gas is delivered to the stack wet or reheated, and whether the stack is also used for hot bypassed gas. These factors appear to affect the performance of lining material more strongly than differences in coal sulfur content, lining application techniques (e.g., surface preparation or priming), operating procedures (e.g., thermal shock), design aspects (e.g., annulus pressurization, stack height, and stack gas velocity), and other factors. Almost all units with bypass capability have a junction in the duct leading to the stack where bypass and scrubbed gases are combined. The exceptions are Winyah 2 and Southwest 1, which bypass gas directly to the stack.

At Duck Creek 1, there is no reheat, and the bypass gas is quenched to minimize stack temperatures. Problems with the mica-flaked-filled polyester lining have occurred in the lower portion of the wet stack. No problems, however, have been encountered in the upper part during 3-1/2 years of service; for 1 year, high-sulfur coal was fired. Quenching the bypass gas may have contributed to good lining performance in the upper part of the stack.

La Cygne 1 is the only operating limestone scrubber that fires high-sulfur coal and reheats all gas (no bypass capability is available). An inert-flake-filled vinyl ester stack lining is beginning to disbond after 2 years of service, as did a previous flake-filled polyester lining. At Sherburne 1 and 2, which fire low-sulfur coal and reheat all gas, unlined Cor-Ten flues have given good service for 3-1/2 years. No stack failures have been reported.

Like most stations, Petersburg 3 reheats scrubbed gas and has emergency bypass capability. This station, which fires high-sulfur coal and has organic stack linings, has experienced lining failure during bypass operation. Two trowel-applied coats of a glass-flake-filled polyester are being tried; results are not yet available. Stations burning high-sulfur coal and not experiencing stack problems have stacks lined with either acid-resistant brick and mortar or inorganic concrete (normally Gunite). The hydraulically bonded concrete mixes that have been used contain calicum aluminate cement as the bonding agent. This type of cement can withstand mildly acidic conditions (pH > 4) and is commonly used in refractory concrete mixes exposed to temperatures of 500°F and greater (which cause portland cement to dehydrate).

Stacks lined with either acid-resistant brick and mortar or inorganic concretes (including a chemically bonded concrete) have experienced no major problems at stations burning low-sulfur coal and providing reheat and bypass. Although the experience time is short, at least four organic lining materials are being used by such stations. Two of these materials are fluoroelastomers (at Apache 2 and 3), one is a glass-flake-filled polyester (at Winyah 2), and one is an inert-flake-filled vinyl ester (at Cholla 2). The lining at Cholla 2, however, has never been exposed to bypass conditions. At Winyah 2, where one-half of the flue gas is always bypassed to the stack, some spot failures have occurred in the lining below the scrubbed gas breeching and above the bypassed gas breeching. These spots are scheduled for repair with the same inert-flake-filled vinyl ester.

Acid-resistant brick linings have exhibited good performance regardless of operating conditions. In spite of this success, however, their application is limited because acid can penetrate brick, mortar, or brick-mortar interface under wet conditions. A possible remedy to this limitation is the pressurization of the stack annulus.

Hydraulically bonded concretes have been successfully used under dry conditions for up to 8 years by stations burning high-sulfur coal. Chemically bonded mixes containing silicate binders are generally considered to be more acid-resistant than hydraulic cement; yet none is in use where high-sulfur coal is fired. Also, the weight of relatively thick (1-1/2-inch) inorganic concrete linings could make them unsuitable for use on steel flues designed for organic linings. The major limitation of concrete materials is the risk of substrate corrosion in the event of acid condensation. Although even cracked material will provide a physical barrier to minimize acid transport to the substrate, membrane backing or some other means of providing secondary substrate protection is desirable.

Organic linings have been primarily based on polyester or vinyl ester resin binders, both of which have deteriorated sometimes when high-sulfur coal is fired and commonly when bypass reheat is used. Resins more resistant to the 300°F bypass temperatures and hot acid conditions appear necessary to lengthen the service life of organic linings. The fluoroelastomer linings used in newer stacks may significantly improve the performance record of organic linings. Whether the new materials will be able to provide reasonable service lives in hot and wet environments (which are typical of stack environments at many stations burning high-sulfur coal) remains to be seen.

Although high-grade nickel alloys have not yet been used as stack liners, they are being seriously considered for wet stack applications by several utilities to avoid loss of generating capacity from unplanned stack failures. These alloys are already being used in the stack breeching area of outlet ducts, and in the bypass duct downstream of the exit damper. As yet, however, they are unproven at stacks that alternately handle wet gas and hot gas.

Pumps

Pumps are used for a variety of services in FGD systems such as slurry feed to the scrubber spray headers, slurry transfer, and clear water transfer. Rubber-lined centrifugal pumps are commonly used for slurry recirculation with generally satisfactory results. Ordinary carbon steel pumps are usually used for clear water transfer. Stainless steel pumps also find substantial usage, especially where small capacity pumps are required. The experience with pumps has varied widely in limestone FGD systems. Some utilities report that pumps have given little problem; others cite problems with slurry pumps as the greatest difficulty in keeping systems operating. Although the reasons for these differences are not obvious in all cases, some causes of rubber lining failure are:

> Poor-quality lining
> Foreign objects in slurry (which cause mechanical damage)
> Cavitation (because of dry operation)
> Overly abrasive slurry (containing large particles)
> Overworked pumps (operating too near maximum capacity)
> Undersized pumps

Foreign objects have damaged the rubber linings in scrubber feed pumps at Huntington 3. Frequent loosening of the pump throat liners occurs at La Cygne 1 because of cavitation when the tank level is not maintained. At Duck Creek 1, the original natural rubber pump linings were replaced with neoprene after numerous failures.

Storage Silos

Because limestone storage silos are subjected to alkaline conditions, carbon steel is a satisfactory material of construction. All limestone scrubber installations have carbon steel silos except Widows Creek 8, which has a concrete structure. Southwest 1 uses a 10-gauge lining of 304 stainless steel in the bottom cone to provide a low coefficient of friction. There have been no problems with the storage silos.

Ball Mills

Most limestone FGD systems have ball mills to grind limestone feed to the desired particle size. Ground limestone is purchased for Cholla 1, but Cholla 2 has a mill. All ball mills have carbon steel shells. Although a few are unlined, most have rubber linings. La Cygne 1 is unique in that the mill has a Ni-Hard lining. About two-thirds of the lining has required replacement, but the reason for the failure of the Ni-Hard is not known. The rubber-lined ball mill at Widows Creek 8 required relining after 4000 hours.

Spray Nozzles

A wide variety of materials, ranging from plastic to extremely hard alumina and silicon carbide, has been used for spray nozzles. Wear, plugging, and installation problems are the only difficulties reported with nozzles. Wear problems have occurred only with plastic nozzles (Duck Creek 1, Hawthorn 3 and 4, and the original Lawrence 4 and 5 systems) and metallic nozzles (Cholla 1, Hawthorn 3 and 4, La Cygne 1, Southwest 1, and Widows Creek 8). Almost 8 years of service, however, has been obtained from 316L stainless steel nozzles at Will County 1. Wear problems have not been reported with any ceramic materials; and in spite of greater care required for installation, ceramic nozzles are commonly used for slurry service in the newer scrubber systems. Both silicon carbide and alumina materials generally provide good erosion resistance. Cost, availability, and design considerations will probably influence the selection of specific materials more than any difference in wear resistance. Stainless steel nozzles appear to be preferred for mist eliminator service, but ceramic nozzles are also used.

Piping

The piping used in limestone FGD systems is required to handle alkaline makeup slurry, recycle and discharge slurry, and reclaimed water. The type of material selected depends to a large extent on the type of service encountered. The piping that handles alkaline makeup slurry does not require the acid resistance needed for recycle and discharge slurry piping, which is subjected to the most severe service condition. The reclaimed water piping is not subjected to the highly erosive conditions encountered by slurry piping.

Rubber-lined carbon steel piping is most commonly used to deliver alkaline makeup slurry and has provided generally good service. Lining wear has occurred in high-velocity regions at La Cygne 1 and Winyah 2, and synthetic rubber linings appear to be giving better service than natural rubber linings in delivering makeup slurry. The longest operating life has been at Will County 1, where the reducers eroded and were replaced after 6 or 7 years. At a few sites FRP has been used, but has suffered from erosion at Cholla 1 and Duck Creek 1.

The slurry recycle and discharge lines are made of either rubber-lined carbon steel or FRP. Problems similar to those encountered in the makeup slurry lines have occurred. Disbonding of the rubber lining occurred at Sherburne 1 and 2, and pieces of rubber plugged spray headers and nozzles. When the lining was intentionally removed, the carbon steel eroded badly at reducers and elbows. The recycle slurry is particularly abrasive at Sherburne 1 and 2 because of the fly ash present in the scrubbing slurry (alkalinity from the collected fly ash is used as scrubbing reagent). Thus, the rubber-lined carbon steel piping has been replaced with FRP piping. At Lawrence 4 and 5, FRP piping has also replaced rubber-lined carbon steel piping, which had eroded. At Hawthorn 3 and 4, however, rubber-lined carbon steel piping was used to replace FRP recycle slurry piping because the latter was too difficult to maintain and replace.

The reclaimed water piping, which carries less solid material, has been made of FRP at the majority of locations. A few cases of failure of FRP piping at joints have been reported, but results in general have been good. Flanged or shop-fabricated joints are preferable to field-cemented joints because pipefitters are not skilled in making FRP joints. Carbon steel is also used as a material of construction for reclaimed water piping; it is unlined at roughly one-half of the stations where it is used and rubber lined at the others. No problems have been reported with these materials.

Spray Headers

The materials chosen for spray headers, which are located inside scrubbers, are primarily FRP and 316L stainless steel. At Cholla 2, 317L stainless steel was selected to obtain the greater corrosion resistance resulting from its additional molybdenum content. No difficulties have been reported with FRP spray headers, but 316L stainless steel spray headers eroded in the venturi scrubber at Widows Creek 8. Rubber-lined and clad carbon steel has been used successfully at six systems, and mat-reinforced polyester applied to rubber-lined carbon steel is used at Huntington 3.

Valves

Valves are used in FGD systems for isolation and control functions. Valve problems are typically related not to materials failures, but to plugging and mechanical problems; and there seems to be a consensus that the number of valves in the system should be minimized. Rubber-lined valves are the most common, although many stainless steel valves are used. Knifegate, plug, pinch, and butterfly valves are the four kinds of valves used in FGD systems.

Knifegate valves made of 316L stainless steel have been used for isolation functions, especially where high pressures require metal. For low-pressure applications, FRP has been used. There have been no reports of serious corrosion of stainless steel valves except at La Cygne 1, where the average valve lifetime has been only 1 year. Trials of polyethylene-lined butterfly valves and valves with polyethylene plugs and seats at this site have yielded promising results.

Rubber-lined plug valves have also been used for isolation functions and have encountered erosive wear at six systems. They have been successfully replaced with pinch valves at Duck Creek 1. Butterfly valves with rubber linings have been used for isolation applications at only a few sites, and no problems have occurred with them.

When used for control functions, rubber-lined plug valves have had erosion problems. Rubber pinch valves have been successful, but periodic liner replacement is required when they are subjected to high-velocity flow. Rubber-lined butterfly valves appear to be giving satisfactory performance where they have been used for flow control. At Duck Creek 1, eroded rubber-lined plug valves have been successfully replaced with rubber pinch valves. At Will County 1, however, the original rubber pinch valves on the spent slurry line lasted only 250 hours and were successfully replaced with butterfly valves.

In general, the performance of valves is site specific and depends on the amount of throttling, the chemical and physical nature of slurry particles, and the pH of the liquid. As indicated, the trend in scrubber design is to reduce the number of valves to the minimum.

Tanks

Tanks have generally been constructed of carbon steel, and many are protected with some kind of lining. The organic linings used in tanks include rubber, flake-filled polyester, mat-reinforced epoxy, coal tar epoxy, and bituminous resin. Where pH is high, the use of unlined carbon steel is common. Concrete and FRP tanks have also found usage. In a few instances (e.g., at R. D. Morrow 1 and 2 and Duck Creek 1), FRP has been used for mist eliminator wash tanks; no problems have been reported.

Pond Linings

No pond problems were reported by FGD system users. Where disposal ponds are used, the preferred lining material is clay where the permeability of the soil is low.

TABLE E-10. ENERGY DEMAND FOR FGD ON A 500-MW PLANT

	Electrical energy, 10^3 kWh	Electrical power, 10^6 Btu/h	Power input, 10^6 Btu/h	Reheat input, 10^6 Btu/h	Total energy demand, percent of coal fired
Low-sulfur coal	7.1	24.2	69.1	Nil	1.5
High-sulfur coal	6.9	23.6	67.4	61.5	2.7

The table shows that use of the untreated flue gas for reheat in the low-sulfur coal case yields about a 50 percent energy savings relative to the high-sulfur coal case.

REFERENCES FOR APPENDIX G

Brova, A. A. 1978. Chemical Resistant Masonary, Flake, and Reinforced Linings for Pollution Control Equipment. Proceedings of Corrosion Problems in Air Pollution Control Equipment Symposium, Atlanta, Georgia.

Crow, G. L. 1978. Corrosion Tests Conducted in Prototype Scrubber System. Proceedings of the Corrosion Problems in Air Pollution Control Equipment Symposium, Atlanta, Georgia.

Fontana, M. G., and N. D. Greene. 1978. Corrosion Engineering. McGraw Hill Book Co., New York. pp. 157-159.

Furman, H. N. 1977. Chemical Engineering Progress. 72(11):92-94.

Gleekman, L. W. 1978. Old Materials for Air Pollution Control Equipment. Proceedings of Corrosion Problems in Air Pollution Control Equipment Symposium, Atlanta, Georgia.

Hoxie, E. C., and A. T. Michaels. 1978. How to Rate Alloys For SO_2 Scrubbers. Chemical Engineering. pp. 161-165.

Inco. 1980. The Corrosion Resistance of Nickel-Containing Alloys in Flue Gas Desulfurization and Other Scrubbing Processes. Corrosion Engineering Bulletin No. 7.

Kensington, K. L. 1978. FRP Applications and Specifications in Pollution Control Equipment. Proceedings of Corrosion Problems in Air Pollution Control Equipment Symposium, Atlanta, Georgia.

Leonard, R. B. 1978. Application of Nickel Chromical Alloys in Air Pollution Control Equipment. Proceedings of the Corrosion Problems in Air Pollution Control Equipment Symposium, Atlanta, Georgia.

Lewis, E. C., et al. 1978. Performance of TP-316L SS and Other Materials in Utility FGD Systems. Proceedings of the Corrosion Problems in Air Pollution Control Equipment Symposium, Atlanta, Georgia.

Mellan, I. 1976. Corrosion Resistant Materials Handbook, 3rd Edition. Noyes Data Corporation, Park Ridge, New Jersey.

PEDCo Environmental, Inc. 1981. Flue Gas Desulfurization Information System. Maintained for the U.S. Environmental Protection Agency under Contract No. 68-02-2603, Task Order No. 6.

Rhodin, T. N., and H. R. Carson. 1959. Physical Metallurgy of Stress Corrosion Fracture. Interscience Publishers, New York. pp. 451-456.

Rosenberg, H. S., et al. 1980. Operating Experience with Construction Materials for Wet Flue Gas Scrubbers. Presented at the 7th Energy Technology Conference and Exposition, Washington, D.C.

Singleton, W. T., Jr. 1978. Protective Coatings from Vinyl Esters. Proceedings of Corrosion Problems in Air Pollution Equipment Symposium, Atlanta, Georgia.

Uddeholm Steel Corporation. 1977. Report on the Corrosion Properties of UHB Alloy 904L. Totowa, New Jersey.

Metric Conversions

This handbook expresses measurements in English units so that information is clear to intended readers in the United States. The following list provides factors for conversion to metric units.

To convert from	To	Multiply by
Btu	kWh	0.0002931
Btu/lb	kJ/kg	2.326
cfm	m^3/h	1.70
°F	°C	(°F −32)/1.8
ft	m	0.305
ft/h	m/h	0.305
ft/s	m/s	0.305
ft^2	m^2	0.0929
ft^2/ton per day	m^2/Mg per day	0.102
ft^3	liters	28.32
ft^3	m^3	0.02832
gal	liter	3.785
gal/ft^3	liter/m^3	0.134
gal/min	liter/min	3.79
gal/min per ft^2	liter/min per m^2	40.8
gr	g	0.0648
gr/scf	g/Nm^3	2.29
hp (mechanical)	kW	0.7457
hp (boiler)	kW	9.803
in.	cm	2.54
in. H_2O	kPa	0.2488
in. H_2	mm Hg	1.87
$in.^2$	m^2	0.0006452
$in.^3$	m^3	0.00001639
lb	g	453.6
lb	kg	0.4536
lb/10^6 Btu	g/kJ	429.9
lb/ft^3	kg/m^3	16.02
lb/gal	kg/m^3	119.8
lb/$in.^2$	kPa	6.895
lb·ft/s·ft^2	Pa·s	47.89
lb-mol	g-mol	453.6
lb-mol/h	g-mol/min	7.56
lb-mol/h per ft^2	g-mol/min per m^2	81.4
lb-mol/min	g-mol/s	7.56
oz	kg	0.02835
scfm (at 60°F)	Nm^3/h (at 0°C)	1.61
ton	kg	907.2

Other Noyes Publications

HAZARDOUS WASTE INCINERATION ENGINEERING

by T. Bonner, B. Desai, J. Fullenkamp, T. Hughes,
E. Kennedy, R. McCormick, J. Peters, and D. Zanders

Monsanto Research Corporation

Pollution Technology Review No. 88

The engineering guidelines contained in this book are a compendium of the available literature on current state-of-the-art technology for hazardous waste incineration. They are intended to be used as a source of information for operational decisions and as a reference in the preparation of permit applications for hazardous waste incineration facilities.

A sizable fraction of the millions of tons of industrial waste material generated in the United States each year is considered hazardous (approximately 57 million metric tons in 1980). Incineration has recently emerged as an attractive alternative to other hazardous waste disposal methods such as landfilling, ocean dumping, and deep-well injection.

The advantages of incinerating hazardous wastes are several: toxic components can be converted to harmless or less harmful compounds; volume can be greatly reduced; heat recovery is possible as a means of saving energy; and incineration provides ultimate disposal, thereby eliminating the possibility of problems resurfacing at a later date. Because of these advantages, incineration may become a principal technology for hazardous waste disposal in the near future.

The various chapters of the book detail waste characterization, current commercial technology as well as emerging technology, incinerator design, and overall facility design. Listed below is a condensed table of contents giving **chapter titles and selected subtitles.**

1. **INTRODUCTION**
2. **CURRENT PRACTICES**
   Commercially Available Hazardous
     Waste Incineration Technologies
     Rotary Kiln
     Liquid Injection
     Fluidized Bed
     Multiple Hearth
     Coincineration
   Emerging Hazardous Waste
     Incineration Technology
     Starved Air Combustion/Pyrolysis
   Air Pollution Control Devices
3. **WASTE CHARACTERIZATION**
   Background Information
   Waste Sampling
   Basic Analysis of Waste
   Supplemental Analysis of Waste
   Analysis Test Methods
   Thermal Decomposition Unit
     Analysis
4. **INCINERATOR AND AIR POLLUTION CONTROL SYSTEM DESIGN EVALUATION**
   Destruction and Removal Efficiency
   Incinerator Evaluation
     Basic Design Considerations
     Physical, Chemical, and Thermodynamic Waste Property Considerations
     Temperature, Excess Air, Residence Time, and Mixing Evaluation
     Auxiliary Fuel Capacity Evaluation
     Combustion Process Control & Safety Shutdown System Evaluation
     Construction Material Evaluation
   Air Pollution Control and Gas
     Handling System Design Evaluation
5. **OVERALL FACILITY DESIGN, OPERATION AND MONITORING**
   Incinerator Facility Site Selection
     and Operation
   Waste Receiving Area
     Typical Operations and Layouts
   Waste Storage Area
     Types of Storage
     Safety Provisions
   Waste Blending and/or Processing
     Before Incineration
   Combustion Process Monitoring
   Air-Pollution Control Device
     Inspection and Monitoring
   Scrubber Waste Stream Treatment
     Inspection and Monitoring
   Continuous Monitoring Instrumentation
     for Gaseous Components
   Plant Condition Monitoring Systems
   Scrubber/Quench Water and Ash
     Handling
   Fugitive Emissions
   Materials of Construction
   Miscellaneous Concerns
   Technical Assistance

APPENDICES

ISBN 0-8155-0877-8 (1981)   385 pages

Other Noyes Publications

HOW TO DISPOSE OF OIL AND HAZARDOUS CHEMICAL SPILL DEBRIS

Edited by A. Breuel

Pollution Technology Review No. 87

This book describes various techniques which can be used to dispose of the collected debris from oil and hazardous chemical spills. It is based on research prepared by *SCS Engineers* and *CONCAWE* (the European oil companies' international study group for conservation of clean air and water).

Engineering constraints and equipment requirements for handling and disposal are evaluated. Debris management aspects considered are storage, transport, treatment, reprocessing, and disposal. Hardware and processing systems are identified and conceptual transport/disposal plans are developed. A literature review and several case studies are included. U.S. and European technology are covered.

Disposal is a complex problem. Spill incidents which have occurred in recent years have highlighted the difficulties in dealing with the large quantities of emulsions and debris which are collected during control and cleanup operations. Close cooperation between authorities and industry is a necessity at every stage, to ensure disposal in an environmentally acceptable, cost-effective, and energy-conserving manner.

Each case is different and no specific rules can be formulated, but guidance is given on the choice of options and their state of development. The condensed table of contents listed below gives **chapter titles and selected subtitles.**

I. OIL AND HAZARDOUS CHEMICAL SPILL DEBRIS DISPOSAL SYSTEMS AND TECHNIQUES
  1. INTRODUCTION
  2. SPILL AND DEBRIS DISPOSAL SCENARIOS
  3. SPILL DISPOSAL EQUIPMENT SYSTEMS FOR OILS
  4. SPILL DISPOSAL EQUIPMENT SYSTEMS FOR OILS AND FLOATING CHEMICALS
  5. DISPOSAL TECHNIQUES FOR HAZARDOUS CHEMICALS
  6. SUMMARY
  APPENDICES

II. OIL SPILL DEBRIS DISPOSAL PROCEDURES—CONCAWE STUDY
  1. SUMMARY
  2. CONCLUSIONS
  3. RECOMMENDATIONS
  4. INTRODUCTION
  5. LOGISTICS
     Ownership
     Oil Quantities and Types
     The Nature of Collected Oil
     Storage of Collected Oil and Debris
     Transport
  6. DISPOSAL METHODS
     Oil Recovery Techniques
     Stabilization of Oily Wastes
     Destructive Techniques

III. OIL SPILL DEBRIS DISPOSAL PROCEDURES—SCS STUDY
  1. INTRODUCTION
  2. SUMMARY
  3. LAND DISPOSAL SITE SELECTION
  4. LAND DISPOSAL METHOD SELECTION
  5. LAND CULTIVATION
  6. SANITARY LANDFILLING WITH REFUSE
  7. BURIAL
  8. MONITORING THE SITE FOR ENVIRONMENTAL PROTECTION
  9. CORRECTING ENVIRONMENTAL PROBLEMS

IV. OIL SPILL DEBRIS DISPOSAL— LITERATURE REVIEW & CASE STUDIES
  1. OILY WASTE DISPOSAL ON LAND: SUMMARY OF LITERATURE REVIEW
  2. CASE STUDIES OF OIL DEBRIS DISPOSAL SITES

ISBN 0-8155-0876-X (1981)   420 pages

Other Noyes Publications

NEW DEVELOPMENTS IN FLUE GAS DESULFURIZATION TECHNOLOGY

Edited by M. Satriana

Pollution Technology Review No. 82

This book covers the latest developments in flue gas desulfurization (FGD) technology, both nationally and internationally. Advanced systems are surveyed, with emphasis placed on those processes which currently seem to offer the best prospects for efficient removal of sulfur oxides from flue gases. Conventional technology is also reviewed.

With the constantly changing state of availability of oil and natural gas, increasing consideration is being given to the use of coal-based energy systems. As coal utilization increases, the potential impact on air quality of SO_2 emissions from coal-fired units will become more significant. FGD is considered to be one of the most commercially developed means of continuous control of SO_2 emissions from coal-fired boilers.

The majority of commercial FGD systems are based on lime/limestone scrubbing. Adipic acid modifications, sodium carbonate, magnesium oxide, and double alkali systems are also used. Emerging systems include copper oxide adsorption, dry adsorption, dry alkali scrubbing and citrate processes.

Included in the discussion of the various FGD methods are process descriptions, operating parameters, development status, problem areas, and case studies of operating systems. A chapter on comparative costs has also been included.

A condensed table of contents, with **chapter headings and selected subtitles,** is given below.

1. **INTRODUCTION**
   Background
   Process Categories
   Status of Operating FGD Systems
   Conventional Processes
     Limestone/Sludge
     Lime/Sludge

2. **RECENT ADVANCES IN CONVENTIONAL TECHNOLOGY**
   Process Chemistry—Scaling, SO_2 Removal, Corrosion
   Emission Control Strategy
   Equipment Design Improvements
   Process Design—Dampers, Scrubbers, Reaction Tanks, Mist Elimination, Reheaters, Solids Separation
   Process Control and Instrumentation
   Construction Materials
   Adipic Acid as Scrubber Additive
   Adipic Acid Degradation Mechanism
   Double Loop System

3. **SURVEY OF ADVANCED SYSTEMS**
   Emerging Processes and Descriptions
     (1) Double Alkali, (2) Wellman-Lord, (3) Magnesia Slurry, (4) Citrate, (5) Lime Slurry Spray Drier/Fabric Filter, (6) Dry Bicarbonate, (7) Carbon Adsorption, (8) Chiyoda Thoroughbred 101, (9) Chiyoda Thoroughbred 121, (10) Conoco, (11) Ammonia Scrubbing, (12) Aqueous Carbonate, (13) Nippon Kokan Ammonia Scrubbing, (14) Electron Beam, (15) Pircon-Peck
   Comparison of Energy Requirements for Selected FGD Systems

4. **DOUBLE ALKALI PROCESS**
   Generic Process
   Concentrated-Dilute Mode Comparison
   Technology Status in U.S. and Japan
   Design and Process Considerations
   Environmental Assessment

5. **MAGNESIUM OXIDE PROCESS**
   General Process Description
   Operating Parameters
   Development Status
   Recent Technological Developments
   Solutions to Problem Areas

6. **WELLMAN-LORD PROCESS**

7. **DRY SYSTEMS**
   Dry Injection/Particulate Collection
   Spray Drying/Particulate Collection
   Dry and Wet Scrubbing Comparison
   Ongoing Activities

8. **CITRATE PROCESS**

9. **COPPER OXIDE PROCESS**

10. **CARBON ADSORPTION PROCESS**

11. **SEAWATER SCRUBBING PROCESS**

12. **TECHNICAL FEASIBILITY AND ECONOMIC EVALUATION OF NINE PROCESSES**

ISBN 0-8155-0863-8 (1981)

326 pages

Other Noyes Publications

EMISSION CONTROL TECHNOLOGY FOR INDUSTRIAL BOILERS

Edited by A.E. Martin

Pollution Technology Review No. 74
Energy Technology Review No. 62

This book describes emission control technology for industrial boilers which, because of their more intermittent power demand, have different problems than those found with electric utility boilers. The book will be of use to industrial executives and engineers, plant managers, boiler operations executives, and other personnel involved in pollution control decision making. It provides a series of technology assessments based on documents prepared for a U.S. Environmental Protection Agency (EPA) industrial boiler study.

The material included covers the current state of the art as well as some alternative technology for emission control in the areas of flue gas desulfurization, NO_x control, particulate collection, and fluidized bed combustion. Coal-, oil-, and gas-fired steam generating power plant emission technology is described in detail and, where available, system performance has been documented.

The summarized table of contents below lists the chapter titles and important subtitles.

1. **FLUE GAS DESULFURIZATION**
   Coal-Fired Boiler Controls
     Lime/Limestone Wet Scrubbing
     Double Alkali
     Wellman-Lord Sulfite Scrubbing
     Magnesia Slurry Absorption
     Sodium Scrubbing
     Processes Under Development
   Oil-Fired Boiler Controls

2. **NO_x FLUE GAS TREATMENT**
   Coal-Fired Boiler Controls
     Selective Catalytic Reduction
       Fixed Packed Bed Reactors
       Moving Bed Reactors
       Parallel Flow Reactor
     NO_x/SO_2 Removal
     Absorption-Oxidation
     Adsorption
     Electron Beam Radiation
     Absorption-Reduction
     Oxidation-Absorption-Reduction
     Oxidation-Absorption
   Oil-Fired Boiler Controls
     Selective Catalytic Reduction
       Fixed Packed Bed Reactors
       Moving Bed Reactor
       Parallel Flow Reactor
     NO_x/SO_2 Removal
     Absorption-Oxidation
     Adsorption
     Electron Beam Radiation
     Absorption-Reduction
     Oxidation-Absorption-Reduction
     Oxidation-Absorption
   Natural Gas-Fired Boiler Controls
     Selective Catalytic Reduction
       Fixed Packed Bed Reactor
     Absorption-Oxidation

3. **NO_x COMBUSTION MODIFICATION**
   Coal-Fired Boilers
     Pulverized Coal
     Stokers
   Oil-Fired Boilers
     Low Excess Air (LEA)
     Staged Combustion
     Flue Gas Recirculation (FGR)
     FGR & Staged Combustion
     Reduced Air Preheat (RAP)
     Load Reduction
     Low NO_x Burners (LNB)
     Ammonia Injection
   Gas-Fired Boilers
     LEA
     Staged Combustion, Air (SCA)
     FGR
     FGR & Staged Combustion
     Load Reduction
     RAP
     LNB
     Ammonia Injection

4. **PARTICULATE COLLECTION**
   Electrostatic Precipitation
   Fabric Filtration
   Wet Scrubbing
   Mechanical Collectors

5. **FLUIDIZED BED COMBUSTION**
   System Description
   Status of Development
   System Performance
     SO_2 Control
     NO_x Control
     Particulate Control

ISBN 0-8155-0833-6 (1981)

405 pages

Other Noyes Publications

FLUE GAS CLEANING WASTES DISPOSAL AND UTILIZATION

Edited by D.L. Khoury

Pollution Technology Review No. 77
Energy Technology Review No. 65

Flue gas desulfurization (FGD) wastes and fly ash, together, are generally referred to as flue gas cleaning (FGC) wastes. Modern fossil-fueled boilers (utility and industrial) produce large quantities of FGC wastes. Those boilers employing conventional coal combustion present a broad spectrum of potential environmental problems, which, due to regulatory constraints pertaining to air and water pollution control, will require focusing on the environmental management of FGC wastes.

This study, prepared under the direction of Arthur D. Little, Inc., assembles and reviews data from ongoing activities relating to pollution control technology for conventional coal-fired combustion sources. Generation of FGC wastes; disposal options including current practice, R&D and field studies; and utilization practice, including technical and economic assessment of current practice and R&D studies, are covered. Emphasis is placed on wastes produced by commercially demonstrated technologies and, where data are available, those technologies in advanced developmental stages.

A **partial table of contents** for this book is given below.

I. GENERATION AND CHARACTERIZATION OF FGC WASTES

1. **INTRODUCTION**
2. **OVERVIEW ON FGC WASTE GENERATION**
   Ash Collection Technology
   FGD Technology
   Categorization of Wastes
   Dewatering of Wastes
3. **PRODUCTION TRENDS AND HANDLING OPTIONS**
4. **CHEMICAL CHARACTERIZATION OF WASTES**
5. **PHYSICAL CHARACTERIZATION OF WASTES**
6. **RESEARCH NEEDS**

II. DISPOSAL OF FGC WASTES

1. **INTRODUCTION**

2. **DISPOSAL OF WASTES**
   Disposal Options
   Land Disposal
   Ocean Disposal
   Options vs Potential Impact Issues
   Site Selection, Design and
      Disposal Practice
3. **REGULATORY CONSIDERATIONS**
   Regulatory Framework
   Groundwater Related
   Surface Related
   State Requirements and Plants
   Ocean Disposal Related
   Stability Related
   Land Use Related
   Air Related
   National Energy Act of 1978
4. **ENVIRONMENTAL IMPACT CONSIDERATION**
   Assessment of Present Control
      Technology
   Summary of Data Gaps
      and Future Research Needs
5. **REVIEW OF MONITORING CONSIDERATIONS**
   Screening Tests for Solid Wastes
   Water Monitoring Methods
   Fugitive Emissions Monitoring
   Biological Monitoring
   Post-Operational Monitoring
6. **REVIEW OF ECONOMICS OF DISPOSAL**

III. UTILIZATION OF FGC WASTES

1. **INTRODUCTION**
2. **UTILIZATION OF COAL ASH**
   Current Utilization
   Ash Utilization as Fill Material
   Ash in Cement and Concrete
   Ash in Miscellaneous Use
   Ash as a Mineral Resource
   R&D Programs
3. **UTILIZATION OF FGD WASTES AND BY-PRODUCTS**
   Nonrecovery Wastes, Recovery Wastes
   By-Product Marketing
4. **REGULATORY CONSIDERATIONS**
5. **UTILIZATION ASSESSMENT AND DATA GAPS**

ISBN 0-8155-0847-6 (1981)   646 pages

Other Noyes Publications

POLLUTION CONTROL TECHNOLOGY FOR INDUSTRIAL WASTEWATER

Edited by D.J. De Renzo

Pollution Technology Review No. 80

This book provides an extensive survey of the reliability and effectiveness of 56 unit operations in industrial water pollution control. These operations include 32 generic wastewater treatment technologies, classified as preliminary, primary, secondary or tertiary, and 24 sludge treatment and disposal technologies.

Each process is briefly described by the type of control equipment required, the major variations of design, flow diagrams, and information on the following: design criteria, common modifications, typical performance, applications and limitations of the process, reliability, chemicals required for operation, residuals generated, and environmental impacts.

Summary tables for most technologies are provided showing the concentrations of various pollutants in the effluents, the minimum, maximum, median and mean removal efficiencies, and the number of data points used to generate this information. Conventional pollutants as well as EPA-categorized "priority pollutants" and "hazardous substances" are covered.

The information reviewed here should be extremely useful to engineers, management personnel, and others involved with regulatory requirements, guidelines, and decisions. Contents are:

1. **INTRODUCTION**

2. **TECHNOLOGY OVERVIEW**

3. **WASTEWATER CONDITIONING**
   Screening
   Grit Removal
   Flow Equalization
   Neutralization

4. **PRIMARY WASTEWATER TREATMENT**
   Gravity Oil Separation
   Clarification/Sedimentation
   Clarification/Sedimentation Using Chemical Addition
   Gas Flotation (Dissolved Air Flotation)
   Gas Flotation with Chemical Addition
   Granular Media Filtration
   Ultrafiltration

5. **SECONDARY WASTEWATER TREATMENT**
   Activated Sludge
   Trickling Filters
   Lagoons (Stabilization Ponds)
   Rotating Biological Contactors
   Steam Stripping
   Solvent Extraction

6. **TERTIARY WASTEWATER TREATMENT**
   Granular Activated Carbon Adsorption
   Powdered Carbon Addition
   Chemical Oxidation
   Air Stripping
   Nitrification
   Denitrification
   Ion Exchange
   Polymeric (Resin) Adsorption
   Reverse Osmosis
   Electrodialysis
   Distillation
   Chlorination (Disinfection)
   Dechlorination
   Ozonation
   Chemical Reduction

7. **SLUDGE TREATMENT**
   Gravity Thickening
   Flotation Thickening
   Centrifugal Thickening
   Aerobic Digestion
   Anaerobic (Two-State) Digestion
   Chemical Conditioning
   Thermal Conditioning (Heat Treatment)
   Disinfection (Heat)
   Vacuum Filtration
   Filter Press Dewatering
   Belt Filter Dewatering
   Centrifugal Dewatering
   Thermal Drying
   Drying Beds
   Lagoons

8. **DISPOSAL**
   Evaporation Lagoons
   Incineration
   Starved Air Combustion
   Landfilling (Area Fill)
   Land Application
   Composting
   Landfilling (Trenching)
   Deep-Well Injection

REFERENCES

GLOSSARY

ISBN 0-8155-0855-7 (1981)

712 pages